# Exploring the Evolutı.ı. of Our Ancestors

This is a captivating evolutionary narrative of the human body, exploring the pivotal traits that make humans unique as a species. It provides a better understanding of why we look the way we do, through an evolutionary morphological lens, by delving into the functional explanations for the unique characteristics of us and our ancestors and the evolutionary pathways that shaped them. It integrates changes in anatomy with functional shifts, but also with underlying genetic and environmental transformations that drove our evolution. The main body of the book focuses around four fundamental themes that have evolutionarily sculpted us into who we are today, ever since the shared origin with the chimpanzee: diet, brain, locomotion and skin. This book not only promises to enrich our understanding of human evolution but also challenges us to reconsider what it means to be human in light of our ancient lineage and ongoing evolutionary journey. It also:

- Provides a complete overview of the major events of human evolution;

- Helps readers understand why our body has been shaped the way it is;

- Integrates genes, anatomy, function, behavior and ecology, creating a more complete picture, written in an accessible text while incorporating many facts and figures building upon both historic and recent literature;

- Offers an up-to-date view of how anthropologists currently see our evolution;

- Focuses on four fundamental changes in the brain, diet, skin and locomotion;

- Explains some aspects of what humans are experiencing today (e.g., why some people are lactose intolerant).

# Exploring the Evolution of Our Ancestors
## On the Human Track

Dominique Adriaens

CRC Press
Taylor & Francis Group
Boca Raton  New York  London

CRC Press is an imprint of the
Taylor & Francis Group, an **informa** business

Designed cover image: Author-designed

First edition published 2026
by CRC Press
2385 NW Executive Center Drive, Suite 320, Boca Raton FL 33431

and by CRC Press
4 Park Square, Milton Park, Abingdon, Oxon, OX14 4RN

*CRC Press is an imprint of Taylor & Francis Group, LLC*

© 2026 Dominique Adriaens

Reasonable efforts have been made to publish reliable data and information, but the author and publisher cannot assume responsibility for the validity of all materials or the consequences of their use. The authors and publishers have attempted to trace the copyright holders of all material reproduced in this publication and apologize to copyright holders if permission to publish in this form has not been obtained. If any copyright material has not been acknowledged please write and let us know so we may rectify in any future reprint.

Except as permitted under U.S. Copyright Law, no part of this book may be reprinted, reproduced, transmitted, or utilized in any form by any electronic, mechanical, or other means, now known or hereafter invented, including photocopying, microfilming, and recording, or in any information storage or retrieval system, without written permission from the publishers.

For permission to photocopy or use material electronically from this work, access www.copyright.com or contact the Copyright Clearance Center, Inc. (CCC), 222 Rosewood Drive, Danvers, MA 01923, 978-750-8400. For works that are not available on CCC please contact mpkbookspermissions@tandf.co.uk

**Trademark notice:** Product or corporate names may be trademarks or registered trademarks, and are used only for identification and explanation without intent to infringe.

ISBN: 978-1-032-96253-5 (hbk)
ISBN: 978-1-032-97122-3 (pbk)
ISBN: 978-1-003-58879-5 (ebk)

DOI: 10.1201/9781003588795

Typeset in Corbel
by KnowledgeWorks Global Ltd.

# Contents

Acknowledgments      *vii*

About the Author      *ix*

Introduction      1
     Once upon a time …      1
     A personal point of view      3
     Four major themes      4

Chapter 1: Fragments of our existence      7
     Fossils as direct evidence      7
     Indirect traces      9
     A battle of opinions      10

Chapter 2: What makes us unique?      13
     The urge to be special      13
     Our body, a sapiens body?      15
     Sapiens is an animal      18
     Sapiens is a fish      20
     Sapiens is a tetrapod      23
     Sapiens is an amniote      26
     Sapiens is a mammal      27

Chapter 3: The equatorial Eden      35
     Apes rule!      35
     Fruit for everyone      42
     Clambering with hands and feet      48
     Tropical paradise      60
     The ape brain      63
     So …      74

Chapter 4: The human cradle      75
     The first hominins      75
     The split      76
     The first one from Chad      77
     *Orrorin*      78
     Ardi      79
     Fruit is not enough      80
     Legs gain importance      81
     Feeling good in our skin?      84

A body to carry around the brain                                    85
So ...                                                              85

Chapter 5: The forest opens up                                      87
Pioneers of becoming human                                          87
Falling back on fallback food ... a hard nut to crack               94
From tree to tree, across the grass                                 100
The sun is shining                                                  117
Big brains not so important yet                                     123
So ...                                                              129

Chapter 6: Fully erect!                                             131
The first tall ones!                                                131
Steak and fries ... almost                                          135
Runners are born                                                    141
Keep it cool!                                                       151
What big brains you have!                                           156
So ...                                                              168

Chapter 7: Conquer the world                                        169
The sapiens track                                                   169
From hunter to farmer to health issues                              179
Walking with big brains comes with consequences                     184
Becoming pale, not yet an option                                    191
Survival through behavior                                           197
So ...                                                              207

Chapter 8: What now?                                                209
Are we truly unique then?                                           209
Are we still evolving?                                              211
What will the future bring?                                         213

*Glossary*                                                          215

*Suggested reading per chapter*                                     221

*References*                                                        227

*Index*                                                             307

# Acknowledgments

This book has traveled a remarkable journey, evolving from a simple idea to share insights on human evolution with a Dutch-speaking audience to an updated edition now accessible to English-speaking science enthusiasts. I extend my heartfelt thanks to the publisher of the Dutch version for facilitating this transition, with special gratitude to Isaac Demey for his invaluable support.

I am deeply grateful to my academic mentor, Walter Verraes, whose guidance has been instrumental throughout this project. My sincere thanks also go to my colleagues Johan Braeckman, Lieven Pauwels, Peter Aerts and Tom Van Hoof for their critical reviews and insightful comments on the earlier version of this book.

A book's success hinges on reaching the right audience, and I am immensely thankful to Charles "Chuck" Crumly for embracing the challenge of bringing this work from a Dutch edition to an English one, aimed at students and science enthusiasts alike. Special thanks to Laura Slater and Kara Roberts for their assistance with the administrative and editorial processes.

Last, I owe a special debt of gratitude to Cindy, Noa, Quinten and Glenn for their understanding and support as I dedicated countless hours to writing this book.

# About the Author

**Dominique Adriaens** obtained a PhD in Biology with research that focused on the functional morphology of the feeding apparatus in an African catfish species. His research focused on African catfish taxonomy and evolutionary morphology. He was appointed as an assistant professor in the Biology Department, advancing to Full Professor in 2012 and Senior Full Professor in 2020. Adriaens became Director of the Museum of Zoology in the Faculty of Sciences and took over the leadership of the 'Evolutionary Morphology of Vertebrates' research group. His research focuses on the evolution of vertebrates, aiming to answer why animals look the way they do by examining how vertebrates have become adapted based on the anatomy of muscles and skeletal systems. His studies have been wide-ranging and include catfish feeding on land, jaw adaptations in Darwin's finches, specializations in African cichlids, tunnel-digging adaptations in African mole rats and the evolution of the prehensile tail in seahorses. Adriaens and his lab have published over 180 articles in international journals. He has supervised more than 20 PhD students, presented at 34 international conferences and universities, and participated in numerous other conferences. He is a Research Associate at the American Museum of Natural History in New York, and he oversees the Biology teacher training program and the faculty's Excellence Program in Sciences. He chaired the Biology program and serves as Director of Studies for the Faculty of Sciences. Dr. Adriaens has diverse teaching experience, including a course on primate ancestors and human evolution.

# Introduction

Man, the wonder and glory of the Universe

—**Charles Darwin (1871, p. 188)**

We are made of stardust

—**Carl Sagan**

## Once upon a time …

Once upon a time … our own evolutionary story could start like this. All the ingredients for a good mystery story are there. We and our ancestors, the main characters, found ourselves in unprecedented circumstances that demanded a constant struggle for survival. And yes, there were losers and winners. This may not sound very scientific, but still, this is appropriate for a nonfiction book about our own evolution and the evolution of our body. Have you ever considered that we are the only species that undertook a quest about how it got on this planet? Or as the evolutionary biologist Theodosius Dobzhansky once put it (Dobzhansky, 1967):

> In giving rise to man, the evolutionary process has, apparently for the first and only time in the history of the cosmos, become conscious of itself.

Mankind has always sought for explanations about its origin, some attempts more exotic than the next. From a belief that life could arise from clay, whether thanks to divine intervention or not, the notion grew in the 19th century that our immediate ancestors were apes swinging in the trees and picking each other's fleas. Charles Darwin was very explicit about this in 1871 in his *Descent of man* (Darwin, 1871), calling the origin of man "*a most interesting problem*". His view even produced spectacle in 1860, during the so-called *Great Debate* between Thomas Huxley and the Bishop Samuel Wilberforce, having someone faint at the thought that we would descend from apes and that this was the result of a completely natural process.

Are we then not much more than glorified apes, walking on two legs instead of hands and feet, covered in clothing instead of fur, and a little smarter? Or is this instinctively an unbridgeable gap? Ultimately, we are capable of accomplishing things against which eating nuts as monkeys do pales sharply. We manage to get gigantic machines stuffed with conspecifics into the air, up to another planet! Through technological ingenuity, we can let the other side of the world know in no time how nice our desert looks like. Ever seen a monkey do that? At the same time, no one will deny the human

DOI: 10.1201/9781003588795-1

traits in a chimpanzee cherishing her young, or two young capuchin monkeys teasingly chasing each other from one branch to the other.

Our story is about a book. Not the book you're holding in your hands, but the body that holds this book: the body of *Homo sapiens*. This is a book that took millions of years to be written and where today we have means to read even the fine print. Not only can we read what it says but we can even reconstruct how this book came to be: it tells our evolutionary story. If you look at the historical literature on our evolution, it is laced with features of a classic epic story (Landau, 1991). The famous fossils or human species were the heroes who faced the ferocious and sometimes epic challenges of Mother Nature when trying to survive. As Misia Landau, an American anthropologist puts it, *"It's my belief that scientists have much to gain from an awareness that they are storytellers"*. Scientists should more be 'storytellers', daring to go beyond merely communicating the facts and figures (as is done and necessary in the scientific literature). 'Tell a story!' is not for nothing the credo for good communication that attracts and holds attention; it is even a trait unique to our species. It actually played an important role in our own evolution, for it was an eminently suitable means of persuading conspecifics to undertake something together (Ferretti & Adornetti, 2021). Gottschall even calls humans Homo fictus, the fiction-man (Gottschall, 2012).

But from when does a narrative of scientific facts become too speculative and subjective? The scientific discipline that focuses on the evolution of hominins, so-called paleoanthropology, is not exempt from this. Stories from the 20th and 21st centuries show that testosterone levels and patriotism within this discipline were sometimes high. For many, their mission was to find the missing link in human evolution. And how great wouldn't the glory be if you could prove that the evolution of man took place in your backyard, and that you were the one who discovered it? The famous forgery and dark period in paleoanthropology, namely the Piltdown Man, attests to this. It must be said, the information available is sometimes very limited, and sometimes one must just dare to put forward an idea. If only to get others thinking and pointing in a direction to gather further evidence. Of course, a nonfiction book about our own evolution is not a fairy tale, but it does make writing an evolutionary story about our own evolution challenging. Matt Cartmill's words are certainly not encouraging in this regard (Cartmill, 1983):

> People who study human origins sometimes claim that their investigations will shed new light on human nature and so help us understand ourselves and predict our future. But this is wishful thinking. A thing is what it is, no matter how it got that way; and human appetites and impulses are what they are, no matter where or what we came from. Knowing their history adds nothing. If it were proved tomorrow that people evolved from cottontails instead of apes, we would still prefer bananas to clover as an ingredient in pie.

# A personal point of view

Human evolution was, is and will remain a hot topic. Bookshelves stacked for meters long could be filled on this topic, but this one is written from the perspective of an evolutionary morphologist. This is what I do professionally: I study the evolutionary morphology of vertebrates – and humans are, after all, vertebrates. I study the morphology (synonymous with 'anatomy') of vertebrate animals from an evolutionary and functional perspective. Simply put, I constantly ask myself the question 'why does an animal look the way it does?' And why does it look different from its ancestors or close relatives? Is it doing something special with that, perhaps?

Like an evolutionary detective, I try to understand how natural selection modified the anatomy of an organism, or a group of organisms, over time, allowing it to function within an ecological context. From that, one can deduce why certain organismal designs occurred or occur, and others do not. As an example, it makes perfect sense that a shark's tail fin is vertical and a whale's is horizontal, as a shark is a fish that moves the spine in a horizontal plane while mammals – and thus whales – move their spine in a vertical plane. Hitting a horizontal fin horizontally through the water does not work well. So, this book follows an evolutionary logic behind our ancestral anatomy, through which I try to unveil our evolutionary antecedents. By illustrating the evolutionary differences between species, we can better understand why our bodies are built the way they are and why they differ from our ancestors. Although evolution involves coincidence, it is not a random fact that we have large brains and naked skin (i.e. no fur), that we walk upright but cannot grasp with our big toe, why ideally we do not have flat feet but chimpanzees do, why we have a wisdom tooth that does not always breaks through or why we do not choke when drinking milk when a baby but do so when we are older.

I have the privilege of having to delve into the literature for a course I teach on 'Evolution of Primates and Paleoanthropology' at my university. At the beginning of my university career, I had to decide whether to take over this course from my then-supervisor, Prof. Walter Verraes. He introduced me to the diversity of fossil species, each with their typical characteristics and a fascinating story, tying them together in an evolutionary narrative. At first, I was not eager to take on that challenge. It seemed a hopeless task to be able to work my way through the gigantic pile of literature in this field, literature that, moreover, changes daily. To give an idea, the *Web of Science* database, which reflects only a portion of research that appears in publications, indicates that in 1980 there were 149 studies published around human evolution (based on the search term '*human evolution*'). In 2001, the year I first taught the course, there were 2952 studies published. By 2021, there were 13,559. We already know a lot about our evolution, and yet there is still a lot we don't. Even how many human species actually existed is still up for debate. Over the past 260 years of research, there have been no less than 206 genera and species of hominins described (Reed et al., 2023). Recently, it was even suggested that the species name '*Homo heidelbergensis*' should be omitted (Athreya & Hopkins, 2021). In what follows, the main protagonists in our evolutionary story and their own evolutionary '*moment de gloire*' will be introduced.

# Four major themes

Where do I want to go with this book? I want to provide the ingredients that form the basis of our own evolution, topped with a subtle sauce of scientific storytelling. Facts may be boring but the story they tell is far from that. That Lucy had an obliquely oriented femur with respect to the tibia is boring. That this was one of the first unambiguous proofs that *Australopithecus* already showed adaptations to walk upright, even though it still had the brain size of a chimpanzee, nevertheless tells a very interesting evolutionary story. I want to inform the reader about what we know, but equally important about what we don't know. I want to suggest possible scenarios that could explain our evolutionary process, giving a better idea of why we look the way we do. But most importantly, I also want the reader to be able to use the insights in this book to approach the information from a critical perspective, including the information in this book!

As you will suspect by now, our evolutionary story is long and complex. Four themes, however, do stand out. Four themes that laid out the major chalk lines of the process that determined what our bodies look like today: our diet, our locomotion on two legs (what scientists call being bipedal), our skin and our brain. Many times, modern humans, *H. sapiens*, are put forward as unique, special, far more evolved than other animals, even compared to our close relatives such as chimpanzees. Are we unique because we constantly walk on two legs? No, a couple of thousand species of birds do that constantly. Are we unique because we have the largest brains? No, there are African elephant fish with larger brains compared to their body size (Nilsson, 1996). Are we unique because of our naked skin? No, naked mole rats or elephants are as stark naked as we are. Digging into the literature also revealed quite some surprises, as assumptions that obtained a life on their own for decades eventually turned out to have no scientific basis whatsoever. Even now you read in books that humans are unique because of their large brains or that we have ten times more brain cells that are responsible for supporting neurons (these are the nerve cells that transmit stimuli) than we have neurons. Is it also a surprise to you that it took until 2004 to show that this is not so?

Who is the reader I am addressing? I hope to reach both the scientist with a good understanding of human evolution and the curious broader audience. This includes those who do not have a direct affinity for science but are eager to know more about human evolution. The book allows to test to what extent the image one has about our evolution corresponds to current facts, visions, hypotheses and theories. This can be useful for teachers who bring aspects around evolution, human anatomy and physiology and, of course, human evolution into their classes but also for medical doctors who can look at medicine and their patients from a different, evolutionary perspective. The reader may be either a staunch humanist or a religious person, wishing to check their view of the world and society against what has been described in the literature. In the end, I want the reader, without any required prior knowledge of human evolution, to have a better view of how we got here on this planet and have grown a more critical view of the process behind it. But most of all, I want the reader to have enjoyed our evolutionary journey story, the story of us, *H. sapiens*.

# 1
# Fragments of our existence

The search for our ancestors plays out like a detective story, laced with numerous unknowns and mysteries. In 2002, for example, the discovery of the fossil remains of the oldest hominin, *Sahelanthropus tchadensis*, briefly turned the world of paleo-anthropologists upside down. Its fossils were found in present-day Chad in Central Africa, where before that only *Australopithecus bahrelghazali* was discovered in the same region, in 1996. What on Earth were they doing there, as everything pointed to an East African origin of human evolution? In 1995, new fossils from Georgia hit the world press. Nothing special, would you think? It turned out that they were estimated to be about 1.7 million years old, hence much older than people had ever thought to find human ancestors there. And what about the discovery of the so-called hobbit in 2004, an Asian early human dwarf species (*Homo floresiensis*) that may have even lived together with *Homo sapiens* on the island of Flores? And what about our skin color? If dark skin protects us from harmful UV radiation, why do Inuit in the far north have a tinted skin even though the sun doesn't shine much intensely there? Do we even know who the first true *Homo* species was? Even that is still up for debate. Despite all scientific technologies and the thousands of studies in the past, we still haven't been able to fully resolve all the questions about our own evolution. However, this should not come as a surprise when you realize what one must work with. Not all fossil remains have been preserved and we will not even find all of those. Because of that, we don't know how much we actually don't know. One can think of paleontologists as being puzzle solvers who fortunately find new puzzle pieces on a regular basis, i.e. fragments that have left a trace of our history. The real challenge starts when trying to put the pieces back together in our great evolutionary puzzle.

## Fossils as direct evidence

Fossils are like puzzle pieces containing the most direct information about who our ancestors were and what they looked like. Yet, that doesn't mean they answer all the questions. Being highly mineralized and therefore less likely to decompose, skeleton preserves the best. But this largely depends on the conditions in which a dead animal ends up being covered with sediment, such as leaves, sand, volcanic ash, as it may be partially preserved or completely destroyed and replaced by another material. If the original bone remains, it is called a subfossil. A 'true' fossil is when the bone has been completely replaced by sediments or minerals, yielding a cast of it. This process of fossilization can occur after as little as tens of thousands of years under ideal conditions.

DOI: 10.1201/9781003588795-2

Over time, a huge number of fossils have vanished through chemical decomposition or erosion. The older the fossil, the slimmer the chances of finding a complete skeleton. Paleoanthropologists – scientists who study hominin fossils – often deal only with scraps of a skeleton, such as from a skull or jaw, or even less. Teeth preserve the longest, thanks to their very tough enamel and dentin. The species *A. bahrelghazali* has even been described based on just a piece of lower jaw and an upper premolar (Brunet et al., 1996). The fossil material of many of our ancestor species is indeed highly fragmented, leaving scientists with educated guesses about how the limbs were built and whether there were differences between males and females.

Variation provides the core of evolution. Being able to map them is essential to biology but remains a challenge when dealing with extinct species. When Donald Johanson in 1974 found the famous fossils of Lucy, his team also found remains of nine other individuals (referred to as the 'first family') (Johanson & Taieb, 1976). Mandibles that differed markedly in size and shape were found, a phenomenon constantly presenting paleoanthropologists with a dilemma: is this variation within a single species or is it across multiple species? This is even more of a challenge when variation within a species is significant, such as differences between males and females, that can even be larger than that between different species. Biologists who study living organisms have access to measurements on a large number of individuals of the same species and can even check whether their DNA belongs to the same species. Paleoanthropologists working with 'true' fossils deal with a cast that has lost all traces of DNA. This explains why certain fossils have been assigned to different species over time. For example, Eugene Dubois in 1896 described the Java man (Dubois, 1896) and Davidson Black in 1929 described the Peking man (Black, 1929), all afterward turning out to be one and the same species: *Homo erectus* (Mayr, 1950). Sometimes paleoanthropologists strike lucky, like in 2015, when more than 1550 fossils from at least 15 individuals were discovered of the new species *Homo naledi* (Berger et al., 2015). These were found deep in the Rising Star cave in South Africa, which was partially submerged, needing trained cavers to bring the fossils to the surface.

Also in time and space, the fossil evidence is extremely fragmented. And yet it can teach us a great deal about how different human species spread around the world. But as said, everything depends on how well a skeleton is fossilized and whether one looks in the right place. Knowing that billions of animals worldwide once died and ended up underground, only a very small fraction eventually fossilizes, of which again only a tiny fraction is found and of which only a few will be identified to which species they belong. Such fragmentation, for example, explains the differences in opinions on whether hominins that left Africa passed through the Middle East – an area called the Levant – or migrated along the coastline (Vyas et al., 2017). Exactly determining what kind of environment our ancestors lived in, and thus whether they showed adaptations to that environment, equally remains a challenge. The time window, represented by fossil evidence and thus when a species lived, is often wide (maybe several thousand to several hundred thousand years). Even if one can reconstruct vegetation within that time window using fossil remains of plants that then and there thrived, it does not imply that a particular ancestor also lived in that particular habitat as suggested by a recent vegetation study (Faith et al., 2021). Even the selection of reference taxa included in an analysis can introduce a bias on the predicted ecological setting and faunistic diversity. By only looking at mammalian fossils from the Eastern African rift valley, it has been shown that this is not representative for the East African mammalian biodiversity as well as the environmental conditions they lived in at a given time window (Barr & Wood, 2024). The same may thus apply for hominin biodiversity in that region.

Age determination of fossils is done using various dating techniques, including using the already known age of other fossils found nearby. The age of 6–7 million years old for one of our first ancestors, *Sahelanthropus*, was defined as such. For a more refined dating, a direct determination of the sediment age in which a fossil was found is done using radioactive isotopes. These physical building blocks at the atomic level are formed in nature but decay and become converted into another building block after some time (referred to as radioactive decay). The decay rate is different for different isotopes, and thus the ratio between the amount of isotopes before and after decay tells us how old a sediment and thus the fossil is. Which isotope to use then depends on the initial estimate using other fossils. For example, carbon dating is used for sediments younger than 50,000 years, uranium-thorium dating for up to 500,000 years, uranium-lead dating for up to a little over a million years and argon-argon dating for sediments of more than 4 million years old. As dating methods are constantly being refined, the age of fossils may also change over time. This was the case for the Java man, long held to be 1.8 million years old, until a new estimate in 2020 reduced it to just 1.3 million years old (Matsu'ura et al., 2020). Such dating techniques allow us to narrow down the time window when human species roamed the globe. And yet again, taking the limited fossil evidence into consideration, when one speaks of a human species having lived during a certain period in a certain area, what it actually means is that fossils are known from that period and that area. It does not exclude that species also occurred earlier or later or in other places. The absence of fossils is not evidence of the absence of a species.

# Indirect traces

Fortunately, researchers are not limited to hominin fossils themselves. For example, they have been able to reconstruct what the environment must have looked like in which humans lived during the Paleolithic – the period that spreads from about 3.3 million to 12,000 years ago – based on fossils of birds (Finlayson et al., 2011). You would be amazed at the ways in which our evolutionary past is all written down. It's just a matter of finding the right glasses to read that information. Fossils of plants and other animals found in the same sediments as hominins tell a lot about the environment of the time. If one finds fossilized pollen grains of lianas, a typical jungle plant, or fossils of small antelopes now found exclusively in tropical forest, one can infer that those hominins probably lived in a forest environment.

Fossils are traces of long ago. We also learn a great deal from traces of today, traces that have been recorded during millions of years of evolution, generation after generation, and transmitted through our genetic code until now. DNA of modern humans has shown that *H. erectus*, our early ancestor from over a million years ago, could not be the last common ancestor of all modern humans (Cann et al., 1987). We even know today complete migration routes of our ancestors based on the genetic traces left in the Y chromosome of males or the whole genome (Krause & Trappe, 2020; Sun et al., 2021). A pioneer in this field is Sweden's Svante Pääbo, who developed techniques to reconstruct the genetic code of recently extinct species (Pääbo, 2014). In their first attempts in 1997, only 379 base pairs (units that build the DNA chain) were obtained, and only from the DNA found in mitochondria (these are located in cells and are responsible for energy metabolism) (Krings et al., 1997). By 2010, this was already 4 billion base pairs of the DNA located in the nucleus (Green et al., 2010), and four years later, almost the entire genome was deciphered (Prüfer et al., 2014). These techniques even lead to the discovery of a possible new species, Denisova humans, and the fact that we still carry Neanderthal DNA in us today (Green et al., 2010; Krause et al., 2010). Today we can even analyze DNA extracted from sediments in caves (Vernot et al., 2021). This genetic information tells us something about the origins of and relationships between species, as well as about the genes that were selected by natural selection, and to what extent these genes explain why certain traits were expressed. For example, we possess multiple copies of a gene – the *SRGAP2 gene* – that is associated with the strong enlargement of the brain (van Straalen & Roelofs, 2017). By the way, I am writing this paragraph in 2021, the year that celebrated the 20th anniversary of the publication of the first version of the genome of *H. sapiens* (Fraser, 2021; Pennisi, 2021; Wible, 2021). In 2023, the analysis of the complete genome of 233 primate species was published (Kuderna et al., 2023). It's that fast.

And this is just the tip of the iceberg of traces about our evolution that can be explored. One gains insights about how our language originated and evolved from fossils, but also the anatomy of a fossil tongue bone from Israel teaches us something about speech in Neanderthals (Arensburg & Tillier, 1991). But also by studying language problems that occur with brain damage, this may teach us something about how language may have evolved evolutionarily (Code, 2021). Have you ever wondered why we manage to communicate fairly well by depicting things when speaking to someone who speaks a different language? Even that is a possible relic of early language, where our ancestors communicated through sign language (Zywiczynski et al., 2021).

# A battle of opinions

Finding the missing puzzle pieces is a challenge, as well as figuring out where to fit them in. This is the never-ending struggle of paleoanthropologists, both literally and figuratively. Often, they have to suffer during excavations under extreme temperatures in arid areas, exposed to sandstorms and mosquitoes spreading all kinds of diseases. Other struggles are like those described by Donald Johanson, the discoverer of Lucy's skeleton in Ethiopia in 1974, who experienced military confrontations and Kafkaesque administration issues upon his return to Ethiopia in 1990 (Johanson & Wong, 2010). Or take Raymond Arthur Dart, an Australian medical student who made one of the most important discoveries of the 20th century in South Africa in 1924, at least in terms of human evolution, being the skull of a child found in the Taung region. He is therefore known as 'the father of the child of Taung'. From the fragments of the skull, which incidentally also contained a petrified cast of the brain, Dart concludes the following (Dart, 1925):

> it exhibits an extinct race of apes intermediate between living anthropoids and man ... I propose tentatively then, that a new family of Homo-simiadae be created for the reception of the group of individuals which it represents, and that the first known species of the group be designated Australopithecus africanus.

He was so convinced that this fossil was a missing link between man and the great apes, naming it *Australopithecus* or 'monkey from the south'. However, not everything turned out as he had hoped for. His eagerness to make that missing link known to the world urged him into too rapid a publication of his findings, making him vilified from multiple sides, including being labeled an active servant of Satan. Dart also made his discovery shortly after the 1920 controversy surrounding Britain's Piltdown Man, a fossil skull and mandible discovered in 1908 but eventually exposed in 1950 as a forgery (Berger & Hilton-Barber, 2000; Gardiner, 2003). It wasn't until 1947 that the scientific importance of his finding was recognized by the international scientific community. Many of the critics of Dart's discovery also went along with the Piltdown story at the time. A battle of egos, it is of all worlds and of all times.

Scientific discourse driven by egos is to some degree still ongoing today. When Michel Brunet and collaborators published their discovery of *S. tchadensis* in 2002 in the renowned scientific journal *Nature*, it was very quickly met with criticism by the discoverers of *Orrorin tugenensis*, a hominin discovered a year earlier (Brunet, 2002; Wolpoff et al., 2002). Brigitte Senut and collaborators had just presented their missing link to the world, which was the then oldest ancestor of humans, dated 6 million years old (Senut et al., 2001). You can imagine their mixed feelings a year later, when 'their' *Orrorin* was deposed from his throne by *Sahelanthropus*, an even older and more primitive ancestor (Brunet et al., 2002). Very recently, Chinese researchers introduced the new species *Homo longi*, based on a skull found 90 years ago but only examined a couple of years ago (Ji et al., 2021; Ni et al., 2021). Other paleoanthropologists quickly expressed doubts about the new species and the proposal that it would be most closely related to us, *H. sapiens* (Gibbons, 2021a, 2021b). A similar discussion is pending on the in 2024 suggested new Chinese and large-brained species, *Homo juluensis* (Bae & Wu, 2024; Wu & Bae, 2024).

11

Does this now mean that all these discoveries of hominins have little scientific value? Of course not. It is precisely this kind of critical eye that forms the core to eventually come to a consensus among paleoanthropologists that can be considered reliable, at least reliable based on the information and technology then available. This is how science works and should work. This is what should always keep paleoanthropologists – and other scientists – alert and critical. This is also the story of the much-criticized hobbit, the tiny *H. floresiensis* from Southeast Asia. Fossil remains of a dwarf human on the island of Flores were attributed to a new human species in 2004 (Brown et al., 2004). Soon this was written off by numerous studies that believed that these were deformed modern humans, deformed by a variety of diseases and impaired growth. Nine years after the description and after many publications putting the existence of this species into doubt, there is no doubt that the hobbit is indeed an early human species that lived on that island about 100–60,000 years ago, and perhaps even much earlier (van Heteren, 2013). In 2019, even a second dwarf species was described from the Philippines, *Homo luzonensis* (Détroit et al., 2019). One cannot say that paleoanthropology, the study of human fossils and their underlying evolution, is an extinct discipline.

# 2

# What makes us unique?

There is no doubt that man occupies a unique place in time and space. No other crea-ture possesses such intelligence from which technologies have emerged that can send a spacecraft fully teleguided to Mars and drive a cart around controlled from Earth. No other being has made arrangements on a global scale to exchange services and products through something we call 'money' (Harari, 2018). No other organism has developed methods to fight deadly bacteria and viruses so that we all don't die from them or designed techniques that allow you to look inside a body without having to make a cut. At the same time, man is also the only creature that overloads his own (and others') ecosystem and threatens to bring it to an irrevocable tipping point. It is the only organism that can provide itself with virtually unlimited food while at the same time parts of its population have inadequate access to food. We are the only ones who have managed to create such a mismatch between evolutionary adaptations to our original environment and the environment in which we live now, resulting in all kinds of societal diseases (Lieberman, 2013). We are the only species of organisms that have evolved into such a complex social animal, while at the same time abusing it on a large scale (corruption, war, ...). But this book is about our bodies. So, is our body so unique and so different from that of other organisms?

## The urge to be special

Man has always wanted to distinguish himself from other animals. The Ancient Greeks, like Aristotle in his *Scala naturae*, placed humans at the top of the natural hierarchy (Praet, 2013). Although Aristotle supported this view on the levels of complexity of organisms, later this was more from an ideological standpoint. If man was created in the image of a God, then it could not be otherwise than that man had to be unique, different and better than other organisms. Charles Darwin and Alfred Russel Wallace were the first to put forward a natural mechanism that could replace a "directing and creating God": natural selection (Darwin, 1859). That this was not well received in the Victorian period was not surprising. Society was clearly not yet ready to come to terms with the idea that man had arisen in an ordinary, natural way like any other organism, and worse, from an ape. This, by the way, is reflected in the name given ever since to the group to which man belongs: the 'primates' derived from the Latin *primus* or 'first ranked'. The scientific name (Primates) was first introduced by Carolus Linnaeus in 1758, when he described humans as the species *Homo sapiens* and placed them within

DOI: 10.1201/9781003588795-3

the Primates (Linnaeus, 1758). It was then logical that that group of animals to which man belonged should be ranked highest.

After the publication of '*On the origin of species*' in 1859, in which he explained the general evolutionary mechanism, it was a logical consequence for Darwin and others to also delve into the natural process of our own evolution. As early as 1863, Thomas Huxley made the first attempt to organize the different groups of primates and thus define the position of humans among them (Huxley, 1863). He was convinced that within primates man was closest to the great apes, and that of the great apes the chimpanzee or gorilla was the most similar. Huxley spoke of '*man and man-like apes*' that belonged to a separate family. Darwin largely adopted that classification in his book '*The descent of man, and selection in relation to sex*', published in 1871. Here he compared all sorts of aspects of human biology with that of other organisms, especially other primates (Darwin, 1871). He compared development and anatomy, trying to determine the origin of all sorts of characteristics. In fact, he went much further. He was convinced that mental traits such as language, morality, consciousness and religion were also the result of natural evolution. This was a very progressive position at the time, even for Wallace it was not acceptable (Wallace, 1858). He believed that intelligence and the ability to speak a language were an expression of a God-provided gift (Browne, 2021). Darwin, strongly convinced of the opposite, even wrote to Wallace "*I hope you have not murdered too completely your own and my child*".

Now, Darwin actually had to make do with very little in his attempt to figure out our evolution. He could only work with information on extant primates because there were hardly any human fossils described in his time (Haile-Selassie, 2021). Only the Neanderthal was already known (although the first fossil had already been discovered in 1829 in Engis, Belgium, it only became clear that this was a new human species after the discovery of fossils in 1856 in Feldhofer, Germany) (Wragg Sykes, 2020). That the importance of this information had not yet fully penetrated Darwin's circles is shown by the fact that in 1864 he even got a skull of a Neanderthal man in his hands without realizing what it meant (he does not even mention anything about it in his book). The skull in question was the one found at the Forbes mine in Gibraltar in 1848. Now, Darwin was severely ill at the time and the fossil was still full of sediment, which might explain why he didn't pay much attention to it. Darwin could thus not yet have known that something like *Australopithecus* or very early *Homo* species would have existed.

Today, fortunately, we have access to a great deal more information. We have numerous fossils of multiple hominins, there are new insights into anatomical and physiological features of modern humans and other extant organisms, as well as DNA from extinct human species. For example, as of 2023, we know that 91% of the human genome is shared with that of at least five other primates and that 11% even matches that of 230 species (Christmas et al., 2023). That considered, it is quite impressive that Darwin managed to make predictions that have since been confirmed. Darwin relied on corresponding patterns in the embryonic development between species and the presence of what he called vestigial organs. To give an example, Darwin explained the presence of dark fetal hair (the *lanugo* hair) in infants as an evolutionary vestige of what forms the fur in chimpanzees. In humans, this hair disappears from the sixth month after birth and is replaced by fine hair. He also saw the small protuberance in the auricle present in some humans, called the 'Darwin tubercle', as a rudiment of the ear tip in other primates. In 5–20% of the human population, another vestigial muscle occurs in the forearm (the *palmaris longus* muscle), which in other primates is important for loco-motion on four legs (Lemelin & Diogo, 2016; Thompson et al., 2001). Some humans can move their auricles, like most mammals, thanks to other vestigial muscles. And so, Darwin drew a great deal of inspiration to correctly conclude that humans possess several unique characteristics associated with walking on two legs, the use of tools, properties of the brain, or the size of canines. Darwin also correctly predicted that the origin of man was to be sought in Africa, and not in Europe (as people have falsely attempted to posit with the Piltdown fossil) or Asia (although the discovery of the Java man in 1891, i.e. shortly after Darwin's book, then casted some doubt on that). Darwin was also visionary in his assertion that it would be wrong to think that the common ancestor of man and chimpanzee, our closest living relative, should have looked like a chimpanzee or some other now living primate. This is a notion that only recently has been increasingly substantiated by fossil and other evidence (Almécija et al., 2021).

Now, it is not surprising that he also occasionally missed the ball completely, such as in terms of the order in which certain evolutionary changes took place, the role of sexual selection in the emergence of certain traits (such as naked skin, height differ-ences between men and women) or the explanation of variation in modern humans. The latter is even the subject of some controversy, to the extent that it had a racist bias (a view that seemed to be standard during the Victorian period) (Killgrove, 2021). Current knowledge of variation among modern human populations (and we already have heaps of genetic information about that) allows one to conclude without any doubt that *"humans are not divided biologically into distinct continental types or racial genetic clusters"* and *"genetics demonstrates that humans cannot be divided into biologi-cally distinct categories"* (official positions of the American Association of Biological Anthropology and the American Society of Human Genetics) (Fuentes, 2021).

## Our body, a sapiens body?

Modern man, *H. sapiens*, looks different from all other animals. The great natural historians of the 19th century already clearly saw that we share great similarities with other primates, such as chimpanzees. The fact that people like Charles Darwin, Ernst Haeckel and Richard Owen were looking for similarities between humans and the great apes at that time, to trace our evolutionary origins, is also not coincidental. Characteristics are simply inherited and maintained even after thousands of genera-tions. Many times it involves characteristics that at some point were advantageous as they helped to survive in an environment, i.e. adaptations. But not everything that was preserved was necessarily an adaptation. As long as it has no detriment to survival or the chance of having offspring, then there is no reason why natural selection should filter it out. Also, something that is an adaptation at time A does not necessarily remain an adaptation at time B. If I have thick fur when living in a cold climate, that is an advantage, an adaptation. But if the temperatures increase substantially, then that fur is no longer so interesting. In fact, it then even becomes a disadvantage as I would be constantly too hot and unable to function well in my new environment.

This is also how we should see our bodies: the result of continuous evolutionary tweaking of an ancestral body. Each time the environment changed, new adaptations arose and old ones were retained (if not detrimental) or simply filtered out. Let's compare it to the history of car models – or to call it in biological terms 'car-bodies'. When one of the first car models was built in 1834, it was a very simple design, powered by a steam engine (Wikipedia-bijdragers, 2022). This relied on the resources available and the needs that had to be met at that time. But, the environment changed, including the discovery of gasoline and the advantages of using it as fuel for an internal combustion engine. But also the requirements imposed by the 'environment', *i.e.* society, were constantly changing (faster, more powerful, more comfortable, safer cars were needed). Adaptations in the automobile body quickly followed themselves to meet new needs. Generation after generation, the car-body changed with performance getting better and better in this new environment. Good features were retained (cars now still run on four wheels and have a steering wheel), bad features were discarded (we no longer use steam) and constant improvements made their appearance change (gasoline engines, better materials for safer bodywork, electric engines, etc.). So, if one wants to understand the logic behind every part in the 'body' of a contemporary car, and why parts are integrated into a car in a certain way, one must go all the way back into automotive history.

The same is true of our bodies, which also did not come into being only when *H. sapiens* appeared on the scene, about 315,000 years ago. In fact, many features of our bodies go back in time much further, long before the emergence of the first hominins (they appeared about 7–8 million years ago) and even long before the origin of the first primates (emerged about 65 million years ago). In fact, some of our characteristics date back to the origin of life about 4 billion years ago. Our DNA is based on the same building blocks that created DNA back then. The fact that our cells have a real nucleus (containing nuclear DNA), that they carry tiny energy machines inside them (in the form of mitochondria with their own mitochondrial DNA), and that they can be equipped with a flagellum for swimming (as in sperm cells) is due to the emergence of nucleated unicellular organisms, called eukaryotes, nearly 2 billion years ago (Zimmer & Emlen, 2013). That our bodies are a cohabitation of numerous, different-looking cells, we owe to the emergence of multicellular organisms. Fossils show that the first real animals already existed 650 million years ago, in the form of sponges. That in our body different types of cells arrange themselves into tissues, forming organs that perform specific functions – much better than having to do so as individual cells – is a process that came into existence 600 million years ago. Our typical 'animal characteristics', such as having to ingest food to obtain energy or engage in sexual reproduction via gametes (sperm and egg cells) originated then. But of course, that does not yet make a human being.

However, one might also ask whether a 'human body', being the body of an individual of the species *H. sapiens*, exists. I suppose you will be raising your eyebrows now, because every one of us has a body and belongs to *H. sapiens*, right? So, what is the problem? Let me rephrase the question: "Does the body of the species *H. sapiens* consist only of cells that genetically belong to that species?" The answer is no. The reason is that genetic information is often exchanged between species and thus pieces or complete DNA of one species end up in the genome of another species. Take for example those energy machines, the mitochondria. These are present in almost every cell in our body but have their own DNA: the mitochondrial DNA or mtDNA for short. Each of our cells thus both have DNA in the nucleus (the nuclear DNA or nDNA) and mtDNA. Well, this mtDNA is actually the DNA of bacteria, because mitochondria are bacteria that were once incorporated into our single-cell ancestors. Their ability to use oxygen to make energy, which a cell can use, turned out to be so interesting that this coexistence is still the rule after millions of years of evolution.

And there are even other bacteria that are part of our body. It is not even off the mark to say that 'our body' is more of an ecosystem of numerous species than the body of one individual of one species. This ecosystem contains about 40 trillion of *H. sapiens* cells (this is an estimate for a human being of about 70 kg) (Bianconi et al., 2013), but on it and in it thrive another 100 trillion microorganisms (a one with 14 zeros behind it) forming our intestinal flora, 1 trillion microorganisms that colonize our skin, and a little less than 1 trillion microorganisms that survive in the vagina, the urethra, on our hair and in our nostrils (Heiss & Olofsson, 2019). So, our bodies carry more cells than we have of our own! And we are not yet talking about pathogens here, but about the entirety of other organisms that live constantly in and on our bodies, and which we therefore carry around daily. In fact, it is becoming increasingly clear that this goes beyond mere carry-on. Substances that some of those bacteria produce affect the course of our development (including the development of our nervous and immune systems) and influence how fully developed organs function, such as our intestinal tract and brain (Cowan et al., 2020). A mother even exchanges bacteria with a baby the moment it is born, which can affect the child's subsequent development (whether a child is born through the birth canal or by cesarean section can thus already lead to a different composition of microorganisms in the child) (McDonald & McCoy, 2019). Microorganisms are also passed on through breast milk which will help determine the baby's gut flora. Bacteria that live in our guts are not only essential for proper digestion, but our diet also determines what types of bacteria are present. For example, it has been found that the composition of the gut flora changes drastically after only four days when we switch from a plant-based diet to a diet of animal products, or vice versa (David et al., 2014).

Different bacteria means different substances being produced that are released into the intestinal tract and affect our cells. This can even influence the formation of brain cells, which in turn adjust behavior (such as feeling hungry) (Adamantidis, 2022). Nowadays one even speaks of a brain-gut connection, where gut function has a direct effect on brain function. Certain mental disorders (such as depression, schizophrenia, Parkinson's or Alzheimer's) have been associated with a disrupted functioning of our gut bacteria (Sanada et al., 2020).

There are also viruses that travel with our bodies, through our DNA, because of the process that viruses go through to multiply. Viruses are not living organisms. Therefore, they need host cells to multiply. A lot of viruses inject their DNA (or RNA) into a cell, and then viral DNA gets built into the DNA of that cell. If that cell does its daily job of doubling its DNA and making proteins, then that cell will make not only its own DNA and proteins but also those of the virus. Hence, new viral particles are born. Once the viral DNA is built in, it can be expressed again at later times, without having to be reinfected with the virus. Take the Herpes virus (Desfarges & Ciuffi, 2012). There are many variants of this virus, but a lot of people deal with the *Herpes simplex* virus. Do you sometimes suffer from fever blisters or cold sores? Then you have Herpes virus DNA in your DNA. No less than 67% of the world's population carries Herpes DNA in them (Wikipedia contributors, 2022). Someone who at some point becomes infected may be symptom-free for a long time, until certain circumstances trigger a new activation (such as a weakened immune system or too much sunlight). Since these people carry this in their cells, there is no way to get completely cured of it either: one drags the DNA with them for life. Of some viruses it is known that their DNA has been carried for millions of years, as it was passed on generation after generation. In some cases, as will be shown later, this can even give rise to an advantage, an adaptation.

# Sapiens is an animal

We, *H. sapiens*, are animals because we have all the typical characteristics of what defines an animal. For example, an animal is composed of multiple cells, all cells having a nucleus containing DNA (eukaryotic cells), an animal is heterotrophic (it must ingest organic food to get its energy), is mobile, is equipped with a nervous system and reproduces sexually. Whether it is a simple sponge, an insect, an earthworm, a snail or a mammal: all possess these properties. For the analogous reason that you call all vehicles with four wheels, a motor and a steering wheel a 'car', we call all organisms that possess these properties an 'animal'. So we are, too.

Animals inherit most characteristics from their ancestors. As a result, you can reconstruct their lineage based on similarities between species. And this can go way back. Apply this to us, and you have the following scenario. We no longer have a large tail, just as in chimpanzees and gorillas, for example. The explanation for this is that we share a common ancestor with those great apes, which also lacked a tail. Both chimpanzees and we inherited that characteristic from that ancestor. That ancestor was itself a great ape. So, you could say that we humans are also apes, because we are evolutionarily 'born' from a great ape. Even though we differ from other great apes in many characteristics, a great ape remains at our origin. Let us go a little further. We have nostrils that face downward, just like the monkeys found in Africa and Asia (for example, a baboon). This is because we share an even earlier ancestor with the monkeys of this region, which also had downward-facing nostrils. From those ancestors then evolved the great apes, which also have those nostrils. And humans evolved from that and still have those nostrils. Hence, we are not only humans and great apes but also monkeys. And so you can go on: we are also primates, we are mammals, we are amniotes (animals with membranes around their embryos), we are tetrapods,

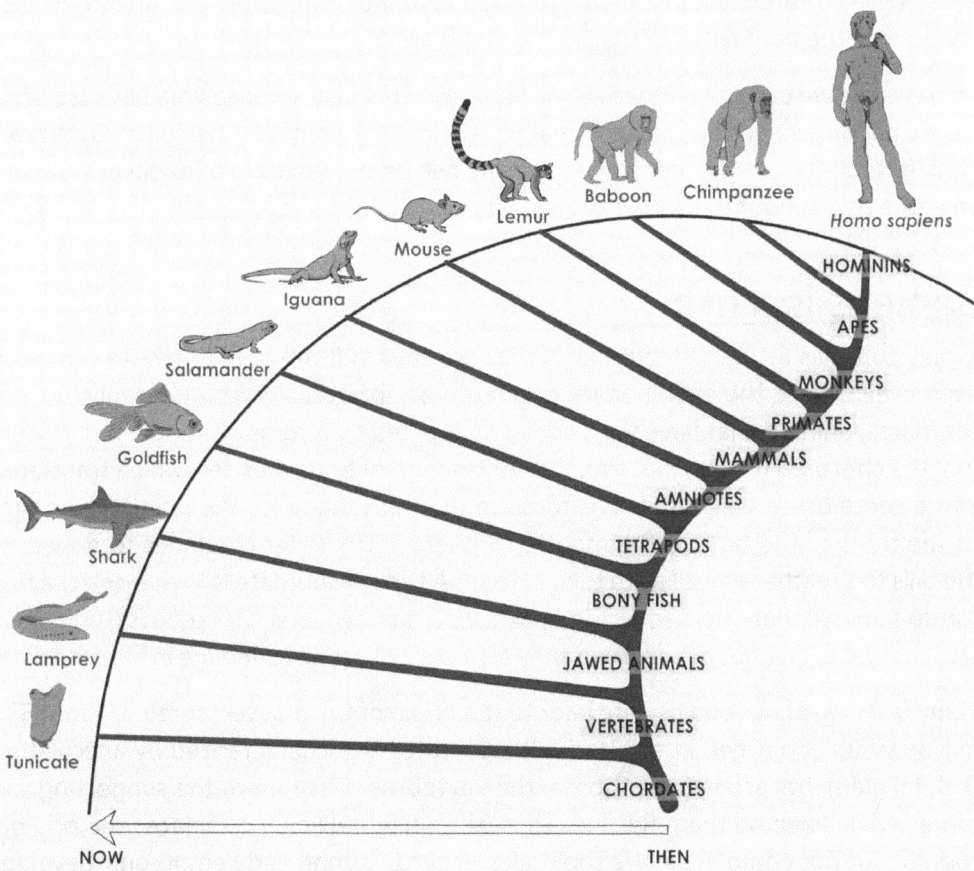

Figure 2.1   Simplified view of an evolutionary tree, showing how *Homo sapiens* belongs to the hominins, but equally to the apes, to the monkeys, to the primates, to the mammals and so on. The arrow points to the timeline, to indicate that all species alive today have had as long an evolution history since the first ancestor, as humans do.

we are bony fish, we are jawed animals, we are vertebrates, we are chordates and we are animals. Put the other way around, we arose from a particular group of animals that gave rise to the chordate animals, within which then arose vertebrates, from which then a group of bony fishes gave rise to land animals on four legs, some of which further evolved into amniotes, within which then arose mammals, including the primates. This method of grouping is done according to the evolutionary logic of descent: if an ancestor belongs to a group X, then all descendants of that ancestor also belong to that group X. Thus, you get, as it were, a branching tree, with the trunk representing group X. Each branch then gets its own name but still belongs to trunk X. So, you have a branch called 'primates', which is itself branched. One of those branches is that of the 'downward-nosed monkeys', which itself has side branches and one of which is the 'great apes'. That branch is then branched further, including one branch to the orangutan, one to the gorilla, one to the chimpanzee and one to the first ancestor of the hominins.

So basically, we can say that we are all fish, yay! This may surprise you, because after all, we no longer seem to possess characteristics that point to a fish past. Or do we? Well, it is fair to say that we are not just any fish on dry land. Neil Shubin even wrote an entire book about it: 'Our Inner Fish' (Shubin, 2008).

# Sapiens is a fish

It may come as a surprise, but our bodies are thus still full of features we inherited from even before the evolutionary origin of vertebrates. Vertebrates evolved from chordates, animals that have their bodies supported by a somewhat rigid but flexible rod, the **chorda**. Through that rod, the contraction of body muscles can be translated into a coordinated wave that runs through the body, allowing the animals to swim. At the front of the body is the mouth, through which water continues to flow over the gills to breathe, as well as to take in food. A typical chordate is the lancelet, a tiny fish-like animal that filters small food particles from the water. Sea squirts (tunicates), also chordates, are not fish-like, yet they also use a pharynx to engage in filter feeding.

Many features of our bodies date back to the origin of the first vertebrates, about 530 million years ago (Chen et al., 1999). 'Vertebrates' are characterized by articulating skeletal elements around the chorda: the **vertebrae**. They make the supporting rod firmer, while keeping them flexible. Thus, the vertebral column replaces the original chorda, but not completely. We too make a chorda during early embryonic development, which then largely is broken down. Only remnants remain at the level of the intervertebral discs. This explains how our vertebrae end up being separated from each other by a disc of soft tissue, which ensures smooth movement between the vertebrae. Such an intervertebral disc consists of two parts: a central core that is soft (the *nucleus pulposus*) and a firmer ring that holds this core in place (the *annulus fibrosus*). Well, that soft core is a leftover chorda we inherited from our fish ancestors. When we walk upright, the vertebrae press against that central core, which thus absorbs the shocks, making the presence of this chorda remnant an advantage. But this can also cause problems. Indeed, if the pressure becomes too large or an improper load is applied, the core may be pushed through that ring, resulting in us having a hernia.

In chordates, the spinal cord lies on the dorsal side, while in other animals, such as earthworms, it lies on the ventral side. Moreover, in the first vertebrates, the spinal cord thickened at the front into three vesicles: the first **brain** was born (Chen et al., 1999). It is therefore no coincidence that during early embryonic development, our brain is still laid out as three vesicles. Eventually they grow into five parts, which is the case in all living vertebrates today. Sometimes you read that the first brains in vertebrates contained only one part, to which a piece was added each time during evolution: a reptile brain was added to the fish brain, to which a mammalian brain was then later added. This theory is still mostly used in psychology, although it has since been shown that this makes no evolutionary sense (Herculano-Houzel, 2016). The idea had grown historically with Ludwig Edinger (late 19th century), and in 1964 formed the basis for Paul MacLean's theory of the so-called triune brain. But as Suzana Herculano-Houzel, an evolutionary brain specialist from Brazil states, *"the triune brain is only a fantasy"*. Not only can all five parts of the human brain in one form or another be distinguished in early fish species but also the main processing center of all stimuli is different in reptiles and mammals (in other words, they are not simply added to each other). Even considering how neurons differ in different areas, brains clearly did not evolve that way (Hain et al., 2022).

Early vertebrates had a hard skeleton made of bone. This offered many advantages and explains why a wide variety of vertebrate species evolved from about 450 to 350 million years ago. Some were even completely covered in a bony armor (Janvier, 2015). But they all had in common that they did not yet have jaws and thus depended on filter feeding. Using a **gill basket,** they sucked up small particles of food. The pharynx is the part that follows after the oral cavity and forms the transition to the esophagus. It is the part of the intestinal tract that is associated with the gills and forms the gill slits, along which water passes the gills, necessary for breathing. The gills are suspended from skeletal elements called the gill arches. Early vertebrates had seven pairs of gill arches, between which there was a gill slit, and thus at least six gill slits in total. Even during our development, four pairs of gill slits still form, something Charles Darwin and Ernst Haeckel had already noticed (Roberts, 2021). But three of these gill slits disappear. One remains and even remains functional! It connects the side of the head to the inside of our pharynx. So, do we still have a gill slit, you may ask? Well yes, our ear canal is derived from the anterior of the six ancestral gill slits. This connects the external auditory canal that runs up to the middle ear, with the Eustachian tube that then connects to our pharynx. So, we still have a pharynx, as well as a gill slit, and even remnants of gill arches! In fact, the cartilages that form our larynx are evolutionarily derived from some of those gill arches, something that came about during the formation of lungs. We still possess genetic programs that allow multiple gill slits to form. If something goes wrong with that, which effectively happens in some people, then remnants of additional gill slits are formed. The re-expression of ancestral genes is an example of what is called an atavism (Stiassny, 2003). The ability to also form gills, however, we have completely lost.

About 450 million years ago, the first vertebrates equipped with **jaws** appeared: a lower and an upper jaw. The emergence of these jawed animals was a very important step in the evolution of vertebrates, and thus our own evolution. Jaws came about because the anterior gill arch became larger and firmer, with more powerful muscles pulling onto it. On top of that, these jaws now started to bear teeth, which allowed them to grab larger food chunks and prey and cut them into pieces. Jawed animals were no longer depending on filter feeding! If you look at our teeth, they are composed of two hard materials: an inner layer of dentin and an outer layer of enamel. Well, teeth with these materials have already evolved from the first jawed animals. In fact, teeth are considered as modified scales already present in jawless vertebrates. Scales located around the oral cavity in these jawless animals came to be placed on the jaws in jawed animals and thus acquired a grabbing and cutting function. So our teeth are derived from scales of early fishes (Rucklin et al., 2021; Smith & Coates, 1998). The jaw bones present in these fish that have bones (not coincidentally called 'bony fish') formed the basis for jaws that would be preserved throughout evolution. As in bony fishes, our upper jaw contains two tooth-bearing bones: the premaxilla which contains the incisors, and the maxilla which contains the other teeth. However, there is a big difference: our upper jaw bones are fixed to the skull, whereas in most fish they can move relative to the skull. This allows them to suck up prey very efficiently, by sliding their jaws forward and forming a funnel. By thus sucking through a straw, so to speak, they can suck up water so quickly that if there is a prey close to the mouth opening, it is sucked up along with the water. The fact that we can no longer move our upper jaw is because it has become attached to the skull ever since fish began to colonize the land, giving rise to the of tetrapods.

But as already mentioned, not all the features that are retained during evolution are adaptations. Some just stayed there because they were not too detrimental. And sometimes our fish past plays out, for example, when we have the **hiccups**. Our hiccups are an inherited conflict that arose in fish that were the ancestors of tetrapods (Shubin, 2008). These fish engaged in both aquatic breathing, in which the pharynx pumps water over the gills in a rhythmic manner, and air breathing, in which an air bubble is occasionally pushed into the lungs. In aquatic breathing, all the muscles of the pharynx are activated in a cyclic pattern. Each time water is pushed through the gills, the opening to the esophagus is squeezed shut so that no water enters the stomach. The stimulation of the muscles is done by two nerves that start from the brainstem: the *vagus nerve* and the *phrenic nerve*. In fish, the brainstem is close to the gills, so the nerves do not have to run far. In mammals, and therefore humans, it is different. We only breathe air. The muscles responsible for that are located posteriorly in the rib cage (like the muscle in the diaphragm, at the bottom of the chest), and thus far from the brainstem. But we come from an ancestor that combined both, something frog tadpoles still do. They alternate rhythmic breathing through the gills with alternate breathing to get air into the lungs. When they send water through the pharynx, they avoid getting water into the lungs by closing off the access to the lungs through a valve: the *glottis*. Therefore, that valve must be closed very quickly as soon as they send water through the pharynx. This is the natural 'hiccup' of these amphibians, which protects their lungs from filling up with water. It is this same movement that also occurs in humans, even in fetuses still in the womb. When we have hiccups, we suck in an amount of air very quickly. About 35,000ths of a second after that, our glottis slams shut, producing the typical hiccup sound. We no longer have natural hiccups because we no longer do water breathing. We get hiccups when those nerves get disrupted, resulting in muscle spasms. This is no big deal and will pass, unless your name is Charles Osborne who had the hiccups for 68 years straight. Now, according to some, it is no coincidence that we have retained that fish and amphibian trait, nor is it a coincidence that we have more hiccups as babies than afterward. Possibly this would have been advantageous to be able to suck up milk without it entering the trachea and lungs (Straus et al., 2003).

And so there are hundreds of more examples of features we owe to our fish ancestors. We hear and have a functioning organ of balance because we have cells in our inner ear that do the same thing as they do in fish: detect displacement of water. Fish use these **hair cells** to detect changes in the environment, such as when a predator approaches. We use this to detect the rotation of our heads. Also, during embryonic development, we establish a **heart** very similar to that of fish: a single tube through which only oxygen-poor blood passes. In humans, and other tetrapods with lungs, this heart undergoes a loop formation (Standring, 2020). Because the heart has to send the blood in two circuits, one to the organs and one to the lungs, the looping gives the heart a double construction: we have a left and a right side where oxygenated blood passes through the left side and de-oxygenated blood passes through the right side. And so, I could go on and on. But let's take a leap, to characteristics we owe to the tetrapods.

# Sapiens is a tetrapod

Tetrapods are not called 'tetrapods' for nothing: they are characterized by four **limbs** with hands and feet, and thus no longer have fins like their ancestors. The first true tetrapods had legs, yet they still lived in the water. Fossils like *Acanthostega* and *Ichthyostega* (lived 365 million years ago), along with fossils of their ancestors, give a very good picture of the evolutionary transformation from fins to legs. Fish ancestors had fins with fin rays, supported by skeletal elements largely the same as those that form our limbs. A consistent pattern in the limb skeleton is that a single base element is followed by a double element. In our arm, this is respectively the humerus, followed by the radius and ulna (radius and ulna lie side by side, in line with the humerus). In our leg, this is the femur followed by the tibia and fibula. These are followed by the wrist bones (*carpalia*) in the arm and the ankle bones (*tarsalia*) in the leg, against which then lie the palm bones (*metacarpalia*) and midfoot bones (*metatarsalia*), respectively. At the very end then lie the phalanges. The number of wrist and ankle bones and phalanges may differ among different groups of tetrapods, but the majority have five fingers and five toes in common. The first tetrapods, however, had more, probably as many as eight. Very soon, five proved to yield a well-functioning hand and foot and thus the fact that we have five fingers and toes is something that came about some 350 million years ago, in early tetrapods. A flexible elbow and knee even originated earlier, before true tetrapods emerged. In the fossil *Tiktaalik* (a lobe-finned fish), the first signs of an elbow and wrist joint were observed, probably to raise the body somewhat and to propel itself forward using the fins while in shallow water (Daeschler et al., 2006; Shubin, 2008). The ability to lean on fins while underwater may even be much older, long before the first legs appeared, as also lungfish have wire-like fins they use to push themselves up and make stepping movements across the bottom while submerged (King et al., 2014). Direct ancestors of tetrapods also had fin rays that were modified allowing a better support of the body (Stewart et al., 2019). The use of fins in those ancestors to lean on may have even accelerated the evolution to legs (Standen et al., 2014). Now, the *Tiktaalik*

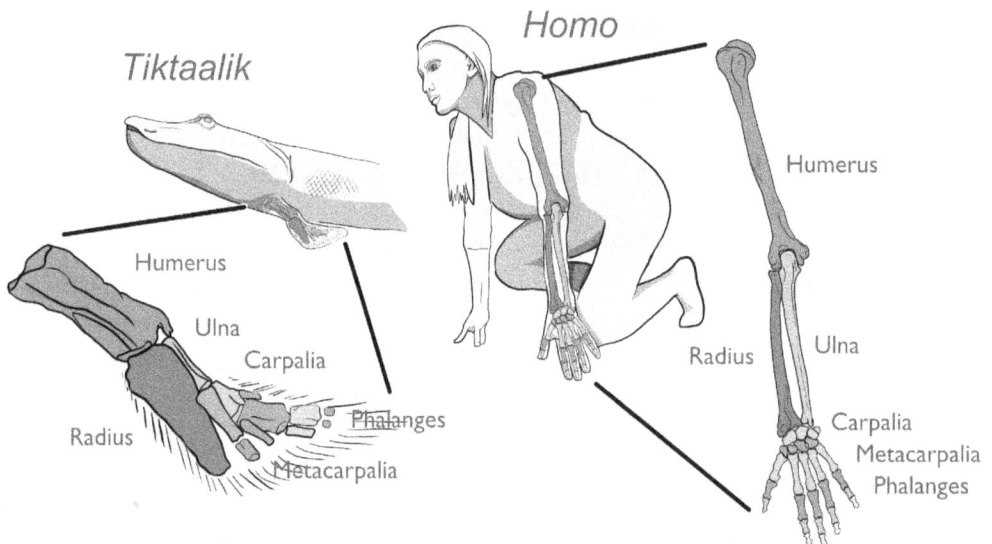

Figure 2.2    Evolution from a fin in fish to limbs in humans. On the left *Tiktaalik*, an extinct ancestor of the tetrapods, with pectoral fins bearing fin rays used to support itself on the bottom underwater. On the right *Homo sapiens*, showing the arm consisting of equal (homologous) skeletal parts as in *Tiktaalik*. (Figure pectoral fin *Tiktaalik* adapted from Shubin et al. (2006).)

elbow joint was not yet as refined as ours. That only really came about from the amniote animals, that is, from the reptiles, about 250 million years ago (Shubin, 2008).

Unlike fish, the skull in tetrapods is solid with most of the bones firmly connected. The upper jaw is thus no longer a mobile bone. What is also new now is the presence of **internal nasal openings** (the *choanae*). These allow that the air that enters the olfactory cavity through the external nasal openings (the *nares*) can be passed into the oral cavity. This is not yet the case in most fish (lungfish are an exception to this). It brings the advantage that breathing remains possible while the jaws are closed. Fish, on the other hand, cannot breathe through their nostrils because they are connected to a blind terminal olfactory sac. So, the fact that we can breathe with a closed mouth we owe to these four-footed ancestors.

For breathing, **lungs** now became the main respiratory organs. Tetrapods were able to colonize land, thanks to their ability to breathe air, rather than relying on water to breathe through gills. Indeed, gills do not work in air: they would collapse, making the surface area along which gases could be exchanged too small. All tetrapods practice air breathing through the lungs, with one exception. A group of very small salamanders (the family of Plethodontidae) are small enough to breathe only through the skin. But our lungs, combined with a way of breathing in which air is drawn into the lungs, thus have their origins from the tetrapods.

Living on land also brings a lot of physical changes, such as the way how **sound waves** propagate. In fact, sound waves travel much faster in water than in air: about 5300 km/h in water, but only about 1200 km/h in air. For fish under water, this is convenient, because the inner ear – where sound vibrations are perceived – is also filled with a liquid. In animals living on land, however, vibrations through the air thus enter at a speed about 4.5 times slower. This brings an impedance difference, because the resistance of the sound waves in air is different than that in the fluids in the inner ear. In tetrapods, this is overcome as vibrations entering through air are amplified in two ways. On the one hand, a thin membrane is formed that easily vibrates with the sound waves, and thanks to a relatively large surface area, amplifies the vibrations. On the other hand, this eardrum is connected to a bone through which these vibrations are transmitted directly to the inner ear. As this bone lies in what is called the middle ear, it is called the middle ear bone. So we owe the fact that we have an eardrum attached to a middle ear bone to this adaptation that originated with the first tetrapods. Now, we are mammals and therefore we have three middle ear bones, as opposed to only one in all other tetrapods (as well as our eardrum has a slightly different origin than that in other tetrapods) (Kitazawa et al., 2015). But one of the three bones comes from that one in the tetrapods, which is the *stapes*. The other two middle ear bones were added much later, as we will see, when mammals came into existence.

In bony fishes, the shoulder girdle attaches to the skull. Hence, they don't have a true neck. We, of course, do have a **neck,** as do all tetrapods. For the first tetrapods to grab prey on land, a neck was useful: it allowed them to move their heads independently of the movement of the legs. Running after prey with a head that was constantly swinging from side to side along with the movement of the front legs was not very convenient. The shoulder girdle had already become detached from the skull before the emergence of the first true tetrapods, as *Tiktaalik* already had a neck (Shubin et al., 2006). Later, the first two vertebrae of the neck also changed shape in tetrapods, making the head even more mobile.

Fish are suspended in the water column, with their body fully supported by the water mass pressing against it. From the moment legs arose and the body was no longer in the water, the pressure became more and more concentrated at the level of the legs and the place where they are connected to the spine via the **girdles**. In fish, the shoulder girdle thus lies suspended from the skull, whereas the pelvic girdle lies completely detached from the spine, embedded in the abdominal muscle mass. In our case, however, the pelvic girdle is firmly connected to the spine, at the level of the so-called sacral vertebrae. Well, that feature, too, is something we inherited from early tetrapods. In early tetrapods, such as *Acanthostega*, the pelvic girdle was already greatly enlarged and directly connected to the spine. So, a pelvic girdle composed of three paired bones (ilium, ischium and pubis) and connected to the sacral vertebrae is certainly nothing new in humans; it was already there 360 million years ago.

## Sapiens is an amniote

Although tetrapods already had some success living on land, the greatest adaptations to land life did not occur until the so-called Amniota. The first amniotes looked very much like what might be considered a lizard-like reptile. From this then evolved all reptiles, but also birds and mammals. Present-day amphibians (such as frogs and salamanders) are tetrapods, but they are not amniotes. They are still very dependent on water: they dry out quickly and they need to be able to lay their eggs in water. As with us, the skin in amphibians consists of three layers (a feature that originated from the first vertebrates): an outer epidermis, a middle dermis and a deeper subcutaneous layer called hypodermis. In amphibians, the **epidermis**, which is supposed to prevent dehydration, is only a few cell layers thick (depending on the species, 2–12 layers thick) (Akat et al., 2022). At the very surface there is a very thin horny layer (*stratum corneum*), a layer in which keratin forms. Keratin is the substance that makes up our hair. The fact that this layer is thin brings them advantages and disadvantages. The advantage is that they can still breathe through the skin (they absorb oxygen from the surrounding water through the skin and release $CO_2$ into the water). This is crucial for amphibians that need to survive in a pond that freezes completely shut at the surface during winter. The downside is that they also lose a lot of moisture through that skin if they move on to dry land for too long. So, they need to stay near the water. They partially compensate for this, thanks to mucous glands in the skin, which deposit a layer of mucus on the skin to keep themselves moist. In Amniota, that problem got solved, thanks to a thicker epidermis, a thick horny layer, a specialized type of keratin and efficient enzymes to form that keratin (Alibardi, 2022). In many amniotes the keratin layers are packed together such that this forms very firm structures. To keep the skin mobile, those sturdy structures are deposited in the form of scales connected by a stretchy skin. Some mammals also have scales, such as the armadillo, pangolin or rat (they have scales on their tails). Primates do not have scales, so neither do we. But we do have a thick epidermis, covered on the surface with several layers of dead, flattened cells, packed with keratin. These are constantly being renewed, as those dead cells wear off and flake off. So the fact that we shed flakes of skin that keep us from drying out too quickly is something we owe to our amniote ancestors.

When we reproduce, we don't have to go into the water to deposit eggs. Amphibians have to because their **eggs** are surrounded by a jelly capsule (you may have held frogspawn in your hands as a child), which dries out very quickly in the air. Amniotes don't have that problem: they have, as it were, brought the water to themselves, at least to their eggs, thanks to an embryonic novelty. During early embryonic development, three germ layers (made of embryonic cells) form in all vertebrates, including humans: an outer ectoderm, a middle mesoderm and an inner endoderm. Each forms specific structures that eventually establish the body's organs. But amniotes do something extra with that: as the embryo forms, those germ layers also start to grow out, away from and around the embryo. Eventually, two membranes will surround a space around the embryo: the amniotic cavity surrounded by an outer chorionic membrane and an inner amniotic membrane. Together they form what in humans is called the amnion, surrounding the amniotic cavity filled with amniotic fluid. By now it will also become clear where the name 'Amniota' comes from: they are the animals with an amniotic cavity filled with fluid, in which an embryo develops in its own aquatic environment. Amniotes could thus deposit their eggs anywhere on land, because they were all provided with their own water pool in which the embryo could develop. To prevent this water from quickly evaporating, most amniotes formed a solid eggshell that covered and shielded everything. In mammals, that eggshell became redundant at some point, as the eggs were no longer deposited outside of the mother but developed inside a mother's womb. Yet, the amniotic membranes did remain. The water breaking in pregnant humans thus implies the tearing of the amniotic membranes, releasing the amniotic fluid, something we inherited from amniotes.

## Sapiens is a mammal

One evolutionary lineage within the Amniota, called the Synapsida, is very relevant to our story. Indeed, it is the group of amniotes that gave rise to mammals, to which we also belong. However, a lot of characteristics that we share with mammals arose much earlier in these synapsid ancestors. Take the characteristic to which the group name refers: the synapsid skull. This is a skull in which at the bottom and behind the eye (at the level of the temporal region) an opening exists between skull bones: the lower temporal window. In ancestors that did not have this window yet, jaw muscles lied tucked away between an outer layer of skull bones (covered with skin) and the inner skull bones (surrounding the brain). But this limits the space in which the muscles can grow and expand when they contract. Skull windows appeared in most amniotes, giving the jaw muscles more space. This was useful for adapting the jaw system to process different types of food. Some amniotes, like most reptiles and birds, have two such windows (an upper and a lower one). In those reptiles, there are two separate windows, whereas in birds, they merge into one large window. In the Synapsida, there is only the lower window, a condition that remains in mammals, including humans. So where is that window in my skull, you might ask? Well, the lower edge is bordered by the zygomatic arch. If you press your index finger against your skull, just above this arch, and then firmly close your jaws, you will feel something moving. That is one of your jaw muscles, namely the temporal muscle. Since you can feel that muscle, this means that it is not covered by bones and is therefore exposed under the skin. So that's where that lower temporal window is, something that originated about 320 million years ago.

Where humans, just like other mammals, differ from other vertebrates is in the way the jaw apparatus is constructed. All the bones that form our jaws have already been inherited from bony fish, such as the two bones that carry the teeth in our upper jaw (i.e. premaxilla and maxilla). If we look at the lower jaw, the functional lower jaw in bony fishes, as well as in tetrapods and other amniotes, consists of multiple bones. In mammals, there is only one bone, the dentary, which, as in those others, carries the teeth of the mandible (see bone 3 in Figure 2.3). During the evolution of those Synapsida, the other bones gradually detached from the mandible and became incorporated into the skull. The loosening of those bones had the advantage that the jaw in early mammals had more freedom to chew, allowing for better chewing and therefore more efficient digestion of food (Schultz, 2020). A finer grinding of food was also possible, thanks to new tooth shapes. Starting from an ancestral dentition with all similar teeth, evolution to a typical mammalian dentition started with the elongation of an anterior tooth a little less than 300 million years ago, resulting in the origin of the canine. This implied a first functional division of labor in the teeth: a canine is primarily for grabbing and killing prey, the teeth in front of it for cutting off pieces of food, and those behind it for further grinding. Cutting off and grinding is improved by a change in tooth shape. The teeth in front became wider and spatula-shaped, creating the incisors. The teeth behind the canine got a wider base, on which eventually several tubercles appear, forming the cheek teeth. Initially three tubercles per molar were arranged in one row, as found in the extinct Triconodonta (literally, 'three tubercles on the teeth') about 200 million years ago (Davis, 2011). When chewing, the tubercles of the cheek teeth in the upper jaw rub against those of the lower jaw, grinding food more efficiently. The posterior cheek teeth, where the biting forces are greatest,

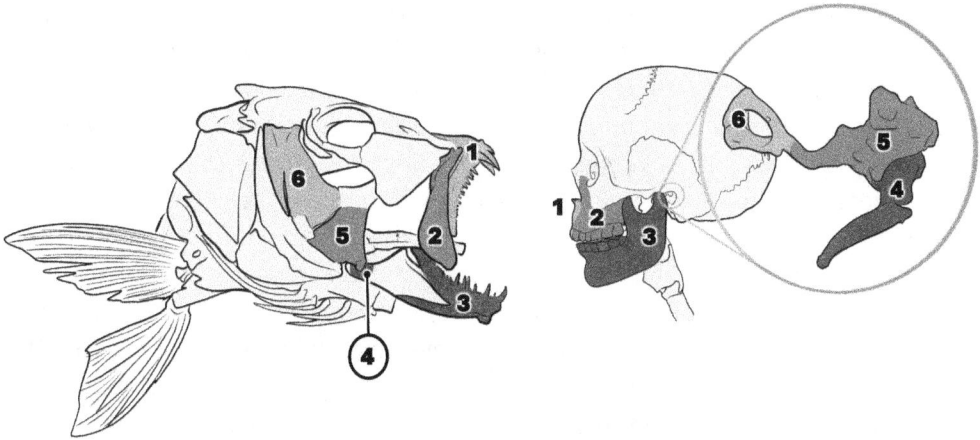

Figure 2.3   Jaw bones (1–3) and middle ear bones (4–6) in *Homo sapiens* (right) and their corresponding bones in a bony fish (left): 1 – premaxilla, 2 – maxilla, 3 – dentary, 4 – hammer (malleus), 5 – anvil (incus), 6 – stapes (middle ear bones 4–6 in humans shown greatly magnified).

became larger and firmer, resulting in a distinction between premolars just behind the canine and molars located behind it. The three tubercles now also changed into a triangular pattern, with one tubercle at each corner, thereby creating tribosphenic molars. Combined with jaws that could move sideways, an even more efficient chewing became possible as the upper tubercles of an upper jaw molar came to lie between the tubercles of a lower jaw molar. A lot of mammals still have tribosphenic molars, just look at the molars in your dog's upper jaw. But in the lower jaw, most mammals have more rectangular teeth, with more than three tubercles, called quadrosphenic molars. In many primates, both the upper and lower jaw molars are quadrosphenic, just as they are in ours.

But then what about those bones that got detached in those Synapsida? And if their role in chewing got lost, how could those animals still chew? Bony fish have a lower jaw where the joint is provided by a bone in the lower jaw (the *articular*) that articulates with a bone in the upper jaw (the *quadrate*). In mammals, those two bones have detached from the jaws and thus no longer form a functional jaw joint. How was this possible? It's not that those animals didn't feed for a few generations while they lay around waiting for a new jaw joint to appear. Also about 200 million years ago, some ancestors of mammals had two different jaw joints at the same time: one that they had inherited from the bony fishes (between the articular and quadrate) and a new one (between the dentary and squamosal). As soon as this first joint lost its jaw function, the other one could take over. And this worked as we still have such a joint: our lower jaw consists only of the dentary that articulates with the squamosal bone in the skull. What about this other jaw joint, you may ask? This got a totally different function: it became a joint between two additional middle ear ossicles. Indeed, the articular and quadrate became completely detached from the lower jaw and came to hang against the middle ear bone that had been there since the tetrapods. This explains why we, like all other mammals, have three middle ear ossicles: the articular became the hammer (*malleus*), the quadrate became the anvil (*incus*), and they came to lie against the stapes (see Figure 2.3). On top of that, another lower jawbone (the angular) forms a bladder-like structure that encloses those middle ear bones (called the *bulla tympanica* in humans). As early as 123 million years ago, this process had occurred (Mao et al., 2020) and hearing and chewing could happen completely separately. For humans, three ossicles proved to have an advantage over one as in other tetrapods, as it allows for refined and adaptable hearing (Bell & Jedrzejczak, 2021). This was an advantage for the creation of language in some lineages.

Brains were around since more than 500 million years ago, but the typical mammalian brain did not emerge until about 200 million years ago. They then became remarkably larger about 100 million years ago, at least compared to other amniotes (Smith, 2022). Improved sensory function, such as smell, sight, hearing and taste, as well as improved food intake, would be at the root of this. About 60 million years ago, when all dinosaurs were long extinct (except for birds, which are also dinosaurs), lots of niches became vacant that could be occupied by mammals. "One man's death is another man's bread" certainly applies to evolution. Mammals then got bigger, but brains first got relatively smaller (Bertrand et al., 2022). This did not last very long, as 50 million years ago the brain was already increasing in size again. But not everything about the brain enlarged in a similar way. Especially areas related to balance, vision and eye movement, as well as control of the head and integration of sensory information (but not the olfactory organ), expanded. Thus was formed the *neocortex*, a new brain structure unique to mammals. Now, it is not that new (the term 'neo' refers to new), because the precursor of this brain area was already present in the other vertebrates. Only there it is very small and located deep inside the brain. In mammals, this is very much expanded and lies superficially in the large brain or cerebrum (hence the term *cortex* which means 'bark', referring to this part lying superficially). This neocortex is very important for higher cognitive functions, also for us. This is partly due to the large number of nerve cells or neurons arranged in six layers (Briscoe & Ragsdale, 2019). In mammals such as cetaceans, ungulates, rodents as well as primates, the neocortex has increased even further in size and complexity. It is therefore not surprising that this is the part that has also grown very much in humans, something that can explain much of our evolution.

When you think of a mammal, you undoubtedly think of some hairy animal. And rightly so, because the presence of **fur** composed of densely packed hairs is a typical characteristic of mammals. Hairs are made by the keratin-producing cells of the epidermis, with the nutrition and energy for this being supplied by blood vessels from the dermis. Hairs are really nothing but an organized accumulation of packets of keratin, packed together into a cylindrical structure that grows through the skin and extends superficially. Now, the fur is older than mammals themselves, going back 125 million years or more. As will be seen throughout this book, the change in our fur is an important aspect of our evolution. We too still have hairs covering our bodies, being large and thick, pigmented hairs on our heads and certain parts of our bodies (in some humans a bit more than others) and small, non-pigmented hairs scattered over almost our entire body. Hair has always been very important for mammals to keep their body warm. In fact, the first mammals were nocturnal (night-active), which allowed them to co-exist with dinosaurs that ruled by day. Fur kept them warm during the night, just as it protects certain mammals today from extremely cold conditions (think polar foxes or polar bears). This is possible because hairs can be straightened via muscles (via so-called *arrector pili* muscles). By straightening the hair, a thicker layer of air is retained, which has an insulating effect. When the hairs lie flat, the air flows closer to the skin and can extract more heat from the skin. Although we no longer have as many hairs that still allow this fur function, we still have the mechanism to straighten the hairs. Like mammals, we also allow those little muscles to contract when we get cold, which gives us goose bumps (the muscles pull the skin slightly up). And just like other mammals, we do this when we want to express certain emotions (a dog does not straighten the hair on its neck to cool down). Even though goose bumps does not really make us feel warmer, it is something we inherited from our mammalian ancestors.

Avoiding water loss and insulating against cold or heat are just two of the many functions of the mammalian skin. The skin regulates the body temperature in other ways as well. For example, the subcutaneous hypodermis contains fatty tissue, which in marine mammals even forms a very thick layer (the blubber layer). This fat forms an insulating layer between the blood vessels that run through the skin, transporting body heat, and the body surface. Also, the small capillaries in the skin are here and there surrounded by small muscle cells. If one of these contracts, the capillary is constricted and less blood flows through it. This prevents too much warm blood from flowing to the surface and cooling the body too much. Conversely, if all blood vessels are fully opened, then more heat flows to the surface, which is useful for cooling. Therefore, it is no coincidence that we look pale when we are very cold (lots of capillaries are closed, so less blood in the skin) or reddish when we are very hot (more blood in the skin to dissipate the excess heat). Many mammals can release excess heat in yet another way, namely through **sweat glands**. Formed from the epidermis, but located deep within the skin, these glands produce a salt-like fluid: sweat. Given that this fluid has a body temperature, it will release that heat when the sweat leaves the body. Minuscule droplets of sweat are released through drainage channels at the surface of the skin, where they evaporate very quickly. In this way, heat is withdrawn from the body allowing it to cool down. The more sweat and the faster it evaporates, the more efficient the cooling. Therefore, air passing by the skin is essential for this sweat to evaporate faster. In mammals, you find sweat glands mostly where there is no fur, because hairs prevent air from getting up close to the skin surface, making evaporation difficult. If your cat leaves a footprint, and she hasn't been walking in the rain, she's probably sweating through the bare skin on the underside of the paws. As you might suspect by now, a lot has happened to our skin during our evolution, something that will be explained further in the chapters to come.

But sweat glands are also special to mammals, and thus also humans, in another way. During evolution, some sweat glands have been transformed into glands that secrete a fluid that is no longer cooling but is nutrient-rich: mother's milk. These glands develop embryonically along a ridge laid out left and right from the armpits to the groin: the milk glands. Depending on the type of mammal, these milk strips will form **mammary glands** in some places and disappear in others. So, you get mammary glands in the groin in horses and cattle, along the abdominal area in cats and dogs, at the level of the thoracic region in primates, or under the armpits as in elephants. In pigs, they run from the chest to the groin; in rodents, such as naked mole rats, mammary glands are even formed along the entire milk strip. In fact, the scientific name of these vertebrates refers to the presence and use of mammary glands: 'mammals' or 'Mammalia' suckle their young with the help of *glandulae mammariae* or mammary glands. Initially, mammalian young licked up milk oozing from an opening of the mammary glands, something still seen in egg-laying mammals, such as platypus. They have mammary glands but no nipples yet. In other mammals, an outgrowth of skin caused nipples to form around a mammary gland opening. This makes it easier for a suckling young to take the nipple into its mouth, taking in milk more efficiently. The fact that mammals have a swallowing reflex and a highly implanted larynx at birth is therefore no coincidence: it allows a baby to suck milk and breathe simultaneously, without constantly choking. This is also true in humans, hence one more of our mammalian characteristics.

I was just talking about egg-laying mammals. So, do mammals lay eggs? Well, mammals such as platypus and echidna do lay amniote eggs, as did their reptilian ancestors. The logical question then is: why don't other mammals lay eggs? That's where viruses enters the story. The DNA of those mammals contains viral DNA from so-called endogenous retroviruses (these carry their own genetic information in the form of RNA, which is translated into DNA in a host cell). That viral DNA itself makes up 8–10% of mammalian DNA (Dupressoir et al., 2012). Like many viruses, they infected mammals millions of years ago and incorporated their DNA. The virus possessed certain genes, so-called *env* genes (short for *envelope glycoprotein-encoding* genes), that allowed the virus to more easily fuse their viral protein coat with the cell membrane of a mammalian cell, while suppressing the immune system's response. This allowed the virus to more easily inject its DNA into mammalian DNA and thus reproduce. But surprisingly, this was also found to bring benefits to the host, making them part of the normal functioning of the body (the term 'endogenous' refers to the fact that they are fully integrated into the functioning of the mammalian body). Indeed, the ability of those *env* genes allowed a foreign organism to more easily fuse its cell membranes to insert itself into the tissue of another organism. I'm not talking about parasites or aliens here, but embryos. Have you ever considered that a baby nestled in a mother's womb is half made up of foreign genetic material? Yep, the DNA a baby inherited from the father is completely foreign to the mother's body, and all logic then says that a mother's immune system would immediately reject any embryo that tried to implant. And yet about 85 million years ago, one of the great success stories of mammals came about, thanks to these endogenous viruses: the formation of a **placenta**. A placenta consists of two parts: a part formed by the embryo and the part formed by the mother. The embryonic part is formed from those membranes that were already there in reptilian ancestors: chorionic and amniotic membranes. But in those mammals, no eggshell is formed around that because those membranes have the capacity to nest in the endometrium of the uterus, thanks to an outer layer of embryonic cells that fuse with each other (that layer is called the *syncytiotrophoblast*) (Rote et al., 2004). The viral genes allow the cell membranes to fuse and nestle in the uterine wall without triggering an immune response. Without those viruses, all mammals were probably still laying eggs, including ourselves!

# 3

# The equatorial Eden

The groundwork has been laid to delve deeper into the more recent part of the evolutionary story behind our bodies. We make a jump from mammals that lived about 200 million years ago to those mammals that are the origin of our evolution as hominins: the great apes. Great apes saw the evolutionary daylight about 20–18 million years ago (Almécija et al., 2021). Our story starts here. For each chapter, we make a time travel through the main periods of our evolution, focusing each time on those four aspects that defined the major evolutionary outlines of our bodies: diet, locomotion, skin and brain. Each chapter first briefly introduces the key players and provides a time line that can help keep track of where, or rather when, we are. We start here with what may be a good point of reference from which our bodies started evolutionary: that of our ape ancestor.

## Apes rule!

We turn here to the apes, but to clarify exactly what these are, a brief introduction to the other primates is necessary. Just what are these 'primates'? And what is the difference between a 'prosimian', a 'monkey', or an 'ape'?

### What's in a name?

The origin of primates is believed to have been around 75–65 million years ago (Finstermeier et al., 2013; Jameson et al., 2011; Wilson Mantilla et al., 2021), arising from a group of tree-dwelling, small mammals that resembled the extinct **Plesiadapiformes**. Plesiadapiformes lacked some of the key characteristics of primates. For example, they still had claws instead of flat nails and lacked some typical primate features in the skull. About 56 million years ago, **true primates** enter the world stage and are said to have even gradually replaced the Plesiadapiformes (they lived until 37 million years ago) (Silcox et al., 2017). Until then, everything took place only in the northern hemisphere, when the Gondwana ancient supercontinent was already split into North America that was close to Eurasia, and South America that was partially separated from Africa. Fossils of true primates appear almost at the same time in North America, Europe and Asia (Rose et al., 2011). One of the first primates belongs to the genus *Teilhardina*, named after the French Jesuit Pierre Teilhard de Chardin, who conducted research on primate evolution. Still today, primatologists, like Thierry Smith of the Royal Belgian Institute of Natural Sciences in Brussels, are

DOI: 10.1201/9781003588795-4

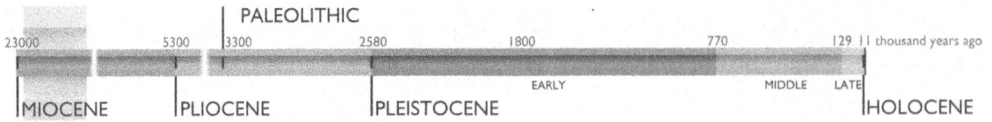

| | | PALEOLITHIC | | | | |
|---|---|---|---|---|---|---|
| 23000 | 5300 | 3300 | 2580 | 1800 | 770 | 129 11 thousand years ago |

| | | | | EARLY | MIDDLE | LATE |

MIOCENE    PLIOCENE    PLEISTOCENE           HOLOCENE

Figure 3.1    Time line showing the period dealt with in this chapter.

conducting paleontological research on *Teilhardina*. He recently discovered that the species *Teilhardina belgica* is one of the most primitive primates. They were small (about 30–60 g) and had hind legs well adapted for grasping (thanks to grasping feet) and jumping. As expected for a primate, they had flat nails on fingers and toes (Gebo et al., 2012). He could show that *Teilhardina* must have originated in Asia and made it to North America via Europe during 25,000 years (Smith et al., 2006). This was possible because at that time the climate was very favorable, as the so-called Paleocene-Eocene Thermal Maximum (55 million years ago) then prevailed.

From then on, things accelerated and diversity increased within two major lines of evolution. One evolved toward **lemurs, galagos and loris**. Lemurs are found only

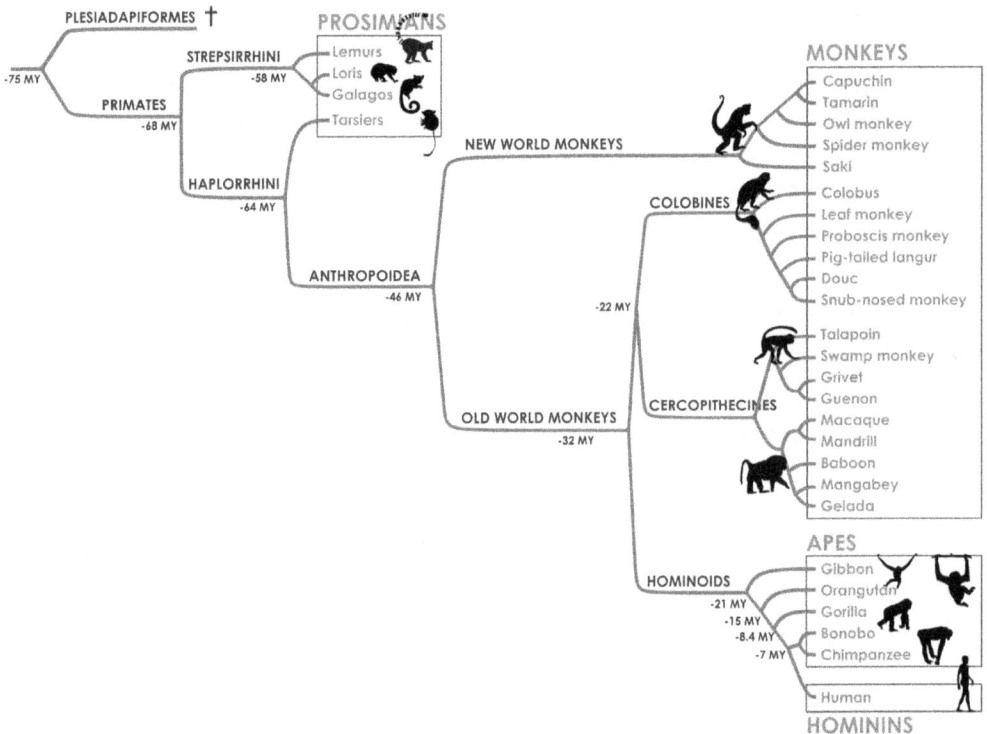

Figure 3.2    Overview of the different groups of primates, and their split from the extinct Plesiadapiformes. Numbers indicate the time (in million years ago) that a particular group originated. (Relatedness and dating based on Chatterjee et al. (2009) and Finstermeier et al. (2013).)

in Madagascar, where they arrived via patches of land that drifted from the African mainland (Wong, 2010). The ring-tailed lemur is perhaps the best known. Galagos or bush babies are very good jumping prosimians from Africa, while lorises are slow crawlers found in both Africa and Asia (Burgin et al., 2020). Most of these primates show adaptations to being nocturnal (night-active). They tend to have large eyes that reflect incident light (just like dogs they have a layer of tiny guanine crystals behind the retina, which causes light to pass through the retina twice, hence increasing the likelihood that the stimulus will be picked up), a well-developed olfactory organ in a long snout and a nose that is kept moist (this allows them to determine wind and thus scent direction). Their scientific name, Strepsirrhini, refers to that special nose.

The other evolutionary line, of which *Teilhardina* was a basal taxon, directed toward monkeys and gave rise to what are called Haplorrhini or 'simple-nosed'. These include the tarsiers, monkeys, apes and the hominins. Tarsiers are special in that they are very small prosimians with enormously large eyes. Like the Strepsirrhini, they are nocturnal yet exhibit anatomical adaptations to a diurnal (day-active) life, as do all other Haplorrhini. They have a light-sensitive area in their retina (the *fovea centralis*), the reflective crystal layer is absent, and vision becomes more important than olfaction (which translates into a smaller and dry snout and relatively larger brain parts involved in vision) (Barton & Harvey, 2000). A shorter snout also resulted in shorter teeth. And yet tarsiers hunt insects very efficiently at night. They can do so thanks to their giant eyes, at least relative to their bodies: it amounts to 4.5% of their body weight, with all sorts of implications on skull construction (Ankel-Simons & Rasmussen, 2008; Rosenberger, 2010). Should the same be true for me, I would have eyes of 3.7 kg. Fortunately, our eyes are much smaller (about 0.03% of our body weight). Occasional switches have occurred between being day- and night-active during primate evolution (Ankel-Simons & Rasmussen, 2008). The first primates may already have been day-active, equipped with all these eye adaptations (Wu et al., 2017). Being day-active also brings evolutionary advantages, as species formed faster than in night-active ones (Scott, 2018).

Haplorrhini other than tarsiers are mainly diurnal. They are generally also larger than the strepsirrhine ones. We now speak of monkeys, apes and hominins, collectively referred to as the Anthropoidea. Among the recent representatives of monkeys, one group occurs in South America and one in Africa and Asia. The former are usually referred to as New World monkeys, the others as Old World monkeys. It may surprise you, but the Old World monkeys are more closely related to the apes than they are related to the **New World monkeys**. The latter arrived in the Americas long before humans did, namely 44–40 million years ago (Kay, 2015). At that time, the still narrow Atlantic Ocean had already separated South America from Africa. The most acceptable hypothesis is that some monkeys ended up on a patch of land containing some vegetation, which had become detached somewhere on the West African coast and floated up to South America. The fact that the oldest fossils from

South America (36-million-year-old *Perupithecus* from Peru) show strong similarities to Old World apes confirms this hypothesis (Bond et al., 2015). Genetic information also supports that the two groups of apes must have split off from each other during this period (Schrago, 2007). Colonization of the New World proved successful. Some gained a prehensile tail (spider monkeys and howler monkeys), others were small monkeys that had claw-like nails (marmosets and tamarins), some became nocturnal again (owl monkeys) and others stood out for high intelligence (capuchin monkeys). What they all have in common is that they have a wide septum between their nostrils, with the nostrils facing sideways. Their name refers to that: Platyrrhini or 'broad-nosed'.

The **Old World monkeys**, on the other hand, had a very narrow septum and nostrils facing downward. Hence, they are called the Catarrhini or 'downward-nosed'. Have you ever wondered why we too have downward-facing nostrils? Herewith you have the answer: we too have a Catarrhini or Old World monkey ancestor, only we don't really look like real monkeys anymore. Those that resemble real monkeys are the leaf monkeys (Colobidae) and vervet monkeys (Cercopithecidae). These include small monkeys that barely emerge from the tree canopies in the equatorial forest (such as West African talapoins that can weigh less than a kilogram), to large monkeys that walk more on the ground than crawl in the trees (such as baboons). The largest among the monkeys are the mandrills, whose males can weigh up to more than 50 kg. These males also have brightly colored faces with red noses bordered by blue nostrils. The red color is caused by more blood flow through the superficial layer of skin, while the blue is due to the way light is reflected off the underlying connective tissue in the skin (Young et al., 2020). That pattern of red and blue occurs in other primates, such as vervet monkeys (a smaller African monkey). Both mandrills and vervets have colored genitalia in common: a red penis and blue scrotum. So, their males are always showing off their sex organ colors. And that pays off, because the redder the color on the nose and sex organ, the more dominant they are in the group (mandrills live in very large groups, sometimes up to more than a hundred individuals) (Setchell & Dixson, 2001). Old World monkeys are very diverse in their physique, species diversity, behavior and ecology. Some have a long-plumed tail (such as the mantled guereza); others only have a barely visible tail (Japanese macaque). Some only eat fruit or leaves; others also eat meat (such as baboons). Leaf eaters, like the leaf monkeys, show adaptations in their teeth and digestive system to grind tough leaves (thanks to molars with parallel ridges on them) and digest them. They have a three- to four-compartmented stomach in which, as in ruminant cattle, there are microorganisms in the anterior parts that can digest cellulose from plant cells (something mammals in general have little or no ability to do themselves) (Matsuda et al., 2022; Milton, 2012). Other leaf-eating primates are more likely to have enlarged blind and large intestines, housing those microorganisms. Where these primates do differ from cattle is that they do not ruminate their food. But now on to the apes, because that's what we were talking about.

**Apes,** what are they? It is not the same as monkeys and yet they are not humans. Biologically, they are a specialized group of Old World monkeys, from which one evolutionary lineage later on gave rise to humans. The scientific name for the group of all apes, including those that became humans, is the Hominoidea. The ending of this word reflects the hierarchical level within biological classification. For example, Hominidae is not the same as Hominoidea. The former is the family of humans and their most closely related great apes, such as orangutan, gorilla, chimpanzee and bonobo. The second includes that family Hominidae, as well as numerous other families of apes, such as the family of gibbons and families of extinct ape species. It may come as a surprise, but although orangutans look very similar, gorillas look very similar, chimpanzees look very similar (they also have limited distribution ranges), yet the genetic variation within each of those great apes is greater than that within *Homo sapiens* (up to 3.5 times greater in orangutans) (Kaessmann et al., 2001). Genetic appearances are thus deceptive (for example, dog breeds, from Chihuahuas to Afghan greyhounds, are 99.85% genetically similar) (Orwant, 2005).

Of all extant apes, **gibbons** are the most primitive and the smallest (they belong to the family Hylobatidae). The evolutionary lineage of the gibbons split off from the great apes sometime between 20 and 14 million years ago (genetic information suggests 20 million years ago, and fossils suggest 14 million years ago) (Alba et al., 2015; Gilbert et al., 2020). At that time, their African ancestors were able to reach the Asian continent. Gibbons became excellent brachiators, swinging below branches using specialized and very long arms (up to one and a half times the length of their legs) and vertebral column, and highly mobile shoulder, elbow and wrist joints (Kivell et al., 2013; Villamil & Middleton, 2024). Gibbons can cover a distance of 10 m from one branch to another, alternately grasping with one arm after the other (Channon et al., 2011). They can also walk on their hind legs on the ground, holding their long arms in the air to maintain balance, but they walk equally well on three or four legs (Vereecke et al., 2006). Even though they represent the most basal clade of extant apes, they are highly specialized brachiators.

The other apes are significantly larger; hence, they are referred to as great apes. The **orangutans** from Sumatra and Borneo (genus *Pongo*) split off from the African great apes about 18–15 million years ago. They also are brachiators, but they more frequently walk on top of the branches with their feet, while holding on overhead with their hands. Males are remarkably larger than females (weighing more than 70 kg versus females weighing about 40 kg) and exhibit specific characteristics (larger canines and large flaps of skin at the level of the jaws). Orangutans have been under media attention recently because of the precarious situation they find themselves in. Genetic research as recently as 2017 showed that a small group of less than 800 individuals, hidden away in an isolated patch of tropical forest in Sumatra, was genetically and anatomically so different from the other two known species that it should be considered a new species (Stokstad, 2017). The ironic thing is that from the moment the species was described, it was immediately given the status of 'critically endangered' due to the destruction of their habitat.

**Gorillas** (genus *Gorilla*) are undoubtedly the most majestic great apes, known for their enigmatic silverbacks. These are the dominant males, who impress not only by their stature (in mountain gorillas, males grow up to 1.95 m tall – when standing upright – and can weigh 270 kg), but also by clapping their fists on their chests. That's not a figment of a Tarzan movie or anything, they do that. In fact, the frequency with which they literally beat their chest is an indication of how strongly they beat their chest figuratively: the larger the gorilla, the lower the frequency (Wright et al., 2021). They do this to impress females both visually and through sound, but also to scare away other males. At least this is true for mountain gorillas, which live hidden in the dense forests of the Virunga Mountains on the border between Democratic Republic of Congo, Rwanda and Uganda. This mountain range is part of one of the African rift formation areas, which will be discussed later. The other species, the lowland gorilla, is found much further west, on the other side of the Democratic Republic of Congo, and has a wider distribution range up to the west coast of Africa.

They are sometimes considered the little 'brothers' of the gorilla, but the **chimpanzee and bonobo** (genus *Pan*) are more closely related to humans than they are to the gorilla. The chimpanzee has a genome that is 98.63% identical to that of humans, of which 99.5% of the coding genes are the same (Rolian, 2014). The lineage to today's gorillas split off about 9–8 million years ago from the evolutionary line that gave rise to a common ancestor of chimpanzees and bonobos, and hominins. The bonobo split off from the chimpanzee only 1.8 million years ago, due to a group of individuals who became isolated after crossing the great Zaire River (due to climatic fluctuations, that river was not always at the same height) (Gonder et al., 2011; Rutherford, 2019). Hence, to this day bonobos live on one side of the river (south) and chimpanzees on the other side (north). The bonobo as a separate species was only discovered in 1928 based on a skull from the collection of the Africa Museum in Tervuren (Belgium). Chimpanzees and bonobos are also brachiators but equally spend a lot of time on the ground. Like gorillas, they do this via knuckle walking: they lean on the leading edge of the first row of phalanges. Thanks to extra reinforcements in the wrist and fingers, and a curved ulna, they can do this without constantly spraining it (which would be the case with us) (Meyer et al., 2023; Sarringhaus et al., 2022).

So, this was an overview of who is who within the club of primates, and how they are related to each other. Of course, there is much more to talk about this, but for our story we now go back in time, to the Miocene, 23–5 million years ago. That was when it was high time for the apes, gradually opening up the evolutionary path to humans.

## Glory days of the apes

There are far fewer apes today than there once were. Today, there are about 20 species of gibbons, three species of orangutans, two species of gorillas (depending on which source you use, two to four species are distinguished), one species of chimpanzees (some make it four species), one species of bonobo and one species of humans. But did you know that quite a few apes even lived in Europe during the Miocene? In 1856, *Dryopithecus* was described, based on fossils found in the French Pyrenees. But apes also lived in present-day Greece, Italy, Germany, Spain and Hungary at that time. Their distribution range went as far as Turkey, the Arabian Peninsula, India, China and Southeast Asia (Almécija et al., 2021). More than 50 genera of extinct apes are known, which is eight times more than what remains today (Almécija et al., 2021; Begun, 2003). Earth was, in a sense, a *planet of the apes* during the Miocene!

The most primitive apes occurred about 20 million years ago in Africa, which was then separated from Eurasia by the Tethys Sea (Begun et al., 2012). These apes had already lost the tail but were not yet brachiators. This was deduced from fossils of *Proconsul*, an extinct ape that walked on branches with four legs about 20–18 million years ago in what is now Kenya and Uganda (Nakatsukasa et al., 2004). Starting 17 million years ago, temporary corridors formed between Africa and Eurasia, leading to an evolutionary spread of great apes in Eurasia. Apes found in Kenya, like *Afropithecus* and *Kenyapithecus*, possibly gave rise to the first Eurasian apes (Almécija et al., 2021; Begun, 2003; Pugh, 2022). That Eurasia was such a success for apes was due to the warm and humid climate that prevailed there at the time, which gave them access to a lot of subtropical forests. This resulted in a lot of diversity, including in body size. The smallest was about 3 kg in weight, but the Miocene also knew the largest among the primates that ever lived: *Gigantopithecus* (Begun, 2003). This one lived about 2 million years ago in China, was 3 m tall and weighed up to 300 kg (Welker et al., 2019).

About 13–12 million years ago, two major branches formed that provided the basis for further evolution into the recent great apes: one branch gave rise to the European apes, the other to the Asian ones. The Asian branch included the extinct *Sivapithecus* from India and the large *Gigantopithecus*, from whose ancestors the line to the orang-utans split off about 12–10 million years ago (Welker et al., 2019). The European lineage is better known, with a lot more fossil evidence: *Dryopithecus*, *Pierolapithecus* and *Hispanopithecus* from Spain, *Danuvius* from Germany, *Rudapithecus* from Hungary, *Graecopithecus* from Greece, *Ouranopithecus* and *Oreopithecus* from Italy (Almécija et al., 2021). So, Europe was the place to be to thrive as an ape. But that story didn't last long. Toward the end of the Miocene, the climate began to reverse, due to mountain formation worldwide (Alps, Himalayas, African Rift Valley), changes in oceanic currents and the expansion of the polar ice caps (Begun, 2003). Consequently, Europe acquired a more temperate climate with more pronounced seasonal variations, resulting in a less rich and diverse vegetation (Merceron et al., 2010). The decline of the apes set in and the European ones became extinct. A few managed to flee Europe in time, possibly because they were better adapted to moving on the ground using knuckle walking. They ended up back in Africa, where the climate was more stable and more tropical forests were available. One of those European apes, *Ouranopithecus*, shows several characteristics that can also be found in gorillas (Ioannidou et al., 2019). This may have been the precursor of the African great apes, whose origins were therefore not in Africa but rather in Europe.

About 10 million years ago, an ancestor (*Chororapithecus*) of the later gorilla (Suwa et al., 2007) may have lived in the Ethiopian region. From that evolutionary line, the branch to *Pan* (chimpanzee and bonobo) and the hominins split off about 9–8 million years ago. Very little is also known of *Pan*'s ancestors, with only a few relatively recent fossils from Kenya (half a million years old) (McBrearty & Jablonski, 2005). Until recently, one was inclined to think that the common ancestor of *Pan* and the hominins must therefore have been chimpanzee-like in appearance. It was no coincidence then that Jane Goodall set out to study the behavior of chimpanzees in East Africa, at the behest of the famous anthropologist Louis Leakey. By studying their behavior, Leakey hoped to discover what the behavior of our early, East African ancestors may have been. Meanwhile, everything points in the direction that extant chimpanzees should not be considered the primitive forms from which humans evolved. No, those great apes underwent their own evolutionary specializations following their separation from the common ancestor, according to their diet and mode of locomotion. That the ancestor of man and chimpanzee was clearly not a chimpanzee is something that even Charles Darwin was already convinced of (Almécija et al., 2021).

# Fruit for everyone

If the last common ancestor of chimpanzees and humans did not look like a chimp, what do we know about our early ancestors? Does the diet of chimpanzees and other great apes tell us about our ancestral diet? And how different is the diet of the great apes from that of other primates?

## *Apes are fruit eaters … and more*

Primates owe their existence largely to the introduction of a new food source: fruit-bearing flowers (Milton, 2012). Fossils of flower-forming plants go back 130 million years (genus *Archaefructus* from China), but flowers are believed to have originated as early as 180 million years ago (Silvestro et al., 2021; Sun et al., 1998). Very soon these plants became ecologically dominant (Friis et al., 2010). The first flowers were small, forming tiny fruits with little flesh (less than 1 cm in diameter). About 60–50 million years ago, fruits became significantly larger, at the same time that primates appeared on the scene (Fleming & Kress, 2011). Coincidence? Most likely not. All indications are that primates saw the evolutionary light of day as insectivores but soon moved up to become frugivores that evolved along with flowering plants (Wu et al., 2017). It makes it no coincidence that they were highly adapted to living in trees, with grasping hands that were also useful for picking fruit. But fruit alone is not enough. As a source of energy (i.e. calories), fruit is very interesting, but nutritionally fruit does not contain all the essential nutrients (certain amino acids, vitamins and minerals are missing). Only 3.2–12.6% dry weight consists of protein (Milton, 1999b). Yet great apes require about 1 g of protein per kg of body weight per day. Great apes solve this by supplementing their fruit diet with animal proteins (Chivers, 1998). For example, chimpanzees use sticks to fish up termites and ants (Whiten et al., 1999). Rotting fruit is also part of the diet – some fruits are also inedible unless they are rotting (Amato et al., 2021). Great apes also eat leaves regularly (almost 90% of the diet in gorillas) (Chivers, 1998). However, mature leaves contain a lot of fiber, making them difficult to digest. Hence, many great apes often turn to young leaves, which are softer and contain more protein. Chimpanzees and bonobos mostly eat fruit (up to 70% of the diet), but regularly also meat (chimpanzees from Gombe, East Africa, eat up to 500 g of meat per week) (Chivers, 1998). This is not exceptional, as other primates also sometimes eat meat (Watts, 2020). For chimpanzees, however, meat is not only of nutritional but also of social importance: it strengthens the bond between groupmates when they hunt together and share meat (Smil, 2002).

Great apes and other primates exhibit many adaptations that reflect them having evolved together with a diet of (among other things) fruits. Some features did not come about solely as a function of eating fruit (such as grasping hands that are also important for climbing trees), but others have clearly become evolutionarily driven in that context. This certainly seems to be the case for adaptations in vision and taste, two senses necessary to find and select the right fruits. That we are talking about a co-evolution, that is, a concurrent evolution of both the plants that produce fruits and the primates that eat them, is evident from several things. For one, plants want to protect themselves against herbivores, because they gain little if they are damaged too much and if their seeds do not get dispersed. But at the same time, those plants need them because they can further disperse their ripe seeds through the fruits. Most plants contain difficult-to-digest (tannins, cellulose, lignin) or toxic substances (alkaloids) in those parts they need to protect from herbivores (such as leaves, unripe fruits). Other parts, such as ripe fruits, can then be more attractive and tastier. The more the ripe fruits are eaten, the more the seeds they contain are dispersed through feces (Milton, 2012). Although some primates, such as leaf monkeys, show adaptations in their stomachs to digest those substances anyway using microorganisms or to neutralize the toxins, that ability is rather limited in great apes (some digestion in the blind gut and colon is possible thanks to bacteria) (Profousova et al., 2011). So, they best avoid that kind of food, which explains why ripe fruits and unripe leaves are preferred. Ripe (wild) fruits contain little indigestible fiber (yet they are only about 50% digestible) and contain many easily digestible carbohydrates (Milton, 1999b). Thus, while they are an ideal source of energy, they are relatively low in protein (on average 6.5% of the total content). Young leaves contain more protein but often also toxins.

Great apes that are equipped with an efficient detection system to locate ripe fruits and young leaves in the dense canopy have an advantage: they must waste less time and energy to get their nutritional fix. Like most mammals, the first primates were dichromatic: they could distinguish colors along a blue-yellow gradient (Dominy, 2004). This is possible because there are two forms of light-sensitive proteins present in the retina: the S-opsin and the M/L-opsin. S-opsin (S stands for *short*) is sensitive to short-wavelength light, corresponding to blue light. M/L-opsin (from *middle/ long*) is then more sensitive to light with longer wavelengths, toward the yellow to red spectrum. In Old World monkeys and great apes, a second variant of the gene emerged that normally produces this M/L-opsin (Wu et al., 2017). As a result, they now make separate S-, M- and L-opsin. This makes them trichromatic and allows them to also distinguish light along a green-red gradient. Now what does this have to do with fruits and leaves? Ripe fruits and young leaves differentiate based on their color, a phenomenon that can be linked to this co-evolution with fruit eaters (Nevo et al., 2018). It is therefore no coincidence that unripe fruits are green and ripe fruits are yellow, orange or red. Old leaves are also green, while young leaves are red. So, thanks to that trichromatism, great apes can easily distinguish ripe from unripe fruits (blue-green-yellow gradient) and distinguish old from young leaves (green-yellow-red gradient). Of course, this system works well only in a diurnal context, where sight is more important than smell.

Once the right food is located, the next challenge is assessing whether it is edible or not. Taste plays a crucial role in determining whether we will eat food or not. This is also very important for survival. It is always better for something toxic to be considered unpalatable, and thus not eaten, than to find ourselves convulsing and foaming at the mouth after having eaten it. On the other hand, to be too picky about what to eat is not good either. A sour and bitter taste is usually an indicator of inedible food. This is not only true for us: the ability to detect something sour, one of the five tastes we can distinguish, was there from our fish ancestors on (Frank et al., 2022). However, a lot of primates find a sour taste pleasant, especially great apes and humans. Some acidity sometimes is also good, as in rotting and fermenting fruit. Filamentous fungi, yeasts and lactic acid bacteria produce acidic substances. While those produced by filamentous fungi are mostly harmful, those from yeasts and lactic acid bacteria are actually nutritionally beneficial: they provide more calories, they contain useful amino acids and vitamins, they make the fruit more digestible, and any toxins are broken down as a result. Great apes even possess an extra copy of the *HCA3* gene, which allows them to detect more easily substances produced by lactic acid bacteria. Thus, great apes and humans have a nose for the right rotten fruit. Now, fruit fermentation also produces alcohol (in the form of ethanol). There are numerous observations of monkeys and other animals that were wasted after eating rotten fruit. However, the ancestors of great apes and humans possessed a modified gene that allows them to break down ethanol up to 40 times better. This allows them to get a lot of energy and nutrients from rotten fruit without suffering from a severe hangover every time.

That eating fruit was essential during the evolution of monkeys and apes is also evidenced by the fact that primates lost the ability to make vitamin C about 65 million years ago, something other mammals can do (Killgrove, 2021). Due to a mutation, our ancestors became completely dependent on food to get vitamin C. Vitamin C is, among other things, essential to produce collagen, a fibrous protein most commonly found in the body where it provides structure and strength (for example, our dermis and bone tissue are full of collagen). Vitamin C is mainly present in fruits, at least certain fruits. Again, this can be linked back to sour tasting, because all fruits that contain a lot of vitamin C are acidic (vitamin C is called ascorbic acid, by the way). It is therefore a great advantage that great apes and humans are not averse to anything that tastes sour.

The ability to recognize foods high in sugar is also interesting. Not only does the smell of fruit tell something about how much sugar it contains, but in addition, primates possess the ability to detect this (Nevo et al., 2022). Larger primates can do this even better than smaller ones (Simmen & Hladik, 1998). This is advantageous for great apes, which are generally larger than other primates, because this larger body also requires more energy. Great apes that get their protein mainly from leaves, rather than animal food, are also able to detect the amino acid glutamate, thanks to special receptors in the taste buds. This has the advantage that leaves taste less bitter as a result.

## Jaws and other adaptations

So, the ancestor of the great apes and humans clearly depended heavily on fruit and showed adaptations to be able to find it. But finding them is one thing; being able to eat them and digest them properly is another.

Being able to properly process food for efficient digestion requires a functioning jaw apparatus. Great apes have a lower and upper jaw, each with 16 teeth and a set of jaw muscles for biting and chewing. Each jaw half contains two flattened incisors, a canine tooth that is higher than the other teeth, two premolars and three molars. Indeed, just like we do. Apes have long jaws and thus a protruding snout that provides ample space to house the teeth. Canine teeth stick out especially in males where they continue to grow longer (Okerblom et al., 2018). It may surprise you, but this has nothing to do with nutrition but is rather related to aggression between males in their fight for females and in maintaining a dominant position within a group. For example, canines in gibbons are similar in size, which can be explained by the fact that they are monogamous and thus don't have to constantly fight for mates (Swindler, 2002). Ape canines also show adaptations that reflect their use during fighting, as they have similar characteristics to canines in predators (Delezene, 2015). Because of the large canines, great apes still have a gap (a diastema) between the outer incisor and the canine in the upper jaw, and between the canine and the first premolar in the lower jaw. When they close their jaws, the canines fit into these gaps, rubbing those of the upper jaw against those of the lower jaw. This mechanism, like that of predators, causes the canines to sharpen (Delezene, 2015). Great apes, including humans, are distinguished from other primates by the presence of five tubercles on the lower jaw molars, with a Y-shaped groove running between the tubercles (Swindler, 2002). In African great apes, the teeth are surrounded by a thin layer of enamel (the outer and hardest layer of the teeth) (White et al., 2015). They also have rather high tubercles on the molars, which is useful for grinding fruit and leaves. The spatulate incisors, in turn, are useful for biting off pieces of fruit flesh.

Like us, great apes mainly use two large jaw muscles when chewing: the masseter and the temporalis (in reality, the jaw musculature is more complex than that). The masseter muscle runs from the side, at the bottom of the lower jaw, to the bottom of the zygomatic arch (the thick muscle you feel on the side of your lower jaw). The temporalis muscle lies medially behind the zygomatic arch and starts from a protrusion on the lower jaw to attach to the side of the skull, in the temporal region. In great apes and in us, the temporalis is smaller than the masseter, but this has not always been so during our evolutionary history. All jaw muscles do two important things during chewing: they produce enough force to cut through or break the food, and they move the jaws in different directions to grind the food. Chewing is an obvious thing; we do it on a daily basis and it seems so simple. Yet a chewing cycle is a very complex event, the result of an interaction between nerves, muscles, the skeleton, as well as all kinds of sensory organs around the teeth, jaws, tongue and muscles. Those sensory organs constantly register how everything interacts with each other. For example, sensors around the teeth (periodontal mechanoreceptors) constantly measure the size and direction of the force applied to the teeth. That information is used to adjust how much more or less muscle fibers need to be activated (Trulsson, 2006). A chewing cycle starts with a phase of rapid closure of the jaws, until the teeth come into contact with the food,

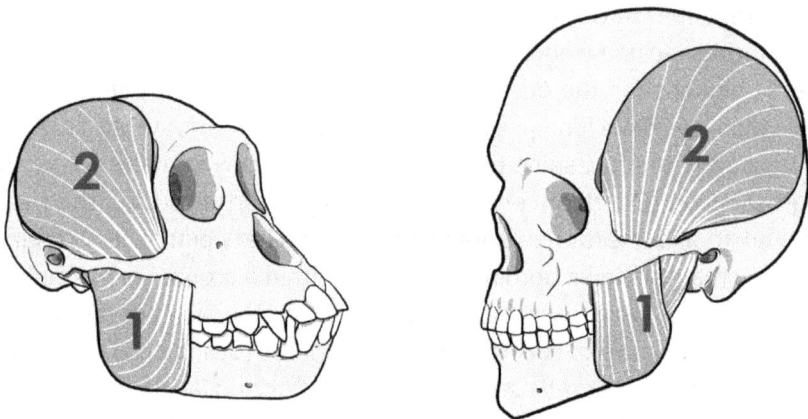

Figure 3.3    Jaw muscles in chimpanzee (left) and human (right): 1 – masseter muscle, 2 – temporalis muscle.

thus firing those sensors into action (Vinyard et al., 2007). As a result, the jaws then close slowly but forcefully. This is the actual biting phase, providing enough force to crush food, but no more than necessary. This is a safety measure. Without this, we would be able to break our teeth and jaw bone purely by the force of the jaw muscles.

So, fruits and leaves are important for great apes. But this comes with many challenges as it requires an adapted digestive system to extract as much energy and nutritional building blocks from the food as possible. When we eat fruit, it is cultured fruit from the supermarket. However, this soft and juicy fruit cannot be compared to the high-fiber and rather tough fruit that chimpanzees have to make do with. Chimpanzees spend on average half the hours that they don't sleep on chewing food (Lieberman, 2013). For us, that's less than half an hour we spend on chewing a day. So, a functioning chewing apparatus is more important in great apes to minimize the energy loss in doing so. Chimpanzees also must take in much more food to get the same amount of energy and nutrients, sometimes up to a kilogram per hour. After 2 hours, the stomach has digested it and new food can be added. Fiber in the food is also harder to digest, requiring additional energy. One would expect that the more fiber, the more time the food takes to pass through the digestive tract, a matter of taking the necessary digestion time. However, the reverse is true, both in great apes and in us. High-fiber food takes a chimpanzee about 38 hours to pass through the intestinal tract. Low-fiber food takes 48 hours! In humans it is 62 hours for food without fiber and 41 hours for fiber-poor food (Milton, 1999b). It seems like the digestive system does not waste too much time on food that is not easily digestible anyway. So not only does high-fiber food contain less digestible food, but the digestive system also extracts less. A double disadvantage, especially for those for whom leaves are on the daily menu (they are very high in fiber)! A challenge for great apes, as they do not have the specializations in the stomach or blind gut like some monkeys, where microorganisms are responsible for digesting those fibers. Fortunately, most great apes eat other things as well. Gorillas do eat a lot of leaves, but they have two characteristics that alleviate the diet problem somehow: they are large (larger mammals require less food per kg of body weight than small mammals) and they are less active than chimpanzees (they require less energy).

The digestive system looks quite similar in great apes. It starts with an ordinary esophagus, followed by a simple stomach in which stomach acid and digestive substances are produced. Then follows a relatively short, small intestine, a small cecum with an appendix and a large colon equipped with pouches. The large intestine comprises more than 45% of the total volume of the digestive tract, the small intestine only 14–19%. So, no particular specializations in the stomach or cecum as in some Old World monkeys, and thus no specializations to properly digest tannins and cellulose or to detoxify plant foods during digestion. Great apes have lost a gene that prevents them from making the enzyme *uricase*. In monkeys, this enzyme is responsible for breaking down tannins. Their saliva does have a special composition that is conducive to breaking down hard-to-digest tannins (Richter et al., 2022). The fact that great apes also lack those extra bacteria means that they are also less efficient in breaking down plant toxins (Chivers, 1998). All indications are that this must have been the general construction of the digestive system of the ancestor of the chimpanzee and us.

# Clambering with hands and feet

Great apes are known as brachiators: they use their hands and arms to hang and swing from branches. But there is much more to it. Great apes already vary greatly among themselves in that respect. They are also well equipped to climb vertically, to walk on the ground on four legs, and are even capable of walking on two legs. All things we can do too, only we only make something out of it when we walk upright on two legs. But we are very good at it. So, our bodies have changed evolutionarily. The question then is the following: did this start with a tree-dwelling brachiator? Or is the story a little more complex after all?

## *Playing the monkey*

We know primates as tree dwellers, scrambling, running, jumping, swinging across the branches with grasping feet and grasping hands. They do so to all degrees, from species that barely leave the trees to those that barely go into trees. Living in trees requires being able to grip well – anyone who fell from a tree as a child will have learned that lesson. But being able to judge where in space the branches are positioned is also essential. A good grip is determined by the limbs; good spatial estimations depend on the eyes and the brain.

Some primates walk on the branches; others hang from them. Some walk on hands and feet; others walk on two legs. Some hang from branches with feet or hands and others even with their tails (such as spider and howler monkeys). Some make small jumps from one branch to another; others manage to jump over more than 10 m using their powerful hind legs (like the indri, a prosimian primate from Madagascar) or swing their arms better than a circus performer (like the siamang, an ape from Asia). You might say that this is not so spectacular; humans own a world record long jump of 8.95 m. The difference is that humans need a long run-up for this, while that indri does it from a standstill. When primates walk on four legs, they also do so differently from other mammals. Most mammals put their hind and front legs parallel to each other: first the right hind leg, then the right front leg, then the left hind leg and then the left front leg. Primates do so diagonally: first the right hind leg, then the left forelimb, then the left hind leg and then the right forelimb (Schmitt & Lemelin, 2002). One also observes this in tree-dwelling marsupials, which are thus not primates, suggesting that the special gait pattern is an adaptation to walking with four legs on branches.

Primates show a lot of variation in arm length relative to leg length. That ratio says something about how they propel themselves. Those champions in jumping that I just mentioned (the indris) have legs that are almost one and a half times as long as their arms. The longer a leg is used to repel, the longer contact is maintained with the ground, and thus, the longer force can be applied to jump. This explains why longer legs allow one to jump farther and higher. It is no coincidence that this has arisen several times independently during evolution (just look at frogs and kangaroos). Long legs also have an advantage when walking on two legs as they allow a greater stride length. If you go hiking with kids, their short legs require them to take many more steps to cover the same distance. Just as leg length determines stride length, arm length determines how much distance can be covered in primates that swing using their arms. It is no coincidence that those siamangs, the champions of swinging, have arms that are almost one and a half times the length of their legs. By the way, the ratio of arm to leg length is called the intermembral index. An index of 100 means that arms and legs are the same length. Typically, four-footed monkeys have an index of around 90 (so arms are about 90% the length of the legs). Jumping galagos have an index that can be less than 60; in siamangs it is almost 180 (Aiello & Dean, 1996; Ankel-Simons, 2007). If you look at great apes in general, you will notice that the arms are relatively long: in gorillas it is 120, in chimpanzees 110 and in bonobos 101. In humans, apparently something else is going on, because it is only 70. You may already guess that there is an evolutionary story behind that! But that's for later.

Most primates have a tail, useful in keeping balance when walking on branches or jumping to branches. In some monkeys, the tail is even a kind of fifth limb for clambering. Some New World monkeys, such as howler monkeys, spider monkeys and woolly monkeys (to a lesser degree also in capuchin monkeys), have such a prehensile tail. Their tails are so strong that they can support their entire body weight. Young animals in particular go to great lengths to dangle their tails (Bezanson, 2012; Garber & Rehg, 1999). A short tail is mostly found in jumpers that have long legs, while species with longer arms that do not jump that much have a longer tail (Sehner et al., 2018). Although apes also have long arms, the tail is so short that it is no longer visible externally. A few tail vertebrae are still present, but they are nicely tucked away between the left and right halves of the pelvic girdle. A tail is just extra weight to drag along when dangling at the bottom of branches.

What do all primates have in common? Gripping hands and grasping feet you might say. That is not entirely correct. Only the grasping foot is typical of all primates; the grasping hand only arose later (Sargis et al., 2007). Now what makes a grasping foot and a grasping hand? When we grasp something with the hand, we are going to rotate the thumb so that it lies along the other side of an object, opposite the other fingers. When a cat wants to grab something with her claws, all five fingers stay along the same side of an object. So the novelty is the ability to be able to oppose the thumb and the big toe: to rotate it so that it cannot only be rotated around an object, but so that the thumb rotates around its axis allowing it to be pinched forcefully toward the other fingers. All primates have an opposable big toe (called *hallux*), but not all primates have an opposable thumb (called *pollex*). This is not to say that the thumb cannot be used to grasp something, just that it cannot always be fully opposable to the other fingers (Pouydebat et al., 2006). Some can open the thumb relative to the other fingers, but only Old World monkeys (including great apes and humans) also rotate the thumb around its axis (Lemelin & Schmitt, 2016; Vereecke & Wunderlich, 2016). Great apes use their hands differently to handle objects. Chimpanzees and bonobos mostly use both hands, but only chimpanzees do so to obtain food (such as via twigging to catch termites in termite mounds) (Bardo et al., 2016). Gorillas make more use of one hand, while orangutans are less mobile in their hands and mostly also use their mouths to grasp (Bardo et al., 2018). They also make less use of their thumb when grasping, which may be because they have the shortest thumbs among great apes (Galletta et al., 2019). From this we can already conclude that we inherited our opposable thumb from our ape ancestor, but we apparently lost the opposable big toe somewhere along the way.

In the event of a crime, criminals can be identified thanks to their fingerprint, a unique pattern created by the skin on our fingers. Evidently, this did not evolve to make things easier for the police. No, this is an adaptation that allowed our primate ancestors to grasp more firmly onto branches. The ridges, which are called dermatoglyphs, are thickenings of the epidermis on the tips of the fingers, as well as the toes, palms and soles of the feet (Montagna, 1972). How much the ridges protrude depends on nerves and water. Indeed, surgeons from the previous century had already noticed that when certain nerves in the fingers were cut, the ridges were much less prominent. You most probably also have noticed that if you've been soaking for a long time in a nice hot bath, the skin on the tips of your fingers looks all swollen. That's because keratin, a protein accumulated in the epidermis, can absorb and retain water. Coincidence or not, the sweat glands also drain at the crest of the ridges (Yum et al., 2020). These are not just annoying features of the fingertips, as swollen ridges offer several advantages for having a firmer grip. First, the ridges themselves form an anti-slip layer, as small bumps on branches get anchored between the ridges. Second, the production of sweat causes a thin layer of water to be present, which softens the keratin, making friction with the branch even greater (Yum et al., 2020). Third, raised ridges also allow that when too much moisture on branches, the moisture is drained more efficiently through the furrows between those ridges, making branches less slippery (Changizi et al., 2011). Same as what grooves do in a car tire. And there are further arguments supporting the idea that our dermatoglyphs are an ancestral inheritance as an adaptation to climbing. First, these ridges are packed with tiny mechanosensory organs, called Meissner corpuscles (at a density of about 12–43 corpuscles per mm$^2$) (Covarrubias et al., 2024). Second, other tree-dwelling mammals can also have dermatoglyphs, where even the prehensile tail in some monkeys has a naked strip of dermatoglyphs where the tail grips branches, something not found in other mammals that have prehensile tails (Organ et al., 2011).

A good grip is important, especially when you're dangling several feet above the ground. But how do you know whether you have a sufficiently firm grip on something or not, and whether you are slipping or not? Numerous sensory organs in the middle layer of our skin (the *dermis*) can detect the necessary information for that. As soon as we touch something with our skin or something moves along it, one or more of the different types of tactile sensory organs are activated, with this information being registered in our brain. And we have lots of them in hands and feet, especially at the fingertips where there are three to four times more present per mm² than in the palm of the hand. In fact, they are so close together that should we touch the fingertips with two fine objects only 2 mm apart, we can perceive them separately. If you do that on the skin of the forearm, they must already be 40 mm apart. Objects anything closer than that and we perceive them as a single touch (Purves & Williams, 2001). Do the test, you will be amazed. Also, in the naked strip of skin in the prehensile monkey tail, there are very many sensory organs (Organ et al., 2011). They can thus also feel with their tails what we feel with our fingers.

Hearing, seeing and feeling ... that's the saying. Hearing is not so important for climbing, unless you hear branches cracking. We already know the importance of feeling. But being able to see well is also vital if you are taking a leap to a branch a few feet above the ground. Grabbing a few inches not far enough or too far can have very painful consequences. We owe it to our ancestors among the Old World monkeys that we are very good at estimating depth. This is called stereoscopic vision or depth perception. Good depth perception has two anatomical requirements: the left eye must see in part what the right eye also sees, and the information coming in through both eyes must be processed simultaneously by both hemispheres of the brain (Jones et al., 1994). Only then does the brain succeed in creating an image that is registered differently by both eyes, which is perceived as a difference in depth. In Old World monkeys, the eyes face more forward so that the visual field of the left eye largely overlaps with that of the right eye. If an object is in that overlapping zone, both eyes will register it but each from a slightly different angle. Prosimians and a lot of other mammals have their eyes more sideways. For example, a rat or an elephant sees nothing of what is right in front of the snout and must turn their head to see it with either eye. Never do they see anything with both eyes at the same time. Predators, such as cats, do have an overlapping field of vision, yet they are not as good at depth perception as Old World monkeys. This is because all information from the left eye goes only to the right hemisphere of the brain and all information from the right eye goes only to the left hemisphere. So, each hemisphere of the brain gets to see only one image. In Old World monkeys, the information from the left eye goes to both the right and left hemispheres of the brain, and vice versa. Just as in our case. Now, stereoscopic vision is not essential to being arboreal. Squirrels are also good climbers and jump from one branch to another and yet they have no or limited stereoscopic vision (Cartmill, 1974). But it does help if you have it.

## Swinging around

So, great apes are **brachiators**, which has nothing to do with gladiators. Although both have very agile and very strong arms, brachiators use them to swing from branches, gladiators use them to beat each other up. An absurd comparison, you might think. Not for David Carrier, a researcher at the University of Utah, who showed that from an upright posture, we can hit much harder with the arms (Carrier, 2011), as well as that our strong hands can absorb the impact of a punch much better (Morgan & Carrier, 2012). Great apes also have short legs, compared to arms. This too is advantageous for fighting against conspecifics, as it lowers the center of gravity and thus increases stability (Carrier, 2007). This aside, back to climbing.

To lump all recent apes together by saying that they propel themselves in the same way just by swaying their arms is to discredit them (Gebo, 1996). In fact, they have a very wide repertoire of locomotory types. Gibbons are true brachiatorial swayers but can

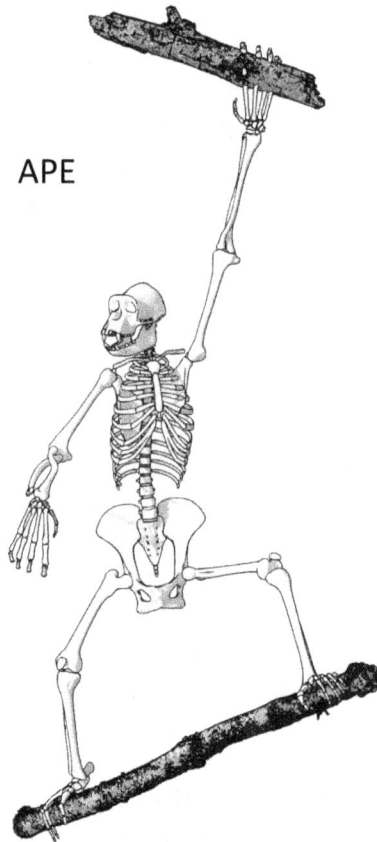

APE

Figure 3.4   Generalized skeletal construction of our clambering great ape ancestor. (Based on 3D model from TurboSquid, by Rembex – https://www.turbosquid.com/3d-models/chimpanzee-skeleton-654529.)

also walk on two legs on the ground. Orangutans scramble by hanging by their hands and stepping with their feet, or walk on four or two feet across the ground. Gorillas swing in trees, do knuckle walking on the ground or also walk on two feet. Chimpanzees do the same, but in a much more active manner. That variation was not always there among apes. In fact, apes walked mostly with four legs on top of the branches. For example, *Proconsul*, an extinct ape from Africa, showed a combination of primitive and specialized ape characteristics. Although they did not yet swing while hanging below branches, they already possessed a more mobile shoulder, wrist and hand (Begun, 2003). Other fossil apes from Asia (such as *Dryopithecus* and *Sivapithecus*) were already arm swingers and already possessed an elbow joint that could be fully extended, just as in recent apes. Indeed, their humerus already possessed a dimple at the back end, into which the elbow protrusion fits when the elbow was extended (Larson, 1998).

So, what typifies such a brachiator (Larson, 1998)? Long and fully extendible arms offer the advantage of being able to reach far into branches, while a fully opposable thumb allows one to grasp firmly around a branch. Compared to primates walking on four legs, apes show strikingly more differences in their arms compared to their legs. It seems that the development of arms and legs became decoupled in apes, allowing long arms to be combined with shorter legs (Rolian, 2014). By the way, the same is true for the hands and feet (Rolian, 2009; Rolian, Lieberman, et al., 2010). The shoulder also became very mobile thanks to all kinds of adaptations, such as a broad and deep thorax that became barrel-shaped (in gibbons) or funnel-shaped (other great apes). In monkeys, it is rather flattened laterally. A broad and rounded chest allows the scapula, which is movably connected to the thorax by muscles, to rotate at a very large angle as it slides along the chest. If we hold our arms stretched out in front and we bring them back, the scapula will slide across the thorax. The rounder the thorax, the greater the angle the scapula can make. Very few animals manage to rotate their arms over an angle greater than 180°. Apes can, and so can we. The scapula also lies more laterally and forward in apes than in monkeys, so the shoulders are also further apart from each other (Gebo, 1996). Combine long arms, a fully extended elbow with short olecranon (bony protrusion at the elbow), a large angle that the scapula makes and widely spaced shoulders, and you get a very efficient grasping system with arms that can cover a great distance while swinging (Spear, 2025). Moreover, the highly mobile scapula is stabilized during large movements thanks to a very sturdy and long clavicle that connects the scapula to the broad sternum (van Beesel et al., 2022). In apes, the shoulder blade is elongated, giving more room for larger muscles that pull the arms back and up. Apes also have increased mobility in the joint between the scapula and humerus, in part thanks to a larger articulatory head on the humerus. And yet everything is held well in place by a typically long clavicle and by an extended *acromion* (a protrusion on the scapula connected to the clavicle) (Spear, 2025). Did you know that a cat has very short and fine shoulder blades, which are separate from the shoulders and sternum? Therefore, they can squeeze through a small opening with no problem.

Although an adult male chimpanzee weighs about 15–20 kg less than an adult male human, a chimpanzee is as much as one and a half times as powerful as a man (Lieberman, 2013). They owe this to their muscles containing more so-called white or fast muscle fibers (amounts to up to 67% of their muscle volume, compared to only 30–50% in humans) (O'Neill et al., 2017). These muscle fibers are not only faster in their contraction, but they also generate more force for a given contraction speed (Fitts & Widrick, 1996). But being able to swing well on branches requires not only powerful arms but also strong yet flexible wrists and hands (Gebo, 1996). Although apes show quite some variation in their wrist characteristics, they do have some traits in common (Kivell et al., 2013). The wrist can move more because it lacks a protrusion on the ulna (the so-called styloid protrusion), something present in monkeys. In addition, apes have cartilaginous structures in the joints (small meniscus, like we have large ones in our knee joint) allowing the wrist to flex more. When an ape swings, it will rotate its body axis as soon as it hangs on with one hand and reaches forward with the other. This involves a lot of rotation in the wrist joint of the grasping hand. This is not a problem thanks to highly mobile joints between the forearm and wrist bones, which allow the forearm to rotate 150° around its axis. Grasping branches is also easier thanks to longer fingers – longer than the thumb – and the fact that the middle phalanges are slightly bent (Matarazzo, 2008). Flexible ankle joints are also useful when walking on branches. Chimpanzees will fold their feet far upward while walking on branches and turn their soles inward to grasp branches on the sides. This brings their center of gravity much closer to the branches and allows for a better balance (Holowka et al., 2017b). On top of that, they have extra mobility at mid-foot level, allowing them to rotate their feet around branches even more (Harper et al., 2021).

Often, the presence of flat nails on fingers and toes is also considered an adaptation to climbing. However, its contribution is probably only limited. Claws are more convenient than nails if the branch to which one is grasping is thick enough. With thick branches, it is enough to anchor the claws in the bark and one can hang on without having to firmly grasp onto the branches with fingers and toes. With thin branches, it is much better to grasp around the branches with a good gripping hand and gripping foot. Claws are rather useless then and may even get more in the way than would be the case with flat nails. Fingertips that are rather broad and soft also are beneficial for a good grip, together with the previously mentioned dermatoglyphs (Toussaint et al., 2020).

You may already have noticed that chimpanzees and gorillas have no tails. None of the apes do and even the primitive *Proconsul* had already lost it. If you are hanging below branches while swinging, you can hardly fall off a branch by losing balance. A tail will do little to change that. For *Proconsul*, it was not so obvious, for it was still walking on the branches. Its chest was still slender though, so the legs were closer together, useful for walking on branches, especially when combined with its powerful grasping hands and feet (Nakatsukasa et al., 2003, 2004). Only very recently has it been discovered why the tail disappeared in great apes. Primates, in fact, possess a specific piece of DNA called an *Alu element*. Now, there are many such *Alu elements* present (together they make up more than 11% of the DNA), but only in primates does the *Alu Sx1 element* occur, located in the *TBXT* gene. This gene allows a tail to develop. Apes, and therefore we too, possess a second *Alu element* (the *Alu Y* element) that plays a crucial role here. When both are present, the protein produced from this *TBXT* gene is shorter than normal. If one adds both *Alu elements* in mice, they also are born with a short to absent tail (Xia et al., 2021, 2024).

So, should all these typical ape characteristics be seen as an adaptation to arm swinging? This is increasingly being questioned. According to some scientists, many of those ape traits are rather adaptations to climbing **vertically upward**, something most apes also do well and a lot (gibbons are an exception here) (Fleagle et al., 1981). In orangutans, vertical climbing is even the most important component during their diurnal locomotion (Gebo, 1996). Strongly muscled forearms are crucial here. Chimpanzees have larger arm muscles than we do, which also produce more force for the same muscle mass (O'Neill et al., 2017; Thorpe et al., 1999). All sorts of differences between humans and great apes indicate that we have largely lost those adaptations, such as the longer shoulder blade in great apes (van Beesel et al., 2022). Also, specific shoulder muscles (parts of the *deltoid* muscle) run differently in great apes than in us, so that in us they raise the arms, while in great apes they pull them down, necessary for vertical climbing. Because of a different attachment of muscles, the arms in great apes are more powerful (such as the large pectoral muscle, the *pectoralis major*) and the hind legs also show clear indications that they are highly adapted to vertical climbing. If you want to climb upright in a tree, you must be able to put your feet on the side of the tree, with the soles of your feet against the bark (DeSilva, 2009). Only then will you have a good grip. Our ankles are not equipped for that, those of great apes are (Lovejoy, Latimer, et al., 2009).

Unlike humans, great apes have a rather stiff lower back, caused by two factors (Machnicki & Reno, 2020). On the one hand, the lower back is shorter with only four lumbar vertebrae (we have five, as do gibbons). On the other hand, they have a longer sacrum that is fixed very deep down, like a wedge, between the left and right iliac wings of the pelvis (the bony ridge you feel at the top side of your pelvis). On top of that, those wings are long and extended anteriorly, as well as they face outward. As a result, the lumbar spine is almost completely trapped within the pelvis, and there is very little free space between the thorax and the pelvis. Hence, a stiff lower back. But this is not so bad for great apes because a stiff back is useful when climbing vertically. This allows them to keep the back nicely straight without having to put too much muscle effort into this (Gebo, 1996).

Chimpanzees can also jump vertically very well: a human jumps an average of 30–40 cm high; a chimpanzee jumps 70 cm high (Scholz et al., 2006). A chimpanzee weighing only 34 kg produces a power of 3000 watts at take-off, which is as much as an adult human weighing 70 kg can produce. That too is thanks to more powerful muscles. By no means do they have it from a long Achilles tendon, a tendon that mammals use during running and jumping. This tendon connects the calf muscles (the *triceps surae*) directly to the heel bone (*calcaneus*) (Myatt et al., 2011). When mammals prepare to jump, they will first go down bending the legs, thereby stretching the elastic Achilles tendon. The resulting energy accumulated in the tendon then is released at take-off, like an elastic band, after which the tendon shortens again. The longer the tendon, the more energy it can accumulate. Surprisingly, however, orangutans, gorillas, chimpanzees and bonobos have very short Achilles tendons. The calf muscles attach almost directly to the heel bone, while other primates, including us, have a long tendon (Aerts et al., 2018). In fact, our Achilles tendon stores 35% of the energy we need to walk slowly (Alexander, 1991). So, don't those great apes suffer from that? On the contrary, it would not be convenient during vertical climbing if the foot were to tilt first due to stretching of that tendon, and certainly not in large animals like these great apes. That gibbons, which are small, have a long tendon and also climb vertically much less seems to confirm this. All evidence suggests that the common ancestor of the apes and humans had a long Achilles tendon, which was shortened in great apes as an adaptation to vertical climbing but retained in the hominins as a function of walking upright (Aerts et al., 2018).

In documentaries, you actually see chimpanzees and gorillas mostly **walking on the ground,** with the less heavy, young animals swinging in the trees (Holowka et al., 2017b). Climbing trees vertically is something they do mainly to look for fruit or hunt monkeys, but if they want to move horizontally, they usually do so via the ground (Gebo, 1996). Chimpanzees also run a lot faster than humans, even faster than elite runners: 40 km/h, compared to 23 km/h in elite runners (we're not talking about sprinting here, but prolonged fast running) (Lieberman, 2013). And yet chimpanzees are not so good at running on the ground. They use one and a half times as much energy as any other mammal walking on four legs, regardless of whether the chimpanzee walks on four or two legs (Rodman & McHenry, 1980). Chimpanzees are better adapted to climbing trees than to walk on the ground, that much is clear. Sturdy arms and agile feet are already useful for this. By applying white spots to chimpanzees' feet, Nicholas Holowka, a researcher at Harvard University, was able to map foot movement in chimpanzees (Holowka et al., 2017a; Holowka et al., 2017b ). This showed that they need to exert more movement while walking than we do, due to the flexible ankle joint. Over the course of our evolution, we have compromised on this flexibility, but this was actually advantageous when we started walking on two legs. In fact, walking on two legs works better with a more stable foot.

Gorillas, chimpanzees and bonobos that walk on the ground on four legs do something very special: they walk on their knuckles. More specifically, they only fold the last two phalanges of the fingers and lean on the leading edge of the first phalanx (White et al., 2015). They don't all do this the same way. Chimpanzees and bonobos lean on the index, middle and ring fingers, while gorillas include the pinky finger in this (Inouye, 1994). Gorillas keep their palm facing backward when walking with the knuckles, while chimpanzees and bonobos keep it turned inward (Simpson et al., 2018). This, together with the observation that it emerges differently during ontogeny in chimpanzees compared to gorillas, supports the hypothesis that they independently evolved using knuckle walking and that knuckle walking was not present in the last common ancestor between humans and chimpanzees (Dainton, 2001; Ito et al., 2024). Were we to try to walk on our knuckles like that, we would constantly flip and sprain fingers and wrists. These great apes obviously don't have that problem, thanks to all kinds of adaptations in their forearms, fingers and wrists. Other mammals that practice knuckle walking (such as anteaters) also exhibit similar adaptations (Orr, 2005). At the level of the wrist, the radius slightly protrudes so that it forms a ridge against which the wrist bones collide if they were to flex too much (Richmond & Strait, 2000). Also, some of the wrist bones (such as the *capitatum* – the wrist bone at the base of the middle finger) have protrusions that provide such protection. And even the metacarpal bones have a ridge, which prevents overstretching of the fingers. This protrusion is also only formed once the young animals begin to walk on their knuckles (Sarringhaus et al., 2022). In addition, the scaphoid (wrist bone located against the radius that supports the thumb, index and middle fingers) in gorillas, chimpanzees and bonobos is fused with another wrist bone (called the *centrale*). We too, no longer have a separate central bone, something we thus inherited from our great ape ancestor. In great apes, this causes a strengthening of the wrist, but this may also be related to vertical climbing (Orr, 2018). Chimps have short flexor muscles in their digits, which is handy for grasping but it limits the capacity to straighten the fingers when the wrist is bent forward during knuckle walking (Ito et al., 2024; Simpson et al., 2018). Gibbons and orangutans do not have all these adaptations strengthening the wrist and fingers. Therefore, you don't see those knuckle walking (orangutans do walk on their flat hands or their fists when walking on the ground) (Simpson et al., 2018). One might wonder why not all great apes walk on a flat hand or fist; surely that would be much easier. Turns out that knuckle walking also has an advantage: it ensures that when the knuckles are placed on the ground, the muscles pulling on the fingers are stretched. This allows for better shock absorption (Simpson et al., 2018). This is especially useful in those great apes, considering they have a shorter elbow protrusion as an adaptation to swinging, making the elbow a less good a shock absorber. The muscles they use to absorb shock in the fingers and wrist are well developed. After all, these are the same muscles needed to grasp at branches. So why don't great apes lean on the underside of their fingers? Most mammals do, however, which makes them digitigrades. The long fingers having difficulty folding upward in great apes could be an explanation. This would be very awkward for digitigrades but is very useful for tree swingers.

## Climbing and walking in trees

Already at the time of Charles Darwin, primates were divided into two groups: the 'pronograde' primates that walk on four legs and the 'orthograde' ones that **walk** upright **on two legs** (White et al., 2015). Guess where humans were grouped. Meanwhile, we know better that splitting into such two categories is just too simple, especially when you know that 70 different categories of body postures have already been described in primates. Moreover, it is hardly defensible to label certain groups of primates as 'orthograde' or 'pronograde'. Where do you place the great apes, knowing that they all occasionally walk on two legs and that even long-extinct Miocene apes showed adaptations to walk upright, even before typical adaptations to arboreal swaying had arisen (Almécija et al., 2021)? This raises a new dilemma: if an orthograde posture arose earlier than swinging below branches, then upright walking could not possibly have originated from an ancestor adapted to swinging. It seems increasingly likely that adaptations to vertical climbing in those early apes gave rise to two evolutionary trends: a far-reaching ability to sway with the arms in recent great apes and an ability to walk more upright on two legs in the ancestors of humans.

When chimpanzees and gorillas walk upright, they do so in a somewhat funny way: they waddle. This is no coincidence, because ultimately, they are primarily adapted to climb vertically, sway, or walk on the ground via knuckle walking. A stiff lower back good for climbing is not useful for walking elegantly on two legs. Long arms and short legs, useful for swinging, are not so useful for walking on two legs. Moreover, the stiff lower back combined with short legs means that when they rise, they must do so by keeping the hips and knees bent (White et al., 2015). They step with a **bent-hip-bent-knee**. The hips and knees are kept in an angled position throughout the entire walking cycle (Harcourt-Smith, 2013). This makes them walk while shuffling but also makes them wobble left and right constantly. But this has a dual cause. Because of the short, angled legs, they must rotate the pelvis much more horizontally to make a large enough step (Thompson et al., 2021). The short legs are also shoulder-width, straight below the body, placing the feet far apart. This makes them have to move their center of gravity over a long distance each time they put their supporting foot down, because they can't lean on that foot until the center of gravity is above their fulcrum (Harcourt-Smith, 2013). Only then can they lift their other foot to swing their leg forward. And so, they keep swinging their center of gravity from left to right and back. You can already guess that this is not very efficient. A chimpanzee uses the same amount of energy whether it walks on two or four legs but uses four times more energy to walk than a human (Sockol et al., 2007). Therefore, you will not see a chimpanzee or a gorilla walking upright if the purpose is to walk quickly somewhere far. Still, there are several reasons why they walk on two legs. Having their hands free to gather as much food as possible is certainly one of them. If one offers chimpanzees many energy-rich nuts, they will be less inclined to walk upright. There is then enough food for everyone. However, if they get a few of those nuts, they will walk up on their hind legs to four times more frequently. The reason? Then they can grab as much as possible and eat it somewhere else undisturbed (Carvalho et al., 2012). The motivation to walk on two legs is also not universal, as it turns out. Chimpanzees and bonobos primarily walk upright to carry food or to get to food that is located high up, but bonobos are also more likely to do so to keep a better eye on their surroundings, while a chimpanzee will also do so to impress conspecifics (by standing upright, they appear taller) (Videan & McGrew, 2001, 2002).

When great apes walk, they put their heels on the ground first, just as we do. Other mammals, however, put their toes down first, because mammals tend to be toe walkers (digitigrades) (Zeininger et al., 2020). Great apes, however, are sole walkers (plantigrades), just like us. But great apes do this in their own way: once the heel has touched the ground, the outer margin of the foot is put on the ground simultaneously (Vereecke et al., 2003). With us, only the heel touches the ground first. Great apes finally put their foot on the ground with the entire sole of the foot touching the ground (Holowka et al., 2017a). In other words, they have flat feet.

When African great apes walk upright, they do so mainly while on the ground. But walking on two legs with an erect posture can also be done in the trees, as orangutans do (Gebo, 1996). Females and young individuals especially do it this way. While using their arms to hold on to branches above their heads, they walk over the branches below using their grasping feet. As it were, they combine walking upright on two legs with swinging with the arms, only now the weight is carried by all four limbs. In other words, they are clambering. Scientists are increasingly convinced that our common ancestor with great apes did not propel itself like chimpanzees do but rather was clambering (Bohme et al., 2019; Ward et al., 2012). The discovery of fossils of apes from Germany and Hungary reinforced this view (Bohme et al., 2019; Williams et al., 2020). Fossil remains of the 11.6-million-year-old *Danuvius guggenmosi* from Bavaria and the 10-million-year-old *Rudapithecus hungaricus* from Rudabánya showed that these apes still used all four limbs to walk on branches but also showed adaptations that indicated upright locomotion on two legs. It is suspected that they even walked on two legs more than recent great apes; only they did not do so to walk on the ground. No, similar to orangutans they were clambering. Scientists sometimes refer to an orangutan phase in our evolution during which our ancestors were clambering (Gebo, 1996). Use of arms, in fact, facilitated walking on smaller and more pliable branches (Thorpe et al., 2007). The evolution within the clade leading to chimpanzees then involved more of a focus on arms for climbing and swinging vertically in trees and running across the ground via a four-legged gait. The evolutionary strategy within our ancestor was then more likely to involve a focus on legs to walk upright on the ground (DeSilva, 2021; Lovejoy, Latimer, et al., 2009).

# Tropical paradise

Our ape ancestor dwelt in tropical rainforest, an oasis of fruit, leaves and trees, to climb, play, sleep, make love, or raise young. A tropical rainforest is tropical, so it is always warm and humid. This creates a stable environment, without excessive challenges leading to overheating, in contrast to what animals living in hot savannahs have to deal with. The body part responsible for controlling body temperature? The skin.

## *Our largest body part*

Should I ask you what the largest organ in your body is, you might think of the brain, or the stomach. Not so. Our **skin** is by far the largest organ: it covers about 1.5–2 square meters and comprises about 16% of our body mass (Bryson, 2019). The skin also has numerous functions: it's a habitat for billions of bacteria, a barrier that (as long as it is intact) is impenetrable to bacteria, fighting bacteria if they do get into the skin, a firm mantle covering the body, forming hairs, holding sensory organs (for detecting vibrations, stretching, heat, cold, injuries or itching), holding numerous sebaceous glands (keeping the skin and hair healthy), protecting the body from harmful UV rays or forming ridges on our fingers for a good grip. Not to mention all the properties of the skin that allow us to control our body temperature (Brant, 2019).

The skin is a very complex organ. Earlier we mentioned the three layers of the skin: the outer epidermis, a middle dermis and a lower subcutaneous hypodermis. The **epidermis** is composed of several layers of densely packed cells equipped with special molecules that firmly anchors their cell membrane and internal cell skeleton (each cell has an internal skeleton of large proteins that give the cell strength and shape) of one cell to that of an adjacent cell. This makes the skin an impenetrable barrier to bacteria (Harris-Tryon & Grice, 2022). The epidermis is also the layer that forms the hair, thus giving rise to the fur. It may come as a surprise, but the epidermis contains neither blood vessels nor nerves. Therefore, you cannot feel pain in your epidermis, nor will it bleed if the epidermis is damaged. If you have a second-degree burn, you will develop a blister. This occurs because moisture builds up between the epidermis and the dermis, loosening the epidermis. When you puncture a blister, you only puncture the epidermis, and it will not hurt or bleed. If it does, you just hit the underlying dermis.

The **dermis** gives strength to the skin, thanks to connective tissue containing lots of collagen fibers. The dermis does have blood flow and is equipped with lots of nerves and sensory organs. Earlier we mentioned tactile organs, which can detect deformation of the skin due to vibration, compression or stretching. It is no coincidence, then, that primates, including great apes, have many such tactile organs in their fingers and toes, such as Meissner's corpuscles (similar organs have been observed in other mammals) (Bolanowski & Pawson, 2003; Covarrubias et al., 2024). These corpuscles lie at the very top of the dermis, that is, close to the surface of the skin. They are sensitive to slight vibrations and are thought to play an important role in primates for touch when grasping branches and other objects (such as fruit) (Verendeev et al., 2015). The number of such corpuscles per mm² varies greatly among primates. For example, chimpanzees have fewer than certain monkeys (such as macaques) and humans. Great apes and humans do have larger corpuscles, which has to do with body size. On our fingertips we carry about 50 such bodies per mm², with a size of about 45 μm (45,000ths of a millimeter). Chimpanzees, on the other hand, have about 37 per mm² of about 40 μm in size. Primates can distinguish between light vibrations and more intense vibrations, thanks to a different type of tactile organs that are sensitive to a different type of vibrations: Vater-Pacini corpuscles lie deeper in the skin and are sensitive to deeper penetrating vibrations, whereas Ruffini corpuscles are sensitive to stretching. The dermis also contains sweat glands, which are formed by the epidermis but grow into the dermis.

The **hypodermis** is the one that contains the fatty tissue. Primates store energy reserves as adipose tissue, which their bodies can draw on when food is in short supply.

If you compare African great apes to humans, some major differences immediately stand out in terms of skin: great apes are very hairy, but we are not. Their skin looks black, but there is a variable degree of darkness in human skins. The hair has to do with thermoregulation, maintaining the body temperature, the skin darkness with protection against UV rays.

## Cooling down with fur

Virtually every primate has **fur**. After all, they are mammals. The only exception you may know is us (Montagna, 1972). Yet, fur is very useful because it protects from harmful sun rays (especially the UVB rays are harmful), but also from radiant heat from the sun (Lieberman, 2013). In fact, fur can shield up to 50% of all radiant heat (Schwartz & Rosenblum, 1981). Hence, fur is not only useful when it is cold but also when it is hot. In a cold environment, the fur helps to better keep body heat inside because the dense hairs prevent the cold air from reaching the skin and thereby withdrawing too much heat from the body. The larger the primate's body, the fewer hairs are implanted per square cm (Schwartz & Rosenblum, 1981). At least that is approximately so, because surprisingly, in a chimpanzee there are as many hairs implanted per square cm as in humans (Roberts, 2021). And yet a chimp looks completely covered in fur and we look naked. This is because chimpanzees have long and thick hairs, while ours are small and thin. Chimps also keep the hair they are provided with when they are born, called lanugo hair. In humans, this falls out (Roberts, 2021).

There is also a connection between fur and sweat glands. In mammals, sweat glands occur mainly in places where there is no hair, such as the palms of the hands and soles of the feet (Groscurth, 2002; Montagna, 1972). At least this is so for sweat glands that are responsible for cooling the body. In fact, there are two types of sweat glands: eccrine and apocrine sweat glands (Baker, 2019). Although they are both called 'sweat glands', it is mainly the eccrine glands that produce sweat to cool down. Sweat is a salty fluid, through which body heat can be given off at the surface of the skin (Folk & Semken, 1991). When sweat evaporates, body heat is withdrawn from the body and a great ape can cool down. **Apocrine sweat glands**, on the other hand, are always connected to hair, through which they release their moisture. In primates, these glands produce a viscous, oily substance containing proteins, sugars and ammonia. They do not do this when they are too hot, but rather under the influence of emotional stimuli. So, this fluid does not help primates cool down, in contrast to other mammals (some even have almost exclusively apocrine glands). Although apocrine glands are formed everywhere, they disappear in many places. How many remain depends on how well developed the fur is. Not surprisingly, African great apes have many apocrine glands: about one-third of all sweat glands are apocrine. They are mainly concentrated at the level of the armpits. As you might expect, that ratio is completely different in humans because we don't have thick fur. Only one in ten are apocrine glands, which are concentrated in the armpits, at the level of the breasts, in the face, the scalp and in the genital region (Best & Kamilar, 2018). Not coincidentally all areas where we humans are often still thickly hairy! African great apes and humans are special compared to other primates, though, because they also have many eccrine glands in places where there are hairs (Montagna, 1972).

So, the **eccrine sweat glands** are the ones responsible for cooling in African great apes and humans. If we compare great apes with humans, we can only conclude that great apes have no real need for a robust cooling system. They have far fewer sweat glands per square cm than a human: about 10 while a human has about 100 (Kamberov et al., 2018). The total is therefore very different, because chimpanzees have proportionally fewer sweat glands for cooling (only 65% of the total number of glands, compared to 90% in humans) and their skin also has a smaller surface area (because chimpanzees are also smaller). A rough estimate is that chimpanzees have 270,000 sweat glands, compared to somewhere between 2 and 5 million in humans. On top of that, a chimpanzee is almost completely covered in fur, which is also not advantageous for cooling off via sweating.

Mammals cool down in different ways. Some pant, cooling through the tongue. Others have large and thin ears that give off heat. And some sweat. All three of these ways have one thing in common: cooling only works if wind passes by those body parts (Carrier, 1984). A dog hangs its tongue out while panting for a reason. Nor does an elephant just flap its ears for no reason. The efficiency of cooling depends on how well the wind gets up to the skin surface. If the wind gets close enough, sweat droplets on the surface will evaporate easily and body heat will be extracted. In African great apes with a dense fur, this is not so obvious. To give an idea, wind would already have to blow at a speed of 18 km/h before a sheep's woolly skin of 4 cm thick will be able to cool the skin.

All indications are that African great apes do not get rid of excess body heat as easily. The logical question then is the following: isn't that a problem? Obviously not, otherwise those great apes would be constantly overheated and passing out. The answer is simple: they do not need to cool down much and quickly because they live in a tropical rainforest where the temperature is not so high and rather stable. On top of that, the humidity in a rainforest is also higher, which means that cooling through sweat wouldn't go so well either. In fact, sweat evaporates much better when the wind is dry.

## The pale chimpanzee with hair

If you look at the naked skin in the face and on the hands and feet of an African great ape, you will notice that it is deeply black. Dark skin results from black-brown pigments or melanin formed by special cells (melanocytes) in the epidermis and dermis (Johnson et al., 1993). In mammals and thus primates, melanin gives the dark color to the skin (Montagna, 1972). How much melanin these cells produce is genetically determined, which provides the color base (Deol, 1975; Jablonski, 2012). But even that can change depending on how much UV rays penetrate the skin. More UV causes more melanin to be produced, hence darkening the skin, even in African great apes. You can for example see the combined effect of genetic control and exposure to sunlight being reflected in the facial skin of the different subspecies of chimpanzee (Kingdon, 1997). Chimpanzee newborns all have pale skin, as clear from their naked facial skin (should you shave a chimp, it would be colored pinkish all over). Where the fur is present, the skin will remain pink as they grow and get exposed to the sun. However, this changes at the level of the hands and face, depending on the extent to which the population is exposed to the sun. The population least exposed is the Central African chimpanzee (subspecies *Pan troglodytes troglodytes*). They live nicely shielded in the dense rainforests of Central Africa. Their faces remain pale. The population most exposed is the easternmost (subspecies *Pan troglodytes schweinfurthii*) found in the border area between the tropical equatorial forest and the East African savannah. Their facial skin turns completely black as they age. The West African chimpanzee (subspecies *Pan troglodytes verus*) retains a pale face and gains only a black band at the level of the eyes.

# The ape brain

We have specialized mammalian brains; that was already clear. But there is more going on in primates, in great apes, and thus in their common ancestor with humans. So, what makes a great ape brain and thus that of our ancestors?

## The primate brain, a special mammalian brain

How the brain is constructed depends on two important factors: what did the brain look like in the ancestor and what does the brain need to do. By the latter I mean the following: what functions must the brain coordinate to have a high enough chance of survival in a given environment? A primate brain, like the brain of other mammals and even other vertebrates, consists of several parts, interconnected by numerous nervous pathways. A mammalian brain stands out because it is large, yet birds also have large brains. So, mammals are not unique in this (Olkowicz et al., 2016). A remarkable example: our brain makes up 2.3% of our body mass, while that of a goldcrest (a tiny bird with a golden stripe on its head) is 7.9%, more than three times the size of ours! Evidently, this must be put into perspective because birds have also become much lighter during evolution, as an adaptation to flight. So, a bird 20 cm tall will weigh less than a mammal 20 cm tall, which makes it logical that the relative weight of the brain is greater (Font et al., 2019). The largest brain, in absolute mass, is found in the sperm whale: they have about 10 kg of brain (Mortensen et al., 2014). They are also very large, yet cetaceans, like primates, have a larger brain-to-body ratio than many other mammals. So, the relationship between the two is not the same for all mammals.

Mammalian brains consist of three major parts: cerebrum, cerebellum and brainstem. The **large brain** (*cerebrum*) consists of two halves, the cerebral hemispheres, which are connected in the middle by the callosal commissure. On the surface of these hemispheres lies the neocortex, the main processing center, characterized by numerous anatomical adaptations. For example, it contains a large number of neurons or nerve cells that are responsible for processing all the information that comes in, as well as storing information in the form of memory and making decisions about what signals will eventually be sent out to what organs, such as muscles. In mammals, neocortical neurons are organized into six layers. In chimpanzees, these make up a neocortex of nearly 1.5 mm thick. In us, it is almost 3 mm thick (somewhere between 2.3 and 2.7 mm) (Pakkenberg & Gundersen, 1997; Xiang et al., 2020). Information enters through a middle layer and is sent to the outer layers (where information processing occurs) and to the inner two layers (from where information is sent to other areas of the brain) (Kolb & Whishaw, 2014).

The thicker the **neocortex** and the larger its surface area, the more space for neurons. A rat has a neocortex with a smooth surface, what is called a lissencephalic brain. Totally different from what our brain looks like. We have gyrencephalic brains, brains full of folds (a fold is a *gyrus*) between which there are grooves (a groove is a *sulcus*). All primates have gyrencephalic brains, but not all like ours. A prosimian monkey is more likely to have few but thick folds. Monkeys usually have more and somehow finer folds. In great apes this is even more refined; in humans this is the most pronounced of all. The more folds, the more room for neurons, so the more intelligent you would think? It is not that simple. The degree of fold formation is actually just a result of the way the brain grows, and especially the rate at which the superficial neocortex grows relative to the interior of the cerebrum. Primate brains are smooth at first, then grooves form as the brain grows. In us, too, the first grooves do not form until mid-pregnancy (Cowan, 1979). As the neocortex grows faster, the slower growing interior pulls against it. As a result, the superficial neocortex comes into folds and you get a large surface area without the full brain having to be proportionally larger (Lieberman, 2011). If you were to fully extend the neocortex of humans, you get an area of about 2230 square cm, the area of 47 by 47 cm square. In a chimpanzee, this is only 735 square cm, about 27 by 27 cm (Xiang et al., 2020). Also compared to other brain parts, the human neocortex is about 50% larger than in other primates (Lindhout et al., 2024).

The **folds** do not form randomly. Certain parts of the neocortex with specific functions come to lie in a particular manner with respect to each other. The relative size of such brain areas also reflects their functionality. Brains are like a mosaic of building blocks, with each building block providing a particular function. If you want a house to meet specific requirements, you are also going to invest in one where the kitchen is bigger, or there is more sleeping space. Brains are very similar; only the size and type of the building blocks are determined by the genome and the environment in which mammals need to be able to function (DeCasien et al., 2017). The neocortex also consists of a few building blocks (Kolb & Whishaw, 2014; Standring, 2020). In front lies the frontal lobe, from where stimuli are sent to the organs all over the body, as well as

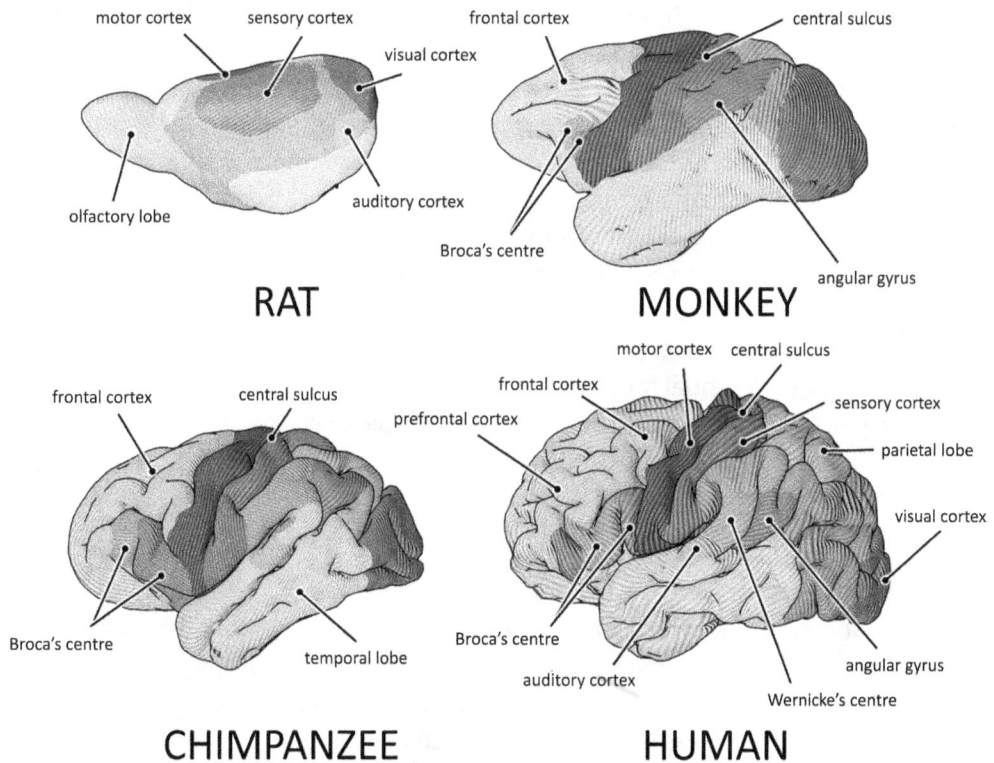

Figure 3.5    Large brain and its lobes in some mammals (view on left side). (Modified 3D model rat and Macaca brain from scalable brain atlas – https://scalablebrainatlas.incf.org/; chimpanzee and human brain from Sketchfab, by Gwuanthro – https://sketchfab.com/3d-models/comparative-brain-anatomy-chimpanzee-human-acf1799d8f9a4242849579f2a782c8df.)

where thinking processes take place, where decisions and plans are made and where information is stored in memory. This lobe runs up to a large transverse groove called the central sulcus. Behind this groove then lies the parietal lobe, where information from many sensory organs enters that is then integrated with that from other lobes (such as occipital lobe and temporal lobe). At the very back lies the occipital lobe, where information coming from the eyes is processed. To the side lies the temporal lobe, where the processing of language and music takes place, among other things, as it receives information from the auditory organ. So, you can start comparing lobes between mammals.

What does this **brain mosaic** look like in primates? The variation reflects the functionality of brain areas in relation to crucial biological functions (DeCasien et al., 2017). Factors that play a role include diet (quality diet with lots of fruit versus a diet of mainly leaves), being active day or night, or living in groups instead of in pairs or alone. Contrary to what you might expect, and to what some researchers also expected, the size of the neocortex is mainly related to the quality of the food, even more so than to the group size in which they live. One also observes this in some New World monkeys (Milton, 2012). It was long thought that group size was limited by the size of the neocortex (Kudo & Dunbar, 2001). Being able to live together does require being able to process a lot of information being exchanged between groupmates. This does play a role but only in second order. The largest neocortex is found in primates that eat quality food and live in large groups. Day-active ones have strongly expanded areas responsible for processing visual information. Obtaining better food is based on the ability to process what one sees. In night-active species, a highly developed olfactory system contributes more to the larger neocortex. Being able to smell whether food is ripe or toxic then becomes very important. But a good nose and brain areas to process everything are also essential for chemical communication. Animals more likely to live alone or in pairs have larger brain areas involved in spatial memory. They also need to be able to look for partners more and know where to find them. Great apes also have a large frontal lobe, especially the prefrontal cortex that belongs to it (Semendeferi et al., 2002; Smaers et al., 2012).

The **small brain** (*cerebellum*) lies behind the cerebrum, and in mammals it is also gyrencephalic and equipped with a left and right hemisphere (with another central piece or *vermis* in between). Mammals have a special cerebellum (Hatten, 2020). Its mass increases much less rapidly with body mass than the cerebrum does, and the number of neurons also increases less rapidly with increasing brain size (Herculano-Houzel & Kaas, 2011). The African elephant is an exception though, where the cerebellum constitutes about one-fourth of the brain mass (Herculano-Houzel et al., 2015). In primates, as well as in humans, the entire brain is about ten times larger than the cerebellum. So, elephants have an exceptionally large cerebellum with exceptionally many neurons. Their communication via infrasonic sounds (sounds at very low frequencies) and a trunk carrying about 30,000 muscle fascicles that must be controlled may explain this (Longren et al., 2023). Indeed, the cerebellum is responsible for motor control, more specifically for complex movements that are automated. When we learn new movements, such as riding a bicycle, we will first control it from the neocortex. So, we think a lot about how to do this, to avoid falling. But once mastered, we just do it without thinking. We can even think about other things at the same time. This is possible because the cerebellum has taken over that function, relieving the cerebrum. Thank goodness, because otherwise a bike ride in nature would be little relaxing. Also, information from the hearing and balance organs is received by the cerebellum, hence the connection to infrasonic sound in elephants. The cerebellum does much more than that, however. They are also important for a variety of cognitive processes and decision-making. In primates, the cerebellum is large in great apes, with especially strong connections to the frontal lobe of the neocortex (Smaers et al., 2012).

Behind the cerebellum then lies the **brainstem**. This is not a prominent part, yet very important. It contains the brain centers involved in regulating the heartbeat and breathing rhythm. From there the brain passes into the spinal cord, from where nerve tracts branch off at the level of the vertebrae to run (mainly) to muscles. At the same place, numerous nerve tracts arrive in the spinal cord, supplying information from all sensory organs scattered throughout the body. Exceptions are the olfactory organ, eyes and inner ear, which are connected directly to the brain by nerves. In the brainstem, as well as in other places in the brain and the spinal cord, a crossing of some nerve pathways occurs, from left to right and vice versa. When our brain sends a signal to move our left hand, this signal departs from the right side of the neocortex (at the level of the motor cortex in the frontal lobe). Similarly, information coming in through the left eye travels to the right side of the brain. This crossing of nerve pathways is not new to primates, and therefore not new to humans. In fact, it possibly dates back to the invertebrate ancestor of the first vertebrates, when during evolution the nerve cord shifted from the ventral side to the dorsal side of the body (de Lussanet & Osse, 2012; Kinsbourne, 2013).

## *Neurons are in control*

The diversity in cell types found in the brain is actually astonishingly high, when you discriminate them based on their genetic profile. Based on a screening of about 7 million brain cells, researchers could identify 5322 different cell types (Yao et al., 2023). Of those, these are the **neurons** that are the main players, yet they require the support of a variety of other cells called glial cells. Neurons are elongated cells, which, like tiny computer parts, receive information, pass it on and store it in a working memory, and use that information to create new information that is sent out as commands. Information that comes in is transmitted as an electrical signal traveling to the tip of a neuron. There, at a widening called the synaptic vesicle, chemicals are being discharged. These are neurotransmitters that will bind to another neuron that is separated from this synaptic vesicle via a gap. This vesicle and gap, combined with the place where neurotransmitters bind to the neuron, form what is called a synapse. Through a synapse, the electrical signal in one neuron is converted to a chemical signal, which is then translated back into an electrical signal in the other neuron. One neuron can transmit signals in this way to numerous other neurons, on average about 1000 neurons (but in special neurons, such as Purkinje cells in the cerebellum, this goes as high as 200,000 neurons) (Chudler, 2020). The number of synapses also varies greatly and is constantly being adjusted. The more stimuli that come in, the more synapses are formed, the more information that can be stored. Thus, memory is determined not only by the number of neurons but also by the number of synapses between neurons (Kolb & Whishaw, 2014). In primates, and therefore great apes and humans, there is even a special type of neuron (Von Economo neuron) that is larger than ordinary neurons and are thought to play a role in consciousness (Evrard, 2018; Nimchinsky et al., 1999).

Neurons handle the transport of **signals,** from the outside world but also from the inside (as in stomach cramps or muscle pain). From sensory organs, stimuli travel to the spinal cord and brain. The distance between a sensory organ or muscle and the spinal cord is bridged by only one neuron each time. This means that some neurons can be very long. The neuron that runs from the tip of your toe to your spinal cord is more than a meter long (depending on how long your legs are, of course). Speaking of a blue whale, a whale that can grow more than 30 m long, they have neurons of more than 20 m long. Some dinosaurs may have even had neurons that grew up to 50 m long (Coyne, 2011). In the spinal cord and brain, neurons provide communication between parts within both. For example, information from most sensory organs reaches the anterior fold of the parietal lobe, a strip called the somatosensory cortex. Before that information arrives there, it has already passed through three parts of the brain (spinal cord, brainstem and thalamus) (Kolb & Whishaw, 2014). From there the information goes, among other things, to the nearby somatomotor cortex, at the very back of the frontal lobe. From here the signals leave for the spinal cord, finally reaching the muscles to trigger contraction. In both the somatosensory and somatomotor cortex, one can identify, as it were, a map of which part of the body is connected to which part of these two areas. For example, nerves from the face will enter at the bottom of the somatosensory cortex and nerves will leave from the bottom of the somatomotor cortex to the muscles in the face. The size of those areas depends on how many nerves enter or leave. For example, the sensory area of the lips and fingertips is proportionally very large because there are a lot of tactile organs there. A lot of brain areas are involved in executing movement. When we reach for an object, we receive signals from the tactile organs that capture information about whether we have a good grip on the object. Based on that, the prefrontal cortex in the frontal lobe will use the information to plan how we should adjust our grip to have a firmer hold on the object. From there, the area next to it establishes the sequence of movements to be performed, then fires up the affected areas in the somatomotor cortex so that the correct muscles are prompted to contract in the correct order and to the correct degree. And so communication occurs between parts of the brain at all levels, depending on their function. The neocortex is also very creative in the ways in which information can be handled, with certain actions also being improved through a learning process (Herculano-Houzel, 2021).

For a long, long time, there was speculation about **how many neurons** our brains count, let alone a good idea of what the situation is like in other primates or other mammals. One often reads in literature that humans would have 100 billion neurons, with another ten glial cells for each neuron. Those 100 billion neurons would be a rough rounding up of a study that suggested there were 85 billion neurons. Rough rounding, no? The ten glial cells per neuron would rely on even less concrete information. In 2004, no one still knew how many neurons were actually present in the human brain! It wasn't until Suzana Herculano-Houzel, a neurobiologist at Vanderbilt University, decided to blend brain tissue into a soup to effectively count the number of neurons (Herculano-Houzel, 2016). She developed a technique to count the nuclei of neurons via a special staining. This allowed her to confirm that humans have 86 billion neurons in the brain. By also analyzing brains of other mammals, she made the surprising discovery that primates have more neurons in the neocortex than other mammals. For example, a 48-kg capybara (a large rodent) has a neocortex of about 48 g, containing 306 million neurons. An Indian crown monkey (a type of macaque) of only 8 kg also has a neocortex of about 48 g, but it contains as many as 1700 million neurons, i.e. 1.7 billion (Herculano-Houzel, 2012). Primates clearly had an evolutionary advantage. Equally surprising was that there are many more neurons present in the cerebellum in primates, although the neocortex is much larger than the cerebellum. In humans, the neocortex contains 19% of all neurons and the cerebellum about 80%, while the mass of the neocortex makes up 45% of brain volume and the cerebellum only 15% (Gabi et al., 2016; Herculano-Houzel et al., 2014)! Thus, we have 16 billion neurons in the neocortex and 69 billion in the cerebellum (Azevedo et al., 2009). This ratio also applies to other primates. Primates also differ from other mammals in that neurons do not get larger as the brain gets larger (Herculano-Houzel et al., 2007). In conclusion, primates have gained larger brains throughout evolution due to an increase in the number of neurons. Admittedly, this trend is true up to a certain level for vertebrates in general, as has also been observed in fish (Marhounova et al., 2019). Only in primates is it much more pronounced. If we also know that the number of neurons, with all their synapses, determines intelligence, then this was a huge advantage for primates and thus also for our great ape ancestor.

All primates follow a similar trend in terms of the number of neurons. At least this is so if we compare it with brain size, but not necessarily so if we include body size. Two great apes are outliers that differ greatly from that trend: the orangutan and the gorilla have abnormally few neurons for their body size, compared to other primates. Yet, they do have the typical primate relationship between neuron number and brain size (Herculano-Houzel & Kaas, 2011). That implies that, although they have a normal number of neurons for their given brain size, they have abnormally small brains for their body size. So, their bodies are simply too big for their brains. And those are the only two where this occurs. Bizarre, isn't it? We'll explore this further when we also take a closer look at human brains.

And then there is that speculation that each neuron would be accompanied by ten **glial cells**. Herculano-Houzel's research showed otherwise. Only in even-toed ungulates (artiodactyls), such as cattle and antelope, are there more glial cells than neurons. Primates have about as many glial cells as they have neurons, although that ratio may differ across brain regions (Fang et al., 2022). In other mammals, it is even less than the number of neurons (Azevedo et al., 2009).

## Olfaction becomes less important

The sense of smell is an important sense; we use it every day, even more than we realize! About 80% of what you think you're tasting is more likely to be something you're smelling. The aroma of food in your mouth passes through the throat to the nasal cavity, so you smell your food more than you taste it. Not surprisingly, when you have a cold with a stuffy nose, your food only 'tastes' half as good anymore. Smelling is also an important part of how our memory works. Who hasn't already experienced certain childhood memories brought about by smelling certain smells we knew as a child? Different areas in our brain are also responsible for so-called olfactory memory, the remembering of smells. That smell is and was important for us and our ancestors is therefore not surprising. We can distinguish odors of close relatives and recognize bodily substances that are part of the immune system (Milinski et al., 2013). It has even been found that we subconsciously have a habit of smelling our hands more after shaking hands with someone (Frumin et al., 2015).

Looking briefly at primates in general, it is striking that prosimians and monkeys have a remarkably long **snout**, at least compared to great apes and certainly to humans. A long snout has the advantage of providing a larger surface area to store sensory cells responsible for olfaction. In humans this area is only 5 cm² on average, which is even less than in a rat that is much smaller (there it is 6.9 cm²) (McGann, 2017). So, animals with larger snouts are, as a rule, better smellers. It is generally said that the first primates were nocturnal animals, where the limitation in vision was compensated for by a very good sense of smell. We still see this in prosimians. Monkeys are more likely to be diurnal, in which vision has become more important. They still have distinct snouts, but comparatively shorter than those of prosimians. In great apes, the trend of muzzle shortening continues. Especially in the evolutionary lineage to humans, this trend has continued furthest, to the situation in modern humans where we no longer have a snout at all. We also see the corresponding shift in importance from smell to vision in the brain. Prosimians still have a large olfactory lobe located in front of the brain where information from the olfactory system enters. In monkeys it is significantly smaller than in prosimians, and in great apes and humans, it is retained as a very small, flattened vesicle at the bottom of the cerebrum (it only makes up

0.01% of the brain mass, compared to 2% in mouse). In contrast, the occipital lobe, where visual information enters, has relatively increased in size throughout primate evolution.

So, you probably think that we must be **bad smellers**. At least, this was long assumed, but research shows it is not that simple. A study in 2017 showed that the olfactory lobe in humans is quite large, at least in absolute terms (about 60 mm$^3$, compared to 10 mm$^3$ in mice) (McGann, 2017). So, we still have a good sense of smell and can smell certain odors even better than rodents or dogs. We only need a sniff that takes as little as 60 ms to be able to identify some odors (Wu et al., 2024). For example, we can better distinguish amyl acetate, the predominant odor of bananas (Gilad et al., 2004). Of the 1000 or so genes that can create receptors to smell substances, only 390 are still active in humans (in mice, there are 1100 active genes) (McGann, 2017). This is where we and our great ape ancestors may have compromised. Some think this is a consequence of the activation of more genes to see colors (Gilad et al., 2004), although there was no decrease in olfactory genes when additional genes for color vision arose in primates (Matsui et al., 2010). In fact, we have retained more olfactory genes than an orangutan or a macaque. But genes, receptors and neurons in the olfactory system and olfactory lobe are only one side of the story. The processing center behind it is another story. The highly developed neocortex in great apes and humans provides a much greater processing capacity for olfactory information. This occurs primarily in an area of the neocortex called the orbitofrontal cortex, which is much more highly developed in humans than in most other mammals. This allows us to distinguish more than 1 trillion different odors from each other, i.e. much more than the 10,000 that people sometimes put forward (Gilad et al., 2004).

In summary, our great ape ancestor did have a shorter snout than a lot of other primates, but it would be wrong to say that it was also a poor smeller. Let us call it a 'different smeller', where visual ability with color vision has become more important.

## New folds

When someone makes a bizarre leap of thought, we sometimes call it a 'special brain twister'. Anatomically, great apes and humans also effectively have special brain **twists**. I am referring, of course, to the folds in the neocortex of the cerebrum. Great apes and humans usually have more folds than other primates. The change was not equal for all lobes of the neocortex throughout evolution. The occipital lobe already became larger as soon as vision became more important in diurnal primates, we already know. Especially striking is the strong fold formation in the frontal lobe, parietal lobe and temporal lobe. The expansion of certain folds has the advantage of providing more space for additional neurons, but also because certain groups of neurons could be given separate functions. Or put another way, more room was provided for neurons to take care of functions in a more coordinated and more extensive manner. About 20 areas are generally distinguished in a mammalian neocortex, which got expanded to about 50 in primates. In humans, there are even about 200 (Dehay & Kennedy, 2020; Kaas, 2019).

Each of the lobes can be further divided into areas, delineated by specific grooves between the folds. For example, **Korbinian Brodmann**, a German anatomist (early 20th century), delineated 52 different areas in the neocortex in humans and some other primates (Brodmann, 1909). This numbering is still used today. For example, area 17 is where visual information enters the occipital lobe, or area 4 where motor stimuli leave for the body. Other areas were further subdivided (e.g., the area where sensory stimuli enter from the body consists of three Brodmann areas, i.e. areas 1, 2 and 3) (Standring, 2020). Brodmann also estimated the size of individual lobes and found that the frontal lobe had greatly increased in relative size, from 11% in a macaque to 17% in a chimpanzee to as much as 29% in humans (Semendeferi et al., 2002). But this turned out to have been a misjudgment. More recent methods, such as magnetic resonance imaging (MRI), allow for more reliable visualization and analysis of brains in 3D. Measurements based on MRI show that, relatively speaking, the frontal lobe is not that much larger than in great apes (the exception being gibbons, which have a smaller lobe), but it is larger than in other primates. If the difference is not in size, then it must be in the folds, in the neuron networks within them, and how those networks are connected to other areas.

One of the specialized areas in great apes and humans is the **prefrontal cortex**, located in that frontal lobe, just before the part where left and right hemispheres communicate. This area comprises about 10% of the total neocortex mass (Ribeiro et al., 2013). The organization of the folds in this area is fairly constant in Old World monkeys and great apes, and thus also in humans (Falk et al., 2018). Yet, in humans this area is mainly enlarged due to an expansion of the white matter, i.e. the parts that connect the prefrontal cortex with other brain parts (Lindhout et al., 2024). Why is this area worth mentioning? This is the area where the most important and complex cognitive processes take place, an area that has played a very important role in our evolution. The development of complex language and the creation of tools might never have emerged if our prefrontal cortex had not become so specialized (Ponce de León et al., 2021). If we rank animals according to the extent to which they use tools, humans are obviously at the top. The chimpanzee ranks second, with their use of sticks to extract insects from crevices, stones as hammers to crack nuts or chewed leaves as drinking sponges (Whiten et al., 1999). Bonobos, gorillas and orangutans use tools much less (Haslam, 2019).

Great apes and humans have extra neocortical folds here and there, which help underlie the development of **language** (Rouault & Koechlin, 2018). Yes, even in great apes there are specific areas that can be associated with language, not only in the prefrontal cortex but especially in the temporal lobe. Now, 'language' and all that comes with it must be viewed very broadly in this evolutionary context. The origin of non-chemical communication, i.e. all communication that does not rely on the diffusion of odorants, presumably lies in body language. This can be inferred in part from the fact that the brain areas that play an important role in language in humans also play a role in handedness in great apes. And as in humans, these areas in great apes are not equal in the left and right sides of the brain. In 1861, French neurologist Paul Broca noted that speech is lost with brain lesions in certain areas of the neocortex (Condemi & Savatier, 2019). He called this the 'speech center', a center we know today as Broca's center. This is located at the bottom, just before the somatomotor cortex (the area from which neurons depart to control muscles). Brodmann designated this center as areas 44 and 45. Meanwhile, we know that when we speak, the motor activity of the vocal cords, tongue muscles and jaw muscles is controlled by this center. This Broca's center is also connected to other language centers, which are responsible for other specific language functions. For example, the temporal lobe contains the auditory cortex, the angular gyrus and Wernicke's center. The auditory cortex is where information enters from sound captured by the inner ear. At that moment, we are aware that we are hearing sound, but no meaning is given to it yet. That doesn't happen until Wernicke's center, discovered by the German neurologist of the same name in 1874. So, this is where the experience and understanding of language originates. But of course, it does not stop there. If we want to give a response based on the meaning of that information, we first appeal to what we already know from the past (including by giving meaning to sounds we already know, i.e. words) and a signal goes to the angular gyrus. This angular gyrus connects Wernicke's center with Broca's center and is also active when we read language (via visual information coming from the occipital lobe) and understand it, when we process numbers and when storing memory (Seghier, 2013). Based on the information from that angular gyrus, the Broca's center is eventually triggered to stimulate muscles so that we produce speech that has meaning, such that a bystander can also understand it (Kolb & Whishaw, 2014).

Remarkably, these language centers are not symmetrically laid out in humans. Carl Wernicke had already noticed that 'his' center was larger in the left hemisphere. There is also a correlation with handedness: in 99% of right-handed people, the left center is larger; in 70% of left-handed people, the right center is larger (Condemi & Savatier, 2019). Actually, I should rephrase this because the center of Wernicke is only present in one hemisphere of the brain. Evolutionarily speaking, this center in humans is an outgrown brain fold called the planum temporale. Great apes also have a planum temporale, on both sides of the brain. And in them, too, this is asymmetrical: in 94% of chimpanzees the left planum temporale is larger than the right (Gannon et al., 1998). Even in orangutans this is noted (Gannon et al., 2005). This is also true for Broca's center (Cantalupo & Hopkins, 2001). Both the planum temporale and Broca's center also contain more neurons on the left side (Spocter et al., 2010). Thus, this asymmetry in our brains most likely originated in our great ape ancestor and formed the basis for language, even without the existence of spoken language. Great apes have asymmetrical language centers and yet they don't speak language? Well, those centers do more than produce verbal language. They are also involved in sign language, i.e. body language with the limbs. So, do great apes use this form of communication? Yes, research on great apes showed that they use sign language to deliberately communicate with each other through more than 80 different signals (Hobaiter & Byrne, 2014; Hobaiter et al., 2022). There exists even a small 'dictionary' for chimpanzee sign language. When a chimp raises his arm, he wants to indicate to a conspecific to pick up an object for him. However, if he waves the arm, it is more likely to indicate that he needs to go away. Sign language is very important for social bonds in chimpanzees and is even more important than vocal communication with those with whom they have a close relationship (Damjanovic et al., 2022). Sign language also goes much further in great apes, as has been observed in the gorilla named Koko, who used more than 1500 learned gestures to communicate with her caretakers, even answering questions asked in English (Patterson & Gordon, 2001). Great apes are also primarily right-handed, just like us. In the brain, Broca's center lies near the area that controls the motor activity of the tongue and lips, which in turn is near the area that controls the hands. And remember that it is the left hemisphere of the brain that controls the right hand. Coincidence that the asymmetry in handedness (left- versus right-handed) is linked to the asymmetry of these language centers, and correlated with handedness and sign language in great apes, and spoken language in humans? Highly unlikely, no?

## So …

So, from all this we can deduce that the common ancestor of the chimpanzee and humans did not look like a chimpanzee and did not move about in trees swinging in the same way as a chimpanzee today does (Almécija et al., 2015; Chaney et al., 2021; Tocheri et al., 2008). This great ape ancestor weighed somewhere between 30 and 60 kg and was slightly taller than a meter. It had a distinct snout with large canines, especially the males. The fingers were longer than the thumb, with curved phalanges. The wrist was strong but still mobile for climbing, as was the thumb. Sturdy forearms had large muscles to fold the fingers while clambering, and the elbow could be fully extended. Ankles and feet were flexible, useful during vertical climbing. The spine had 7 cervical vertebrae, 13 thoracic vertebrae, 4 lumbar vertebrae, a sacrum with 6 vertebrae and 3 short caudal vertebrae (Thompson & Almecija, 2017; Williams & Pilbeam, 2021). The ancestor lived in the rainforest and presumably walked upright on the branches with arms and legs, both with four and two feet. Fruit was the main food. The skin was pale but strongly hairy, with a limited ability to cool via sweating. The brain was a normal size for a primate of that body size, about 300 cm$^3$. They had a well-developed neocortex with many folds, with certain areas involved in sign language already being asymmetrical. Just as anthropologists have had to do, we will further refine and modify this picture once we include the first hominins in the next chapter.

# 4

# The human cradle

We now know how the foundation of our bodies was laid by our common ancestor with the great apes. Let's make a taxonomic step, from great apes to hominins. Expressed in terms of evolutionary trees and branches, we are now right after the split, whereas in the previous chapter, we were right before it. We are at the very beginning of the branch that moved toward humans, away from the branch that gave rise to modern chimpanzees and bonobos. This seems like a big step, but evolutionarily speaking, it is no bigger than that from one great ape to another. We now also move in time, from the Miocene toward the Pliocene, to the period between 9 and 5 million years ago (Carroll, 2003). When exactly this branch split off, we are not entirely sure. Fossils point toward 8–7 million years ago, genetic information rather toward 9–5 million years ago (depending on which source one follows). But this is unimportant to the story of our bodies. What is important is that significant changes occurred during this period resulting in the evolutionary birth of the first hominins. Unfortunately, fossil evidence from that early period is so far rather limited. Much information about the physiology of those first hominins is therefore still unclear.

## The first hominins

The evolution of humans is sometimes presented as if it were a gradual process, a simple line-up of adaptations that followed each other at a fixed pace. However, this is not the case at all. Not only did one branching follow another but the pace also changed constantly. The start of our evolutionary lineage was actually quite slow. On top of that, our evolution also comprises changes that did not involve true adaptations but rather changes that were simply brought about because of coincidence. These traits did not disappear from the population because they were not detrimental enough (Harris, 2010). In many ways, the first hominins were still very similar to our great ape ancestor and lived in a largely similar environment (Andrews, 2020). This is not surprising either because, as said, we made only a small step from just before the branch split till just after it. This also explains why a lot of scientific discussions quickly arise when new fossils are described of species that lie close to that split. It is then not so obvious to identify whether they belong to a species before or after the split, and in which branch after the split. This happened with the discovery of the hitherto oldest hominin, *Sahelanthropus*, where some researchers were initially convinced that this was another great ape (closely related to the gorilla), and not a hominin (Wolpoff et al., 2002). More on this later.

DOI: 10.1201/9781003588795-5

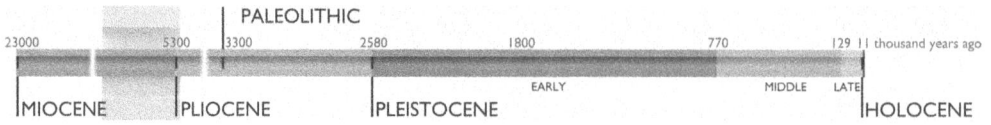

Figure 4.1   Timeline showing the period dealt with in this chapter.

# The split

What do we know about that split? The genome tells us, based on the number of genetic differences between recent great apes and humans that accumulated over millions of years, that the split occurred somewhere between 9.3 and 6.5 million years ago (Moorjani et al., 2016; Prado-Martinez et al., 2013). We also know that this occurred in Africa, something Darwin had already predicted in 1871 (Darwin, 1871). Initially, it was thought to have taken place in East Africa, relying on the

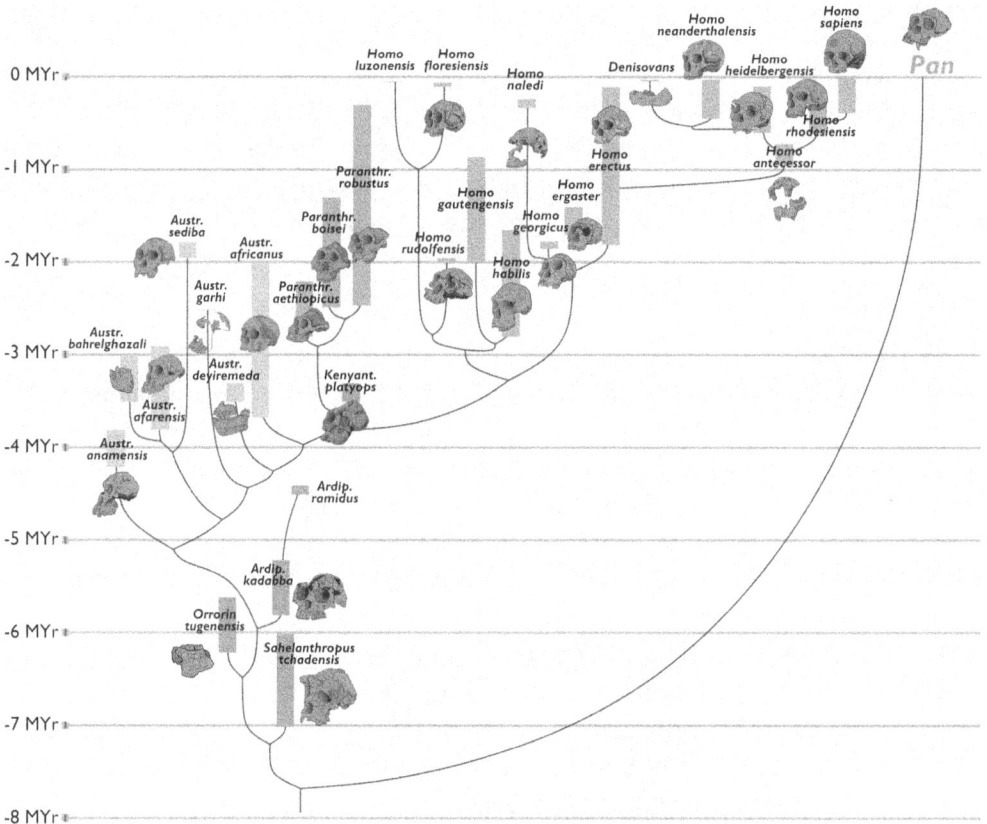

Figure 4.2   Hominin evolutionary tree, from the split with the lineage leading to chimpanzee and bonobo (*Pan*). From the latter evolutionary lineage, there is virtually no fossil evidence. Fossils just don't preserve very well in a tropical rainforest (bars indicate the time window of available fossil evidence).

evidence that East Africa was changing into a dry environment by then, with its rainforest that had turned into a dry savannah. This gave rise to the 'East side story', not the Broadway musical but our own East African origin story (Coppens, 1994). The discoveries of the first hominins and some later forms has since, however, partially contradicted this view (Domínguez-Rodrigo, 2014). The origin of the human lineage may well be Northcentral Africa, rather than East Africa (Sekhavati & Strait, 2024).

Indeed, the discovery of the first hominins tells a slightly different story. If we throw together all the information about body size, brain size, construction of the jaw apparatus, mode of locomotion and developmental rates of all hominins, four major groups can be distinguished (Collard, 2003; Lieberman, 2013). The oldest group, comprising *Sahelanthropus*, *Orrorin* and *Ardipithecus*, are small hominins (less than 130 cm tall and weighing less than 50 kg) with small brains and a sturdy jaw apparatus, who clambered through the branches with their hands and feet, and presumably had a fast development and became sexually mature early, just like great apes. They lived in African woodlands – rather than tropical rainforests – from central to Eastern Africa during a period from 7 to 4.4 million years ago. Their diet was presumably more diverse than that in great apes, making them less dependent on fruit (Almécija et al., 2021).

## The first one from Chad

Since the discovery of the first *Australopithecus* in 1925 by the South African anthropologist Raymond Dart, until the early 1990s, everything pointed toward our evolution originating in East Africa, east of the rift valley. From there it unfolded in East and South Africa. A first shock in the anthropological world came in 1996, when fossil remains of a lower jaw were found from some *Australopithecus* in Chad, i.e. Central Africa, all the way on the other side of the rift valley (Brunet et al., 1996). Not only was this in the wrong place from what one would expect but also Chad was a forested area during the period when that species lived there, about 3.5–3 million years ago. Not only did this seem to disprove the hypothesis that human evolution had occurred exclusively on the eastern side of the rift valley, but also the then prevalent belief that that origin was accompanied by a savannah that had replaced forests and woodlands. So how this *Australopithecus bahrelghazali* had gotten there was a mystery. Possibly this was a branch that had simply migrated from East Africa into Chad, leaving the East Side Story hypothesis intact.

And then came Toumai ('hope of life' in the local Daza language), a very well-preserved fossil skull attributed in 2002 to a new hominin, *Sahelanthropus tchadensis* (Brunet et al., 2002). You can guess from the name where it was found ... indeed, Chad (Toros-Menalla, more specifically), about 2500 km west from the rift valley. One might think, another hominin that wandered into Chad from East Africa and thus the East African origin may remain valid. Could be, were it not for the fact that Toumai must have lived between 7.2 and 6.8 million years ago, i.e. during the period when the split of our hominin ancestor had occurred. After much debate as to whether this was a hominin or a great ape, it is since a consensus that this is the oldest fossil evidence of our earliest hominin ancestors (including based on the analysis by which the distorted fossil was corrected via computer reconstruction) (Guy et al., 2005; Zollikofer et al., 2005). One of the features pointing toward it being a hominin is a snout that is shorter than in great apes and canines that are smaller. However, not many fossils have been found of this species. Along with the skull, another small piece of mandible and a few teeth were found, and later, two mandibles and part of a femur (Brunet et al., 2005; Macchiarelli et al., 2020). In the same week that I wrote this paragraph, the scientific journal *Nature* published the discovery of a left femur and two ulnae assigned to *Sahelanthropus* (Daver et al., 2022). Among other things, it was possible to deduce that *Sahelanthropus* weighed about 44–50 kg (Daver et al., 2022).

The discovery of Toumai caused a second astonishment among scientists. If this is such a primitive hominin, so close to the split, one expected it to find a skull very similar to that of great apes, demonstrating only primitive features. This was true when looking at the base and back of the skull, as well as brain size (amounted to about 360–390 $cm^3$, so very similar to that of a chimpanzee) (Lieberman, 2022). But to a great surprise, the facial skeleton showed highly specialized features, features that only appeared 3–4 million years later during human evolution, in *Australopithecus*: a shorter snout, no gap between the teeth in upper and lower jaws indicating small canines, and a thicker enamel layer in the teeth. From then on, the realization grew that our evolution should no longer be graphically seen as a tree but rather as a shrub. For a long time, this was thought of as a tree, where during early evolution one species along a line (represented as a trunk) gradually passed into a younger species and only later, from about 3 million years ago onward, a crown formed with branches of all other species. If at the very bottom *Sahelanthropus* clearly could not be fitted onto that trunk, because of those specialized features, then it had to be a separate branch. So, our evolutionary tree branched from the very beginning, about 7 million years ago, and continued to do so. It is better to say that we are dealing with an evolutionary shrub rather than an evolutionary tree.

So far, no fossil remains of the spine or limbs have been found, which makes it a preliminary guess as to the size of *Sahelanthropus*. But all indications are that it was similar in size to chimpanzees and other early hominins.

# *Orrorin*

As of today, *Sahelanthropus* still represents the oldest fossil evidence of hominins remains. The time window of the first hominins has been brought forward several times over the past years. That also happened in 2000, when Martin Pickford found a piece of a mandible, some teeth and a piece of an upper arm and femur that turned out to be about 6 million years old (somewhere between 6.2 and 5.6 million years) (Senut et al., 2001). He had found them in the Tugen Hills in Kenya, not far from Lake Baringo. But this did not happen without a struggle. Pickford, who had been researching in Kenya since the 1970s, clashed with the well-known anthropologist, Louis Leakey, then director of the Kenya National Museum and active in the Tugen Hills. Coincidence or not, but Pickford in 1985 no longer received the necessary permissions to conduct excavations in the Tugen Hills, permissions normally issued by … the Kenyan National Museum (Butler, 2001). Under the banner of the Community Museums of Kenya founded by Pickford and together with Brigitte Senut of the Musée National d'Histoire Naturelle of Paris, and with the help of Kenya Minister of Research and Technology, they were able to resume their excavations back in 1998. Only to be shut down again in March 2000 after police raided those Community Museums of Kenya and even arrested Pickford while he was excavating in the field. He ended up being jailed for five days under the charge that he was illegally excavating. I think this feud is still not settled.

But, finally, in 2001, Pickford and Senut made the new oldest hominin known to the world: *Orrorin tugenensis*, aka the 'millennium man'. A fragment of a femur showed that it was a small hominin, about 1.1–1.4 m tall and weighing 39–46 kg (Daver et al., 2022; Nakatsukasa et al., 2007; Senut, 2020). This is very similar to a male chimpanzee. Although the fossil evidence was limited, some important conclusions could be drawn (which of course involved some discussion). Attachment sites of muscle on the upper arm indicate adaptations to climbing trees, while the femur indicates walking on two legs. Other fossils from the same location indicate that *Orrorin* dwelled in evergreen forests (Senut, 2020).

# Ardi

The first discovered hominin more primitive than *Australopithecus* (this is the genus to which the famous Lucy belongs) was found just before the New Year in 1993, in the bed of the Awash River in Ethiopia (White et al., 1994). Tim White and colleagues had found a lot of fossil remains there that were very old, being 4.4 million years old (somewhere between 5.55 and 3.85 million years) (White et al., 2009). Until then, the oldest hominin fossils were those of *Australopithecus*, just over 3 million years old. Initially, they assigned them to the genus *Australopithecus*, just like Lucy. But it soon became clear that this was about something else after all, something more primitive for which a separate genus had to be created: *Ardipithecus* (White et al., 1995). The species was *Ardipithecus ramidus*, with the species name derived from the word for 'root' in the language of the local Afar people. *Ardipithecus* was thus the new root of hominin evolution.

The search yielded more than 110 pieces of skeleton, including many parts of the same skeleton of a young woman of about 120 cm and 50 kg. This skeleton was named Ardi. So, Lucy gained an old 'sister'. In fact, Lucy and her kind lived in the same area as Ardi (Lucy was found at Hadar in Ethiopia), but more than a million years later. When Ardi lived there, it still was a forest and woodland but already more open than a dense tropical rainforest (Negash et al., 2024; White et al., 2009). Remains of skulls, mandibles, teeth, arms, pelvis, hands, feet, ribs and vertebrae were found. These were examined very carefully, as they did not want to draw any hasty conclusions. Broken and compressed fossils were virtually restored. For example, the skull was compressed to less than 4 cm thick (Suwa, Asfaw, et al., 2009). With CT scanners and software, they were able to return them to their original positions. It took until 2009 for the researchers to publish their findings (White et al., 2009). And what publications. Ardi can be considered as important in how we look at early human evolution today as Lucy was in 1978. The picture of what our ape ancestor and the first hominins looked like and what they were capable of was recreated (White et al., 2015). So no, our great apes' ancestor did not look like chimpanzees.

The wonder of Ardi's old age was still reverberating, but Ardi still left a gap in the fossil evidence from the split away from the chimpanzee lineage, somewhere between 9 and 6 million years ago to 4.4 million years ago. In 2001, this came to a rather abrupt end. Not only was the 6-million-year-old *Orrorin* discovered but additional fossils were found in the same year from the area where Ardi was found. And now these were remains of a mandible, teeth, arm bones, clavicle, finger and phalanges that were dated 5.8–5.2 million years old (Haile-Selassie, 2001). So, it is almost as old as *Orrorin*, and yet slightly different. The fossils showed similarities to Ardi. Although first subsumed under the same species as Ardi, as a separate subspecies, this became a new species in 2004: *Ardipithecus kadabba*. The discovery in 2002 of the more than 7-million-year-old *Sahelanthropus* then further filled this gap.

# Fruit is not enough

Although we don't know much about the first hominins, evidence suggests that their diet changed. They also no longer lived in a dense tropical rainforest, but rather in patches of forest and woodlands in between which grasslands began to form (Cerling, Wynn, et al., 2011). Recent evidence suggests that already 6 million years ago, East Africa was for about 80% covered with forest, with the onset of a gradual shift toward grasslands (with so-called C4 grasses) (Negash et al., 2024). The landscape slowly began to open up, bringing them into contact with other foods. Their teeth were also not the same as the chimpanzee's, indicating that fruit stood no longer as central in their diet. Ardi no longer had the typical incisors of a fruit eater and had more substantial jaw teeth to accommodate other foods (Lieberman, 2013; White et al., 2015). *Orrorin* used his teeth differently than chimpanzees, also indicating a different diet (Senut, 2020). The enamel layer was also not very thick, admittedly thicker than in chimpanzees but thinner than in later *Australopithecus*. The position of the zygomatic arch shows that the jaw muscles were not yet as highly developed as in *Australopithecus* (Suwa, Asfaw, et al., 2009). Thus, they were not yet heavily dependent on hard food, something that would be typical for survival in open grasslands. Rather, the first hominins were omnivorous: they ate a little of everything that was not yet too hard (Dominguez-Rodrigo & Pickering, 2017; Suwa, Kono, et al., 2009). Isotope studies on teeth confirm this: Ardi's diet comprised 90% forest plants and only 10% grasses (Suwa, Asfaw, et al., 2009; White et al., 2009). This is similar to chimpanzees but different from *Australopithecus*, in whom the diet consisted of more than 30% grasses.

It is also notable that the canines are less prominent in first hominins. Similarly, the gap between the teeth into which the canine slides when the jaw is closed is smaller (Haile-Selassie et al., 2004). *Sahelanthropus* and Ardi had already lost the sharpening mechanism for canines (i.e. they no longer rub against each other), but in *Orrorin* this was still somewhat possible (Haile-Selassie et al., 2004; Senut et al., 2001). However, this has nothing to do with diet but rather reflects a sexual difference in great apes, where males typically have large canines. In first hominins this has gradually disappeared, first in the upper jaw and later in the lower jaw. Variation in sexual differences in canines in *Ardipithecus* is similar to that of modern humans but is more pronounced in the canines in the lower jaw (Suwa et al., 2021).

# Legs gain importance

People have long racked their brains about how we went from climbing apes, through swinging great apes, to finally becoming upright walking human beings. Arthur Keith suggested as early as 1923 that this had occurred in four stages (Keith, 1923). First, there were the apes that walked on branches on four legs (such as Old World apes), then came the small, swinging apes that hung upright by their arms (such as gibbons) that progressed to large, swinging great apes (such as chimpanzees), and then shifted from that vertical position to hominins walking upright onto two legs. Other scenarios are where there was a transition phase from walking on four legs to slightly erect via knuckle walking, from which a fully erect posture then became possible (Kelly, 2001; Richmond & Strait, 2000). Both scenarios are now cast aside. Vertical climbing combined with erect clambering with hands and feet would have allowed the transition from a horizontal posture to a vertical posture (Fleagle et al., 1981; Kivell & Schmitt, 2009). Although there are some similarities between vertical climbing in chimpanzees and the way humans walk on two legs (Gebo, 1996), we also know by now that what chimpanzees do today is not representative of what our great apes' ancestors did then. Thus, this scenario of a vertically walking and clambering great ape ancestor becomes more prominent. The first hominin also lived in a highly forested environment. Therefore, it would not be surprising if they still spent a lot of time in trees (White et al., 2009).

But what do the fossils really tell us? Do they give us a glimpse into the past of how first hominins moved around? Fortunately, we have fossil remains of the limbs, but the skull can also tell us something about posture. For example, in *Sahelanthropus*, the foramen magnum – the place where the spinal cord enters the skull – had already shifted more forward than in great apes (Lieberman, 2013). Hence, the spinal column was already positioned more below the skull rather than connected to it at the back, and thus the body was already more erect (Zollikofer et al., 2005). Is this also proof that they were walking around on two legs on the ground? Not necessarily. First, an anteriorly located foramen magnum is not always linked to bipedal locomotion (primates that have it positioned slightly more forward do not necessarily walk more erect on two legs) (Landi et al., 2020). A first-found fragment of a femur of *Sahelanthropus* showed no obvious signs indicating bipedal locomotion (Macchiarelli et al., 2020).

Initially, it was thought that this situation changed from *Orrorin* and *Ardipithecus* on. However, the discovery of a better-preserved femur of *Sahelanthropus*, described in 2022, suggested that bipedal locomotion occurred as early as the first hominins (Daver et al., 2022). Although a reanalysis of these remains casts doubts on the bipedal nature of *Sahelanthropus*, it cannot be ruled out at this point that early hominins had femurs somewhat facilitating bipedal locomotion, like those in later *Australopithecus* but not yet like what we see in humans within the genus *Homo* (Cazenave et al., 2025; Richmond & Jungers, 2008). In *Homo sapiens*, the femur has a large articulatory head that fits into the hip joint, a short neck (connecting the head to the rest of the femur) and a long shaft that has about the same thickness everywhere. In great apes, the head is small, the neck short and the shaft also the same thickness everywhere. In these first hominins and in *Australopithecus*, the head is small, the neck is long and the shaft is thicker at the top than at the bottom (Senut et al., 2001). So, what does this tell us about being bipedal? Well, a longer neck allows for greater leverage with which to pull on the femur. Muscles on the pelvis can therefore pull harder on this femur, which is important when standing on two legs. If we stand on two legs and lift one leg, our pelvis will tilt downward at the lifted leg because of body weight. The lateral glute muscles of the supporting leg then pull on the femur which keeps the pelvis stable. So we don't sag every time we lift a leg, thanks to this system. Now, the longer the neck of the femur, the more force the muscles can exert on the pelvis and stabilize it, just as we can lift more weight with a wheelbarrow with longer handles. This system for pelvic stabilization is only needed when one is bipedal. So why do hominins have a thicker part at the top of the shaft, you may ask? Well, when those muscles stabilize the pelvis, they exert a lot of bending forces onto that shaft. The longer the neck, the more bending forces. On top of that, the feet were still quite far apart, which makes those bending forces even greater (Lovejoy, Latimer, et al., 2009). A shaft that is thicker at the top can absorb those high forces better (Richmond & Jungers, 2008). This system also works better if those lateral glutes attach to a pelvis that is wider at the top. In great apes this is narrow, in *Homo* and *Australopithecus* it is wide. Well, Ardi and *Orrorin* provide evidence that early hominins also had a wider pelvis (Lovejoy, Suwa, Spurlock, et al., 2009). Ardi also no longer had its lumbar vertebrae tucked away between the pelvic protrusions, which made the lower back not as stiff as in great apes. Ardi thus already possessed a slightly flexible lower back, useful when walking upright (Lovejoy, Suwa, Spurlock, et al., 2009). A special bone in the foot (the *peroneum*) shows that Ardi could brace her ankle while supporting herself, as in other bipedal hominins. In contrast, great apes lack this bone. The slightly longer midfoot was also already more rigid in Ardi, also useful for walking on two feet, but disadvantageous for grasping branches with the feet (DeSilva, 2009; Lovejoy, Latimer, et al., 2009). The first adaptations to walking upright on two legs show themselves much earlier than had long been thought, i.e. already shortly after the split from our great ape ancestor.

Does this mean that by then our ancestors had already left the trees to walk fully upright on the ground? Probably not. Fossils still show many adaptations to climbing. The ulna of *Sahelanthropus* is slightly curved, as in climbing great apes. That flexion allows the muscles to maintain more tension in the arm while the elbow is folded when climbing (Daver et al., 2022). *Orrorin* still had curved phalanges, useful for grabbing branches (Haile-Selassie, 2001; Senut, 2020; Senut et al., 2001; Tocheri et al., 2008). Ardi's shoulder joint was built providing an advantage while walking four-legged on branches but was no longer equipped to do much climbing with the arms (White et al., 2009). The hands were also no longer as sturdy and the thumb was shorter than in great apes (White et al., 2015). However, they still had a flexible wrist (Lovejoy, Simpson, et al., 2009; White et al., 2009). While the upper part of the pelvis in Ardi showed adaptations to walking upright, the lower part was well-equipped for the attachment of muscles important during climbing (Lovejoy, Suwa, Spurlock, et al., 2009). Ardi walked upright but still had a grasping foot, with a big toe as far open as in great apes (Lovejoy, Latimer, et al., 2009; Prang, 2019). The big toe was long, though, which was already slightly more advantageous for the push off during walking (O'Neill et al., 2022; Prang, 2019; White et al., 2009). Presumably, the Achilles tendon was also long (it is very short in great apes) (Aerts et al., 2018). The push off was still using a foot held completely flat on the ground (White et al., 2015).

So, what can we conclude about the mode of locomotion among our early hominin ancestors? It is best to take *Sahelanthropus* and Ardi, we know the most about them. Everything indicates that they climbed trees and walked on the ground. They did not swing in trees but walked on four legs or upright on two legs, scrambling with their hands. On the ground, they walked alternately on four feet or two legs, although recent research suggests that *Sahelanthropus* might have used knuckle walking (Dainton, 2001; Lovejoy, Simpson, et al., 2009; Meyer et al., 2023; White et al., 2015). However, walking upright was not yet efficient, in part because of the gripping foot. Ardi is said to have stepped in her own special way.

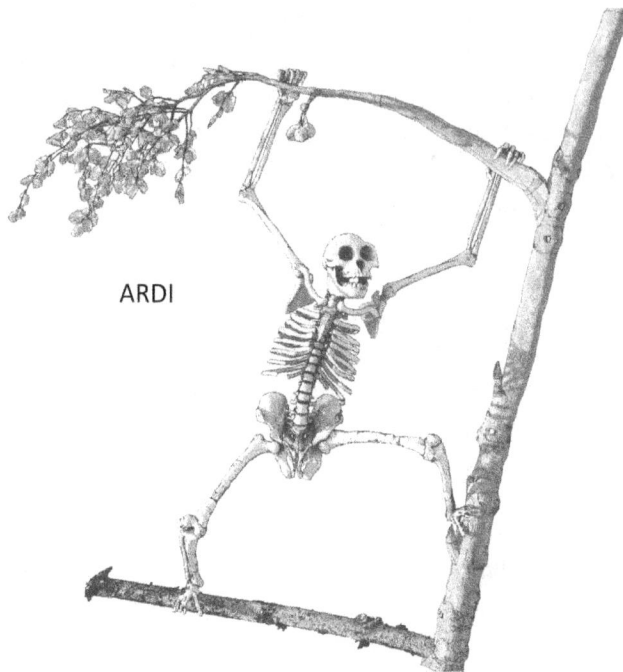

Figure 4.3 Skeleton of a clambering *Ardipithecus*. (Modified from 3D model of Lucy of Brassey et al. (2018).)

## Feeling good in our skin?

We already know a lot about early hominins based on their skeletons, but we know nothing about the soft organs. For example, we have no direct evidence that can tell us anything about how the skin was built and the role it played in cooling. All we can surmise is that which accords with the simplest scenario. We know that chimpanzees have pale skin full of long hair. We also know that they do not have that many sweat glands to cool off through sweat, but that this is OK for them while living in tropical forest.

Of the first hominins, we know that they still lived in forested areas, but that already contained more grassland. So most likely they were already more exposed to direct sunlight and a functioning cooling system became more important. That they showed adaptations to also walking more upright already suggests that they also started to explore those grasslands, looking for new types of food.

Such changes in the environment put a great deal of selective pressure on the skin, which should not only provide cooling but also protect against harmful radiation from sunlight. So, it is not so speculative to think that natural selection had kicked in, favoring hominins with better cooling and more shielding skin. But what this specifically involved remains speculation for now. So, let's keep this in mind for a while, and pick it back up when we have concreter evidence of evolutionary changes in skin.

# A body to carry around the brain

We also know very little about the brain of the first hominins. Still, Toumai and Ardi allowed us to lift a small corner of the veil, thanks to the rather well-preserved skull. Although flattened, computer reconstructions allowed to estimate the volume of the brain cavity, and thus the size of the brain (White et al., 2015). Thanks to Lucy, a later hominin, we know that relative to body size, the brain was then still similar in size to that of chimpanzees. Not surprisingly, this was also the case in *Sahelanthropus* and *Ardipithecus*. Toumai had a brain capacity of about 320–380 cc (Brunet et al., 2002). Ardi, with her 120 cm body size and 50 kg body mass, had a brain volume somewhere between 280 and 350 cc (Suwa, Asfaw, et al., 2009; White et al., 2009). This is similar to the chimpanzee and about 25% smaller than that of Lucy and relatives.

Considering that primates show a general trend in the number of neurons in the brain in relation to brain size, we can predict that Ardi must have had about 21 billion neurons (Herculano-Houzel & Kaas, 2011). Based on that, we can even estimate how much energy Ardi's brain consumed, which is 3.4 W (this is the amount of energy consumption per kg of brain over time) (Aiello & Wheeler, 1995). This is not much. In chimpanzees it is 4.5 W, in Lucy it was 5.8 W and in us it is 17–24 W!

The first hominins had a slightly shorter snout than the great apes, and thus a smaller olfactory organ. But as we already know, a shorter snout doesn't say everything about how efficiently the olfactory organ works. And what about the different parts of the neocortex? Had Ardi also implicated the left-right asymmetry in areas in handedness and language? For now, we can only speculate that it probably did because our great ape ancestor probably had this as well.

# So ...

In the meantime, we do know quite a bit about our early hominin ancestors. But many question marks also remain. The finds of *Sahelanthropus*, *Orrorin* and *Ardipithecus* confirm that we may have to abandon the traditional view that our great apes ancestor looked like a chimpanzee. Ardi shows more similarities to early Miocene apes than to recent great apes, including the chimpanzee (Lovejoy, 2009; Lovejoy, Suwa, Simpson, et al., 2009). Putting all this information together, we can predict that our great ape ancestor weighed about 35–60 kg, with arms about as long as the legs. The hands were equipped for grasping but not so much for swinging. Grasping feet were useful for climbing but also for walking on branches with four or two legs. The phalanges were still long, useful for gripping branches, but the midfoot was already a little stiffer. They were not doing knuckle walking, but they could already walk slightly better on two legs on the ground. The lower back was already a little freer, thanks to a shorter pelvis. The spine consisted of 33–34 vertebrae. Brains were about the same size as those of chimpanzees, so somewhere around 300 cc. Considering they were still forest dwellers, fruit was probably still central. The skin was pale and completely hairy.

Therefore, the transition to the first hominins was not so substantial. Rather, it came with subtle differences that are put forward that typify the transition from our great ape ancestor to the first hominins. For example, the pelvis had already expanded more sideways, advantageous for muscles to provide more stability when walking on two legs. The shorter pelvis now also allows more flexibility in the lower back, useful when walking on two legs. So, the first hominins clearly showed the first adaptations to walking upright. That they did still rely heavily on living in trees, we can deduct from their occurrence in forested areas, but now already more combined with areas of grass fields where walking upright was perhaps more convenient. The prehensile foot also points toward life in trees. Expanding into a wider habitat with grasses presumably translated into a broader diet, where fruit gradually forming a more limited proportion. However, the jaw apparatus was not yet equipped to handle hard foods. Whether the gradual move to more open grassland had also brought about changes in the skin or brain remains to be guessed at for now.

# 5

# The forest opens up

In 1974, the world was introduced to Lucy, also known as 'A.L. 288' (code stands for 'Afar Locality, fossil 288') (Lovejoy, 1981). Admittingly, Lucy sounds better. So did Donald Johanson must have thought when he and his team discovered the fossil on a sandy hillside in Hadar, Ethiopia (Johanson & Wong, 2010). A piece of elbow protruding from the sand eventually gave rise to the uncovering of a skeleton that was about 40% complete and was about to shock the world. A beauty of a fossil, given the apt Ethiopian name 'Dinkinesh', which stands for 'you are beautiful'.

It is this fossil that really introduced the world to the second group of hominins: the australopithecines (Collard, 2003; Lieberman, 2013). These were small hominins with small brains, increasingly faced with a new ecological challenge: the desiccation of the environment and the opening up of the rainforests. Walking upright on two legs and chewing hard food has now become the norm. They met that challenge successfully as numerous species roamed over East and South Africa from about 4.2 to 1.3 million years ago. About 2.8 million years ago, they even co-existed with humans of the genus *Homo*. This chapter focuses on the period when they were, so to speak, autocrats among the hominins. We are now well into the Pliocene.

## Pioneers of becoming human

Australopithecines can be considered the pioneers of our humanity. They laid the evolutionary foundation of what will later characterize us as *Homo sapiens*. But again, these australopithecines should not be seen as a straight evolutionary line toward *H. sapiens*. About 2.5 million years ago, our sapiens evolution struck out in a different direction. More on that later, first get acquainted with the australopithecines. They are represented by a dozen species of small hominins – weighing less than 50 kg and less than 1.3 m tall – with adaptations to walking on two legs but also to climbing the trees. Characteristic was their sturdy jaw apparatus, combined with brains that were similar in size to those of great apes (Collard, 2003).

Excluding the genus *Kenyanthropus* (only one species known), the australopithecines are represented by two forms: the slender and the robust forms. This has nothing to do with a slender body figure but has everything to do with their jaw apparatus. Although quite sturdy in both forms, that of the slender *Australopithecus* pales to that of the robust *Paranthropus*. I briefly introduce the most iconic species.

DOI: 10.1201/9781003588795-6

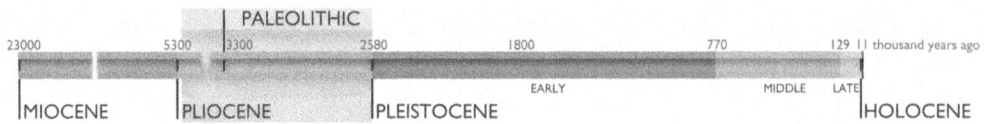

Figure 5.1   Time line showing the period dealt with in this chapter.

## The ape from the South

*Australopithecus* stands for 'monkey or ape from the South'. It was Raymond Dart, a South African anthropologist who first introduced the name in 1924 (Dart, 1925). He assigned this name to a fossil skull he had found, giving it the species name *Australopithecus africanus*: the 'African monkey from the South'. Since then, several species have been added to this genus, but where for some fossils it is not even clear yet whether they represent a separate species or not. Fossil evidence does suggest that there are still unrecognized species, so the number of generally accepted as valid species (about seven species) may well increase in the future (Melillo et al., 2021).

The oldest and most primitive species is *Australopithecus anamensis* (lived 4.3–3.8 million years ago in East Africa), described in 1995 based on fossil jaw, leg and arm fragments from Kanapoi and Allia Bay in Kenya (Leakey et al., 1995; Leakey et al., 1998; Lewis et al., 2024). In addition to some primitive features, as found in Ardi, a shift can clearly be observed in the jaw and limb structures, among others. For example, there is no longer a large cavity present at the base of the upper arm bone, into which the elbow protrusion would fit when the arm is fully extended. Also, the molars are significantly wider than those of Ardi and the teeth have a thick enamel layer. The tibia already shows a wider proximal surface that helps form the knee. All indications that the first australopithecines already showed adaptations to walk upright and possessed a powerful jaw apparatus (Harcourt-Smith, 2016; Wood & Boyle, 2016). This is indicative of the transition in how they interacted with the environment to get around and feed. That environment was dense forest with shrubs, interspersed with patches of grassland (Bobe et al., 2020). They shared this with large herbivores, which may have been partly responsible for the slow opening up of the dense forest. Anamensis lived together for at least 100,000 years in the same area with the species to which Lucy belongs: *Australopithecus afarensis* (Haile-Selassie et al., 2019). That one will get a little more attention further on.

For some species, fossil evidence is highly scarce, such as for *Australopithecus garhi* from Ethiopia and *Australopithecus bahrelghazali* from Chad (Asfaw et al., 1999; Brunet et al., 1996). Some fossils are suspected to belong to yet unnamed species, such as the most complete australopithecine skeleton: the 3.67-million-year-old Little Foot from Sterkfontein, South Africa. Little Foot started with four foot bones being discovered in 1978 in the so-called Silberberg Grotto, tucked away in a miner's shed and untagged, only to be 'discovered' again in 1994 (Dawkins & Wong, 2004).

Still later, other remains were found, eventually yielding a 90% complete and very well-preserved skeleton (Clarke, 1998, 2002; Clarke & Tobias, 1995). The construction of the cervical vertebrae, foot, arm, shoulder girdle and inner ear show that they walked upright but were still adapted to climbing trees (Araiza et al., 2021; Beaudet, Clarke, Bruxelles, et al., 2019; Beaudet, Clarke, de Jager, et al., 2019; Beaudet et al., 2020; Carlson et al., 2021). It is believed to be a different species from the other South African *A. africanus*, and yet it has not been given its own species name.

New species are also being discovered rather recently, such as the South African *Australopithecus sediba* and the East African *Australopithecus deyiremeda*. Of the latter, only a few pieces of lower and upper jaws are known (Berger et al., 2010). A lot more fossils of *A. sediba* (lived about 2 million years ago) from the Malapa cave in Gauteng, South Africa were described in 2015 (Haile-Selassie et al., 2015). Also for this species, everything indicates that they walked on two legs, but still adapted to climbing trees. However, sediba did have relatively short fingers (less useful for climbing) and a pelvis that may have allowed more efficient upright walking. The lower back showed a curvature (the lumbar lordosis), as in later humans, useful for upright walking (Williams et al., 2021). The thumb was relatively long, though, compared to the other fingers, which some believe to indicate adaptations that facilitated manipulation of objects, such as tools (Dunmore et al., 2020).

## Child of Taung

Back to the fossil that started the 'Australopithecus' story: the skull that prompted Raymond Arthur Dart to create *A. africanus*: the 'African ape from the south'. A long story of controversy surrounds this fossil, called 'the child of Taung'. Dart was an Australian, medical anatomist who had only just arrived in South Africa in 1923, at the age of 29, to take over the department of anatomy, when rather by accident he became the center of a turbulent period in the anthropological world (Berger & Hilton-Barber, 2000). He is even said to have declared that he despised bones and fossils ... little did he know how one fossil was going to further define his life.

At the Buxton mine, a limestone quarry just outside the village of Taung, dynamite was regularly used to clear the mine. Fossil skulls of baboons had previously been found after such blasts, something that caught Dart's attention. In November 1924, as the dust cleared after yet another blast, a miner's attention was drawn to a fossil that looked somehow different. He thought it was a fossilized skull of a Bushman. A doctor, seeing the fossil, suggested he should inform Raymond Dart about it. And so it eventually landed on Dart's lap in Johannesburg. Although he had to act as best man for a wedding that day, he couldn't resist studying the fossil and was amazed about the small, fossilized skull whose fossilized brain was exposed. His enthusiasm almost caused him to be late for the wedding, but equally caused him to rush back to the fossil immediately after the reception. Using his wife's knitting needles, he carefully began removing some debris (the fossil still shows traces of those knitting needles)

(Holloway et al., 2014). About 23 days later, the fossil was cleaned up and ready to be described.

The teeth were the first thing that caught his attention: they looked like human teeth, as well as the fossil contained deciduous teeth. So, it was a child, hence the name 'Taung's child'. He estimated the brain volume to be 520 cc, thus too large to be that of a young baboon or chimpanzee. The occipital opening was directed more downward, rather than backward as in four-footed primates. So, this child had to have walked upright. And the teeth thus looked more like those of a human than those of great apes. This convinced Dart that it had to be a human ancestor. *A. africanus*, an ancestor of man was born. This unique scientific discovery had to be made public, and for Dart, the sooner the better. In 40 days, his manuscript was ready and in January 1925 it was published in the renowned journal *Nature* (Dart, 1925). His rush to get it published was Dart's scientific downfall, however. For example, there was a lot of criticism of the sloppy descriptions, including misidentifications of some important grooves in the brain (grooves that could identify whether it was a brain of great apes or hominins) (Falk, 2009). Also the fact that he did not allow others to study the fossil at first, did not argue in his favor (Berger & Hilton-Barber, 2000). In addition, Dart defied the zeitgeist of, on the one hand, the religiously inspired view that humans were superior to other animals in terms of intelligence and on the other hand, the race among anthropologists to claim the ancestor of man to be from their own region. Dart's African fossil with a small brain already did not fit this image. During the same period, a belief prevailed in Europe that the origin of man had to be located there, which supposed to have been confirmed by the discovery of the fossil skull of Piltdown Man, 'discovered' in 1912 (Lowenstein et al., 1982). This fossil, found at Piltdown in the United Kingdom, with a brain volume similar to that of a modern human, did fit this story much better. Even Arthur Keith, a then renowned anthropologist, went along with this story. Until about 30 years later, this story began to crumble as the first doubts surfaced and finally in 1955 the forgery was demonstrated: it was a skull of a modern human and a lower jaw of an orangutan, whose teeth had been somewhat filed off. A recent genetic analysis of fossils even showed that the human part came from two different individuals (De Groote et al., 2016).

Meanwhile, attention also went to the new discoveries of fossils of a new human species in China, the Peking man, discovered in 1929 (Black, 1929), causing Dart's discovery to be ignored for a long time. He was even portrayed as a traitor to the Creator and an accomplice of Satan. This slowly caused him to slide into depression. Yet his findings were defended by another anthropologist who would eventually achieve fame: Robert Broom. He published another study on the fossil in 1925, defending the thesis that Taung's child was a link to our own evolution (Broom, 1925). In 1938, he himself described a new fossil species, which bore similarities to Taung's child (Broom, 1938). At the first *Pan African Prehistory Conference* in Nairobi in 1945, where both Dart and Broom were able to present their research, acceptance of Dart's findings grew. The renowned Louis Leakey then stated that he no longer had any doubt that those South African fossils were precursors to man. After a few more intermediate steps, the verdict finally fell when Donald Johanson presented his Lucy in 1974 and showed that it too was similar to Taung's child. Raymond Dart was still able to witness this (he died in 1988), Robert Broom was not (died in 1951).

There is no longer doubt about *A. africanus* being a true hominin species that lived in South Africa, sometime from 3 to 2.4 million years ago, and that Taung's child was indeed a two- to three-year-old child belonging to this species (Falk et al., 2012). Like other australopithecines, the species shows a sexual difference in body size: males were about 138 cm (about 67 kg) tall and females 115 cm (about 33 kg) tall. Jaws had quite large teeth, forming a sturdy jaw apparatus. Brains were still small, about 460 cc (Wood & Boyle, 2016). This species seems to be just a bit more specialized than the other well-known one: *A. afarensis*.

## Lucy in the sky with diamonds

Lucy ... what's in a name. And yet the fossil A.L. 288 became world famous because of this, thanks in part to the Beatles. In 1967, they had released their single 'Lucy in the sky with diamonds'. In 1974, this song was still popular even in the tents of Donald Johanson's team in the middle of the Ethiopian sand hills in Hadar, in the Afar region. Had the Rolling Stones been playing on the radio as they celebrated the discovery of a largely complete skeleton of an early hominin, it might have been called Angie. But so it became Lucy, a primordial mother of the hominins who dwelled in the East African rift valleys 3.2 million years ago, at a time when volcanoes were active and large herbivores and predators inhabited the ever-opening grassy plains, punctuated by ever-shrinking bushes. This landscape was the result of a geological activity that had begun about 20 million years ago, with the East African plateau shifting relative to the Central African part. This produced multiple, gigantic fault lines, with deep gorges flanked by long mountain ranges, slowly but dramatically altering the East African landscape. Rivers began to flow in different directions and valleys began to fill up. About 5 million years ago, this was transformed into large and small rift lakes, such as Lake Tanganyika, the second deepest lake in the world (1470 m deep) and Lake Victoria, the largest among the reef lakes (only 83 m deep but with a surface area 1.8 times that of Lake Tanganyika). Also, the mountain ranges ensured that the moist, western winds dried up completely before reaching over them. East Africa no longer received the annual precipitation that it was accustomed to and the climate gradually began to become drier. An originally lush rainforest gradually gave way to a forest savannah, where forests were interrupted by patches of grass. In time, this became a grass savannah, where vast patches of grass form the transition between small patches of forest. This East African habitat forced australopithecines to become adapted ... or disappear.

With Lucy, a new species was known, *A. afarensis*, a hominin similar to the previously known *A. africanus* from South Africa, yet also different (Johanson & Taieb,

1976). Discovered at Hadar in Ethiopia, but fossils have also been found in Kenya and Tanzania. The fossil time window indicates they lived there about 3.7–3 million years ago. It may come as a surprise, but they were not the only hominins back then. In fact, their time window overlaps with that of the Central African *A. bahrel-ghazali* (which may have been a westward-migrating population of afarensis), the Ethiopian *A. anamensis* and *A. deyiremeda*, the South African *A. africanus* (Haile-Selassie et al., 2019) and the Kenyan *Kenyanthropus platyops*. Presumably split off from anamensis in East Africa, afarensis may also have spread to Central Africa and into South Africa. According to one scenario, the migration to South Africa, which eventually gave rise to the species africanus, may have been driven by the evolution of the tsetse fly. Indeed, with the desiccation of the climate, the flies were driven more to coastal areas, freeing up a corridor between East and South Africa (Sperber, 1983).

Not only is afarensis the most popular among australopithecines, it is also the species we know the most about. Where it was once thought that there might be a larger and a shorter species, it is now clear that these reflect sexual differences, even more pronounced than in africanus (Melillo et al., 2021; Wood & Boyle, 2016). By the way, it was once thought that Lucy should actually be called Lucifer or Brucy because of being a man. This hypothesis has, however, been refuted (Tague & Lovejoy, 1998). Males were significantly taller than females. Lucy was about 1.05 m tall and weighed somewhere between 20 and 25 kg (some say she was taller, up to even 1.22 m) (Brassey et al., 2018; Helmuth, 1992; Lordkipanidze et al., 2007; Porter, 1995). Males were previously 1.5 m tall and weighed about 65 kg (Wood & Boyle, 2016). Compared to the closely related but more primitive anamensis, afarensis had already evolved slightly further toward a more substantial jaw apparatus, but slightly less substantial than that of africanus (Ward et al., 2001; Wood & Boyle, 2016). From this, scientists infer that this species had a different diet than the first hominins, which is also evidenced by the environment in which they lived and had to survive. That afarensis walked upright, we know thanks to Lucy. Various adaptations in her body leave little doubt about that and fossilized footprints also confirm it. Somewhat less unequivocal is how efficiently they were already able to walk and run on two legs. There are no fossil remains of the skin, but the new environment in which they lived allows for some educated guesses about how they had to deal with heat. And what about the brain? Fossilized skulls show that the brain was still small, like that of chimpanzees, and that it showed a sexual difference as well. Male brains were about 500 cc in volume, while this was about 380 cc for female brains (Wood & Boyle, 2016).

## Lions among the hominins

I already briefly mentioned Robert Broom, the South African paleoanthropologist who defended Raymond Dart and Taung's child. In his search for evidence to support Dart's discovery, he discovered a skull in 1936 that he first attributed to the species name *Plesianthropus transvaalensis*, but which later turned out to be an adult *A. africanus* (Broom, 1938). This is not yet the nearly intact skull known as 'Ms. Ples', which Broom only discovered in 1947. In 1938, Broom made a new discovery, albeit with the help of a schoolboy, Gert Terblanche, who had found a skull and a piece of mandible in Kromdraai, not so far from Sterkfontein. The boy was unaware of the scientific value of that finding and used rather crude equipment to remove it from the rocks, thereby damaging it. Through an intermediary, Broom got his hands on a tooth and a piece of the upper jaw and immediately realized that this was an important discovery. Immediately he began his search for that schoolboy, or as Broom stated, "*I had to hunt up the schoolboy*". First to the boy's house, where his mother informed him that Gert was in school and that he had found four more beautiful teeth. Then to his school, Broom writes "*I naturally went to the school, and found the boy with four of what are perhaps the most valuable teeth in the world in his trouser pocket*".

Once the fossils were in his hands, it quickly became clear to Broom that this was a new species, which, although similar to africanus, had remarkably larger teeth, a robust jaw apparatus and accompanying skull. Sufficiently different to consider it a new genus: *Paranthropus* (Broom, 1938). The species became *Paranthropus robustus*, the one with the robust skull. Later skulls with similar robust features were also found in East Africa. In 1959, Louis Leakey dug up a nicely preserved skull in the Olduvai Gorge in Tanzania (Leakey, 1959). This skull also had a very robust jaw apparatus, very large molar teeth (canines and incisors were relatively small) and a remarkable structure that also can be found on a lion skull: a bony ridge running down the center of the skull roof from front to back. What the connection is with lions, we'll see later. The new species was christened *Zinjanthropus boisei* but soon saw exposed its similarities to the South African species: it became *Paranthropus boisei*. Earlier, French paleoanthropologists Camille Arambourg and Yves Coppens had found a hefty mandible in Ethiopia (first assigned to a new species *Paraustralopithecus aethiopicus*) that also showed strong similarities to the South African robustus (Arambourg & Coppens, 1968). It became *Paranthropus aethiopicus*. Three species with a very similar, very robust jaw apparatus, two in East Africa and one in South Africa. Studies suggest that these robust australopithecines have a common origin and thus one migrated from East to South Africa (Gunz, 2012; Herries et al., 2009). Other studies still question this (Constantino & Wood, 2007). In East Africa, they lived between 2.8 and 1.3 million years ago, in South Africa from 2.5 to 1.2 million years, and possibly even as recent as only 300,000 years ago. At the same time, *Homo* species were also roaming East and South Africa. In 2024, fossilized footprints were discovered in East Africa (in Koobi Fora), indicating they were from the same period and at the same place, but generated by two different hominin species: *Homo erectus* and *P. boisei* most likely crossed paths then (Hatala et al., 2024). During the period between 2 and 1.94 million years ago, when South Africa underwent major climatic changes, *H. erectus* even co-occurred with both *P. robustus* and *A. sediba* (Herries et al., 2020). Three species of three different genera of hominins in the same area at the same time. Which paleoanthropologist wouldn't love to go back to that period and place!

Compared to *Australopithecus,* the sexual difference in *Paranthropus* was slightly less pronounced. Although estimating body size is somewhat more difficult, given that very little limb or spine skeletal material was found, it is estimated that males weighed about 40–50 kg and females about 32–34 kg (Wood & Richmond, 2000). They also walked on two legs, which could be inferred, for example, from the ossification in a femoral head of a robustus (Cazenave et al., 2021). But, like *Australopithecus,* they also showed adaptations to climbing trees. The brain volume was also not yet very large, being somewhere between 450 and 530 cc (Falk, 1985; Holloway, 1972; Wood & Richmond, 2000).

So, the pioneers of our humanity have been introduced. Now let us take a look at those 'lion jaws' and how they took advantage of their changing environment.

# Falling back on fallback food ... a hard nut to crack

The initial contiguous roof from tree canopies under which our earliest ancestors dwelled slowly began to open up. With it, the protection from high temperatures also started to fade away, making them more and more exposed to direct sunlight and associated harmful radiation but also it made the oasis of succulent fruit gradually becoming more and more fragmented. The growing challenge for australopithecines to protect themselves and obtain a new kind of food eventually became an evolutionary opportunity.

## *Plants adapt*

About 16 million years ago, during the Middle Miocene, seawater temperatures gradually decreased (Barrett, 2003; Lieberman, 2013), and the entire globe cooled significantly, with substantial fluctuations between warmer and colder periods (Zachos et al., 2001). By 6 million years ago, this began to manifest itself clearly in the African vegetation, and more pronounced in East Africa. Forests slowly retreated and became divided into patches of forest, separated by growing grasslands (Cerling, Wynn, et al., 2011). Desiccation, however, did not proceed in one direction but was a fluctuating process and was site dependent. For example, there was an increase in the ratio of forest versus grass about 3.6 million years ago (with about the same amount of forest as grassland), followed by a second period of desiccation with an expansion of grasslands. Plants simply adapt to climate. Many tree species thrive somewhat less well in dry environments, and if they can, they usually form hard fruits that are better protected from desiccation. Grasses adapt better to dry environments, as do plants that have thick roots in which they can accumulate water.

Plants adapt biochemically, physiologically and anatomically to survive in dry areas (Basu et al., 2016). They form leaves and fruits that lose less moisture, which usually makes them tougher. There are far fewer tree species that are sufficiently adapted to survive in open savannahs, and those that can do so produce fewer fruits than those in tropical forests (Domínguez-Rodrigo, 2014). They grow during periods when there is sufficient water and bridge dry periods in the form of seeds or underground roots. Special roots allow them to hold more water even if the soil is very dry. The way they engage in photosynthesis can also differ, allowing a distinction between three types of plants in East Africa: C3 plants, C4 plants and CAM plants (Malone et al., 2021). All three use carbon (via $CO_2$), which comes in two forms or isotopes: carbon-12 and carbon-13. Shrubs and trees found in groves and forests are C3 plants that primarily use carbon-13. Grasses and sedges are C4 plants that use much less carbon-13. The third type, CAM plants (CAM stands for 'crassulacean acid metabolism', referring to the family of Crassulaceae or succulents), lie somewhere between C3 and C4 plants in their carbon-13 consumption (they tend slightly more toward C4 plants). Translating this in relation to climate drying in East Africa, this means that there has been a shift from C3 plants to C4 plants in the environment where the first australopithecines roamed. Why is this relevant to note? Obviously, because it indicates that a change in diet was inevitable. But it also allows us to look back in time at how australopithecines effectively changed their diet. Indeed, the two isotopes, carbon-12 and carbon-13, accumulate in the body as they enter through food, where they are stored in mineralized tissues, such as bone or enamel. By analyzing the ratio of the two isotopes in tooth enamel, one can find out whether the diet of the tooth owner consisted mainly of C3 or C4 plants, i.e. leaves and fruits, or more of grasses.

## Other plants, other food

If the food changes, then those who depend on it must adapt to survive. That's just how evolution works. We can therefore expect that australopithecines underwent adaptations in everything that had to do with food. Of course, this is not to say that this transition happened overnight, or that it was all or nothing. Australopithecines did still eat fruit, but an increasing proportion of their daily meal will have had to consist of other foods, foods they had to find in forested grasslands. The diet thus became more diverse (Lovejoy, 1981). This transition was already underway in the first hominins, such as *Sahelanthropus* and *Ardipithecus* (recall their heftier teeth compared to chimpanzees) and slowly continued in the australopithecines (Lieberman, 2013). The presence of carbon isotopes confirms that grasses and sedges were an actual part of their diet. This was still limited in early anamensis, but increased systematically in later species (Cerling et al., 2013). Some sedges also formed underground but edible roots that were supplemented by herbs, leaves and fruits (Grine et al., 2012). Could be a coincidence, you might think? Probably not. In chimpanzees, isotopes found in their teeth never indicate a diet of grasses (i.e. C4 plants), even when grasses are available in the environment.

Getting adapted to these changes occurred at three levels: finding and gathering food, grinding food and digesting food. Finding food in a habitat that became more open and drier inevitably meant longer foraging times until enough was found, both in quantity and quality. Even today, hunter-gatherers travel on average 15 km/day in search for food (Lieberman, 2013). And that's not by car to the supermarket but on foot and scanning everything in the area, looking for any useful food. Current hunter-gatherers do have tools that already make it easier for them to outwit animals, such as slings, bows and arrows, or guns. Australopithecines did not. They certainly did not have the luxury to be picky and let opportunities pass by. Presumably, plants still made up more than 90% of their diet (Wrangham et al., 1999a). Subterranean roots were probably a very important part of this, especially during periods of scarcity in other foods, such as fruits. Roots are also more abundant in a dry forest and grass-lands (up to 40,000 kg/km²) than in rainforest (where only 100 kg/km² are found). There are also many more species of plants that form these roots in dry areas. For example, it is mainly lilies, pumpkin and pea species that form such food- and water-rich roots. On top of that, they also contain a lot of proteins, even up to 40%, enough to meet daily needs (Wrangham et al., 1999b). Even today, there are mammals living in the African savannahs that specialize in finding such underground roots and tubers, such as African mole rats (Van Daele et al., 2007). Getting at them too is yet another challenge. Imagine trying to get these roots out of dried-out soil with bare hands. Australopithecines may have used sticks or bones and horns of animals to dig out roots (Bunn, 1999). Being already adapted to standing and walking upright would have made this easy to do so, as standing on two legs allows efficient crouching and using both hands to dig out roots.

Eating a raw carrot or turnip is something you literally must sink your teeth into. So, it's no surprise that the jaw apparatus in australopithecines shows every indication that they were good at that, sinking their teeth into hard foods. Larger molars and jaw muscles, such as a temporal jaw muscle running high up onto the skull, made the jaw apparatus ideally suited to deal with this. Nevertheless, this also came with limitations, as processing hard foods requires a lot of energy, whereas they do not provide that much energy. Such hard food was thus quite limited in the final amount of energy it provided to the body of australopithecines. Rather, it was fallback food: something to fall back on during periods when nothing better was available (Leonard & Robertson, 1997).

Whether there were also many modifications to the digestive tract to aid digestion of this new food is, of course, difficult to deduce from fossil evidence. After all, intestines rarely if ever fossilize. Nevertheless, it is suspected that australopithecines possessed a relatively long digestive tract, like that of herbivores that ingest foods that are difficult to digest. The fact that australopithecines still had relatively small brains also fits the hypothesis that more energy was invested in forming and maintaining a large digestive tract (Gibbons, 2007).

## Relying on fallback food

Australopithecines had a heftier jaw apparatus, as an adaptation to a more diverse diet that at times was a tough fallback food. In 1962, it was argued that early human evolution showed adaptations to surviving periods of abundant food availability and periods of food scarcity (Shubin, 2008). Perhaps this also applied to australopithecines, which would have come in handy. Now, the underground roots are not so bad. Even today, they make up a third of the diet of hunter-gatherers (Lieberman, 2013). Ultimately, they contain a lot of moisture, proteins and especially a lot of starch. Starch consists of very long molecules of sugars joined together. When fruits were few, roots were an important source of sugars and thus energy for australopithecines (Laden & Wrangham, 2005). But the yield is relatively limited because it requires a lot and powerful chewing, as well as these large molecules are difficult to digest. Having large molars (about 50% larger than in chimpanzee) and a thick enamel coating is therefore very likely no coincidence in australopithecines.

We don't chew for no reason. Chewing is costly. Measuring oxygen consumption in people while chewing gum, in a way that only the consumption by the chewing muscles is measured, shows that the energy consumption of the body increases by 10% (for soft gum) to 15% (for hard gum) (van Casteren et al., 2022). If one extends this to australopithecines chewing on carrots and apples, this energy consumption would have been as high as 20%. Chewing is also not so trivial a thing to do but is a complex event coordinated by multiple muscles, such as jaw and tongue muscles. Saliva production during chewing also consumes energy. On average, we spend 35 minutes a day chewing (Organ et al., 2011). In great apes this is considerably longer (in gorillas up to 80% of daily active hours) (Ross et al., 2009). In total, this requires about 1% of all the energy we consume in a day. If you chew high-energy foods, it pays off. But if you depend on fallback food that requires a lot of energy but yields little, then this can impact survival (as has been observed in some mammals). So, australopithecines had every interest in having a robust jaw apparatus. The way they chewed had also presumably changed, as australopithecines chewed more with the hind teeth than was the case in early hominins. This is advantageous, as their jaw muscles could exert more force on the hind teeth (White et al., 2009).

But of course, their survival did not depend exclusively on this hard food. Other primates such as capuchin monkeys, baboons or chimpanzees also seek their proteins elsewhere when it is insufficiently available: through meat (Dominguez-Rodrigo & Pickering, 2017). Logic suggests that australopithecines will not have done otherwise. Eventually, animals were present in grasslands. But is there any evidence for this? Bones of grazers showing cutting marks already strongly suggest this occurred more than 3 million years ago (McPherron et al., 2010). According to some, this is insufficient evidence, as the marks may just as well have been caused by displacement of the fossils in the soil (Dominguez-Rodrigo & Pickering, 2017). But also the discovery of stone tools with sharp cutting edges from about the same period seem to confirm that australopithecines processed cadavers to obtain meat (Harmand et al., 2015). All in all, it seems highly unlikely that australopithecines would not have been able to obtain meat during periods when they were most in need of it. At the end of the day, they must have gathered enough energy through food to carry on for the next few days.

How much energy is consumed depends on many factors. The more active, the more energy needed. Australopithecines that traveled long distances in search of food therefore consumed a lot of energy. A primate at rest has an average energy consumption equivalent to 69.1 times its body weight, to the power of 0.761 (Leonard & Robertson, 1997). For a 70 kg human, this corresponds to 1752 kcal/day. As soon as we move, this obviously increases. Even just sitting upright requires 20% more energy than in a flat rest. If we stand up, this increases by 40%. For light labor this is 120%, walking 260%, moderate labor 280% and heavy labor 450%. Hunter-gatherers of the Ache tribe from Paraguay, who sometimes travel 19 km/day, consume about 3200 kcal/day, this compared to 1600 kcal when at rest. A chimpanzee consumes about 1500 kcal (including 1000 kcal at rest) and does not cover more than 5 km in a day. Australopithecines were somewhere in between, where because of the sexual difference in body size there was also a big difference in energy consumption. For Lucy's species, a male's energy consumption was estimated at 1700–2400 kcal/day (of which 1200 kcal at rest) and a female's at 1200–1300 kcal (of which 900 kcal at rest). In early *Homo* species, such as *Homo habilis* who still had a small stature, this was similar (2000–2700 kcal in male and 1300–1400 in women). Once the stately *H. erectus* appeared on the scene, this energy consumption increased significantly to very similar levels to ours, *H. sapiens*. This fluctuates somewhere on average between 2300 and 3200 kcal/day in men and 2000 and 2100 kcal/day on average in women (at rest it is 1600 and 1400 kcal/day, respectively).

## Lions that feed on grass

Australopithecines thus relied more on hard fallback food, something their powerful jaw apparatus bears witness to. As for their diet in forested grasslands, it leaned more toward that of baboons than that of chimpanzees (Sponheimer et al., 2022). But what about those robust *Paranthropus*, who exhibited traits similar to lions? They possessed molars whose grinding surface was nearly double that of *Australopithecus* and jaw muscles that were up to three times larger, resulting in twice the bite force. Compared to us, they could even bite 4.5 times harder (Eng et al., 2013). The teeth also had a much thicker enamel coating, advantageous for dealing with greater forces. These teeth were implanted in a gigantic mandible, and the skull bore a substantial zygomatic arch, which ran wide and far anteriorly. This provided more attachment surface for a larger masseter jaw muscle and more space for a larger temporalis jaw muscle to pass through. The most striking feature, also present in lions, is the bony ridge in the center of the skull roof. This too has much to do with the gigantic temporalis jaw muscle because in *Paranthropus* it ran to the very top of the skull and was attached to that ridge. That ridge allows more muscle fibers to attach, allowing greater bite force. Also in lions, the entire cranial roof is covered by a very thick temporalis jaw muscle. Possibly a larger ridge also played a role in social communication, to make more of an impression. This is already seen in gorillas and orangutans, where the ridges in males are remarkably larger than in females (Balolia et al., 2017).

From this it would be logical to infer that *Paranthropus* must therefore have been a carnivore, as were lions. Robert Broom and Louis Leakey, the discoverers of two of the robust species, were also initially convinced that this was the case, in part due to the presence of stone tools in the same areas that could be used as cutting tools (Leakey, 1959; Sponheimer et al., 2022). Leakey was particularly convinced that they processed hard food, which is why he called 'his' species the 'nutcracker'. The fact that the canines are rather small in *Paranthropus* also points more toward a diet that did not consist of soft meat. Teeth can tell us a lot about the diet. In South African robustus, the wear patterns indicate eating nuts and other hard foods, which is confirmed by the isotope values in the enamel that show that their diet consisted for 75% of C3 plants (shrubs and trees that form nuts and berries) (Sponheimer et al., 2022). So, this was presumably a true nutcracker. A variable diet comprising C3 plants was also the case in South African *Australopithecus*, with recent isotope analyses also showing that meat formed a minor part of their diet (Ludecke et al., 2025). But this does not seem to hold true for the East African boisei that Leakey had discovered. Indeed, isotope analysis of the enamel brought a big surprise: their diet consisted of 80% C4 plants (grasses and sedges that form culms and grains) (Cerling et al., 2013; Cerling, Mbua, et al., 2011; Towle et al., 2021). The wear pattern also did not match with what you would expect from a primate eating hard foods. Everything combined indicates that Leakey's nutcracker was actually a grazer, feeding mainly on leaves and culms of grasses and sedges. This is similar to extant Gelada baboons, which, as the cow among primates, feasts on grass in the meadows of the Ethiopian highlands (Sponheimer et al., 2013). Sand grains present on the grass may then explain the wear pattern on the teeth (Towle et al., 2021).

All this suggests that the slender *Australopithecus* and the robust *Paranthropus* made use of their environment in somewhat different ways. The robust forms were more specialized herbivores, while the slender forms had a broader, more omnivorous diet (Falk, 1990; O'Brien et al., 2023). The transition from an omnivorous to this grass diet also presumably occurred gradually, with grasses initially serving as a fallback food during some periods and eventually becoming the regular diet (Malone et al., 2021). This eventually resulted in this robust jaw apparatus, which appeared about 2.4 million years ago (Wynn et al., 2020). And this may also explain why they eventually became extinct, namely due to changes in the environment and available vegetation (Patterson et al., 2022). Even within robust species, variation in climate can be associated with changes in skull morphology (Potts, 2020). The evolutionary history of quite a few plants and animals repeatedly showed that being overly specialized is not always a long-term solution.

# From tree to tree, across the grass

Transitioning to an environment of alternating forests and patches of grassland came with an impact not only on australopithecines their diet but also on the way they had to move around. Where the first hominins still moved mainly via erect clambering in trees, using hands and feet, australopithecines already walked more upright on the ground, i.e. on two legs, but at the same time still adapted to climb trees. Being able to do both is useful when moving around in and between groves: scrambling in the woods, walking between the woods.

## From forest to forest

An equatorial forest that slowly opens up means that ever greater distances must be traveled to find sufficient food (Carvalho et al., 2012). Moving in trees is best not done in the same way as moving on the ground between trees. Having to move more between trees separated by grassy areas imposed a selective pressure to move differently. Mammals living in grassy areas move on four legs, with front and back legs performing similar functions. Our ancestors, however, come from clambering primates that moved upright in trees. Their arms for grasping had a different evolutionary history than their legs for support. This might explain why a shift from clambering upright in trees to moving on four feet in more open grassy areas did not occur. The evolutionary strategy of our ancestors built on what originally allowed them to clamber like this in trees, to move through further refinement to an efficient way of walking upright on two legs across the grass.

This is where our evolutionary story differs from that of the chimpanzee. Rather, their ancestors further refined their specialization as arm swingers (brachiators) in trees. To propel themselves on the ground required another way that was more reconcilable with such a brachiator body, such as knuckle walking. The behavior of present-day chimpanzees seems to confirm this. By comparing the foraging behavior of chimpanzees from different habitats, it became clear that those living in a more open vegetation, with fewer trees spaced more widely apart, do not necessarily move more on the ground (Drummond-Clarke et al., 2022). On the contrary, those more likely to live in denser forests moved more frequently on the ground. Thus, there does not seem to be a relationship between forest density and the extent to which they search for food on the ground. Presumably, the increased likelihood of encountering predators in a more open area is a barrier for chimpanzees to search much for food on the ground.

Climbing out of and into trees also requires much more energy than moving around in a tree, which may also have been a selective driver. Chimpanzees also walk mostly on two legs when moving around in trees, not on the ground! In 80% of the times they walked upright, it was in trees.

All this suggests that the ancestors of chimpanzees completely played the card of brachiation, i.e. swinging under the branches using the arms. This also fit within their environment of a dense forest. Australopithecines, on the other hand, took a different path as forests became less dense and grassy areas more extensive. They took the path of walking well on two legs, relying on what they had already inherited from their tree-clambering ancestors. Where it used to be thought that the transition from an equatorial forest to a completely open grass savannah, about 3–2 million years ago, was the impetus for walking upright, everything indicates that this evolutionary process had already started much earlier, already from the time when an equatorial forest was still transitioning to forested grasslands. This explains why much of the fossil evidence of our earliest ancestors and australopithecines comes from areas then forested. In other words, the so-called savannah theory as people used to see it needs to be adjusted (Domínguez-Rodrigo, 2014). While scientists used to see this as adaptations to life in an open grass savannah, they now consider rather that seasonally fluctuating combinations of forest and grassland were the driving force in this phase of our evolution.

## Why erect?

That it sounds logical that we moved from upright clambering ancestors to upright walking australopithecines is one thing, but that does not explain why it also turned out that way. In evolution, novelties only arise if they also have no disadvantage, and preferably that they have an advantage. So, what might the selective pressure have been as to why our evolutionary line took that path? Was walking upright in slowly outcropping grasses advantageous, knowing that other mammals do well there on four legs? There are several arguments for thinking that it was indeed so (Rakovac, 2021). Hypotheses about what would have stimulated upright walking are numerous: being taller to reach food hanging higher more easily, freeing up hands (allowing food and children to be carried or tools to be used), sticking out above grassy areas (useful for spotting predators or making more of an impression on predators), to being able to wade through puddles or even swim (Lieberman, 2013). Energy consumption is also put forward, where it would be more energetically efficient to travel longer distances between bushes on two legs than on four legs. And then there's the advantage that walking upright in a hot grassy savannah allows us to cool off better (more on that later).

Many of these hypotheses also sound logical, which can quickly lead to scientific storytelling. But how do you test such them when those ancestors are no longer around?

You can hardly ask them why and when they chose to walk on two legs. And fossil skeletal remains don't tell everything about behaviors during locomotion either (Almécija et al., 2021). The closest we get is by looking at how chimpanzees do this. Problem: you also can't have them fill out a questionnaire about why and when they prefer to walk upright and when they don't. But you can present them with challenges to see which ones trigger upright walking more. For what kind of actions does a chimpanzee decide to do that and is it advantageous? That could go a long way toward helping us. That's exactly what Elaine Videan and William McGrew did in 2001, with chimpanzees and bonobos kept in captivity (Videan & McGrew, 2001). To test four different hypotheses, the chimpanzees were put into four different experimental setups. The *carry hypothesis* states that freeing the hands to carry food (and other things) is the driving force behind walking upright. To test this, a large amount of food was provided at once, rather than smaller portions several times a day. If it turned out that they were walking on two legs more frequently than usual to grab large portions of food and walk away with it, it would support this hypothesis. The *vigilance hypothesis* tests whether vigilance to look above something is the driving force. Here, the glass walls of the animals' enclosures, through which visitors can observe the animals, were partially covered. Just as humans like to look at great apes, great apes like to look at humans. If their field of vision is obstructed and they start walking more upright to look above the covers, this confirms this hypothesis. The *display hypothesis* states that making one appear larger has an advantage to scare off predators or conspecifics. We see this kind of impressing behavior in a lot of other animal species as well. To test this, branches were placed in the enclosure. Walking upright then offers the advantage that they can drag branches around and use them to scare others away. And finally, the *foraging hypothesis* was tested, which states that as is (incorrectly, though) put forward for giraffes, standing upright is advantageous for getting hard-to-reach food (Cavener et al., 2024; Simmons & Scheepers, 1996). To test this, bags of food were suspended at 1.5 m high. After many hours of observation, it could be concluded that three of the challenges gave rise to more upright walking. Only the foraging hypothesis was not supported, which makes sense given that great apes can climb trees smoothly to get to hard-to-reach food. Other research also shows that monkeys, such as macaques, walk more upright when it is dark (Manaka & Sugita, 2009). They can then use their hands to feel where they are walking.

Whether grasping food, carrying food or dragging branches, it makes sense that freeing the hands had a clear selective advantage by walking upright. Handling tools and making them were also among the advantages. Even Charles Darwin saw this as the driving force behind the emergence of upright walking and even linked it to the small canines in humans: why need canines to defend yourself when you have weapons and hands to wield them (Darwin, 1871)? Australopithecines could then have already used tools more easily to dig up roots, for example, or for other purposes (Galletta et al., 2019). Australopithecines already had some features in their hands that indicate they could perform a precision grip, albeit not yet as sophisticated as ours (Marzke, 1997). But, unlike Darwin, it is now believed that carrying food was the impetus for freeing the hands. Australopithecines were less likely to encounter large amounts of food when hiking in grassy areas than they were in a forest where a fruit-loaden tree provided more than enough food on the spot. On top of that, it was much riskier to eat food in an open area once found because of exposure to predators. Once food is found, it was certainly beneficial to collect and carry it along until one could consume it in a safer place. Only at a later stage would this then have made it possible to refine hand use, finally allowing even tool use (Lovejoy, 1981).

## Lucy tells a story

Lucy's discovery in 1974 suddenly casted a very different light on how adaptations to walk upright got translated into our australopithecine ancestors. Lucy provided the first unequivocal evidence that these ancestors were already showing anatomical adaptations to walking on two legs, but in the process had not yet completely lost the ability to climb trees. Meanwhile, we already have a lot more fossil evidence from several species that give a better picture as to how far those adaptations went. In addition, scientists increasingly obtained access to very complex computer simulations that allow them to virtually test what these extinct ancestors were capable of. Relying on computer models of the musculoskeletal system of modern humans, one can modify these models to see what the impact would be if, for example, Lucy had moved in a way that we do (Nagano et al., 2005). All of this provides us with a better picture of why our ancestors took the evolutionary path of walking upright on two legs, and eventually started running.

**Climbing and walking**

Australopithecines represent the evolutionary transition from almost exclusively tree-climbing hominins to almost exclusively two-legged humans. This is not only because they are intermediate in time but also with respect to the adaptations in their bodies. We'll talk about the adaptations to walking upright later. First, let's see what adaptations they retained for climbing.

Referring to the difference in the evolutionary history of arms versus legs, arms have always been central to climbing trees. Not surprisingly, adaptations to climbing are found primarily in the arms and shoulder girdles. Long arms are advantageous for climbing and clambering. Lucy still had rather long arms, but they were already shorter than the legs (legs are 1.16 times the length of the arms) (Heaton et al., 2019). Chimpanzee legs are shorter than the arms (0.97 times the length of the arms), while our legs are significantly longer than the arms (1.5 times the length of the arms). Also useful for climbing are long and curved fingers, with the more primates climb, the more curved the phalanges can become (Kivell et al., 2022). Lucy and other australopithecines also had curved phalanges. But there were also differences among australopithecines. For example, the South African sediba showed more adaptations indicative of arm swinging and the shoulder girdle in an afarensis child was also very ape-like (Alemseged et al., 2006; Rein et al., 2017).

To put it bluntly: Lucy had arms to climb and legs to walk (Gebo, 1996). This is broadly true, but obviously does not fully hold true. For example, the fingers are shorter than in those of chimps (compared to the thumb), making hands less useful for grasping around a branch (Tocheri et al., 2008). Also, the way the fibula articulates with the foot is more similar to that of climbing great apes than that of walking humans, and Lucy still had curved toe phalanges, useful for grasping branches (Marchi et al., 2022; Suwa, 1984). Now, that Lucy and other australopithecines exhibited numerous adaptations to walking on two legs can also be inferred from features other than the legs alone, such as the position of the occipital foramen, the anatomy of the inner ear or the organization of the spine. Even fossilized footprints prove this.

The occipital foramen is the opening at the back of the skull along which the spinal cord connects to the brain. Considering the spinal cord running along the vertebral column, it shows the extent to which the spine was behind (as in quadrupeds) or below (as in

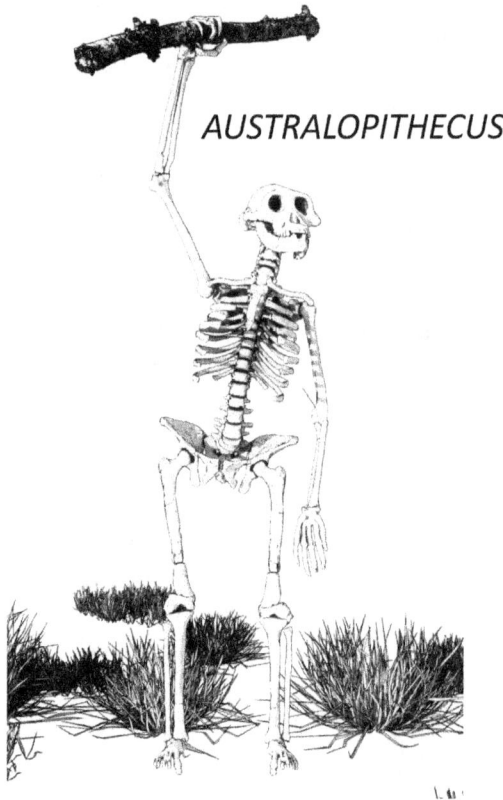

*AUSTRALOPITHECUS*

Figure 5.2   Australopithecine skeleton in forest and grassland environments. (Based on 3D model of Lucy of Brassey et al. (2018).)

bipeds) the skull (Russo & Kirk, 2017). In Lucy, however, it was difficult to determine this because the skull was too deformed during fossilization. But other fossils clearly showed that the spine was below the skull, in a way that was even very similar to ours (DeSilva, 2021). This was also the case in the first hominins, with a position clearly different from that in great apes (Ahern, 2005). But does that tell us anything about the extent to which they walked upright, i.e. constantly or occasionally? Unfortunately, this characteristic is not reliable enough to infer as such, and other characteristics need to be looked at as well (Landi et al., 2020).

What can the inner ear tell us about walking upright, you might ask? Well, the inner ear, in addition to being our hearing organ, is also our balance organ. It determines how our head is positioned in relation to the gravitational field and constantly senses the degree to which we tilt and rotate our head in three-dimensional space. The latter happens thanks to three half-moon-shaped canals in the inner ear, filled with a fluid and each equipped with a sensor. That sensor detects movement of that fluid within that canal, just as fish use this kind of sensor on their bodies to detect displacement of the surrounding water. Only here it is rather the channel that moves relative to the fluid, while the fluid stays behind when we turn our heads (just as water in a glass does not move with it when you turn a glass around). Two of these channels are in a vertical plane, one in a horizontal plane. It is the latter that detects horizontal rotations of our head. In great apes, in whom the skull is connected to the spine more at the back and who also incline their skull more, this horizontal canal is more inclined (Jeffery & Spoor, 2004). Our canal is more horizontal, which is consistent with the fact that we also keep our skull more horizontal because of our upright posture. In australopithecines, this is already more like the situation in humans. Also in australopithecines, the two vertical canals are larger than in great apes, but smaller than in humans (Urciuoli et al., 2020).

An erect posture has a major impact on how the spine must bear the entire body weight and how that weight is transferred to the limbs. In quadrupeds, the weight is spread more along a horizontal spine and that weight is then distributed among four supports, the front two that are distant from the hind two. In bipeds standing upright, not only is the weight carried along a vertical spine but also all this weight is distributed among only two supports that are close together. While quadrupeds, and therefore great apes, absorb their weight by keeping the spine stiffer to prevent flexion, in bipeds all vertebrae collide under the influence of gravity. While in quadrupeds about one-fourth of the body weight ends up in the joint between the spine and a girdle (shoulder or pelvic girdle), in bipeds half the weight ends up in a hip joint (called the sacroiliac joint). Comparing the spinal column of australopithecines with that of great apes and humans can teach us a lot about how that column was loaded during locomotion. At the start of our evolution, in the common ancestor with the chimpanzees, the spine comprised 33 vertebrae: 7 cervical, 13 thoracic, 4 lumbar, a sacrum with 6 vertebrae and 3 short caudal vertebrae (Thompson & Almecija, 2017; Williams

& Pilbeam, 2021). We still usually have 33 vertebrae, but all indications are that one of the ancestral thoracic vertebrae has become transformed into a lumbar one (i.e. we have 12 thoracic vertebrae and 5 lumbar ones), and one vertebra of the sacrum has become a caudal vertebra (i.e. we have 5 sacral vertebrae and 4 caudal ones) (Saladin, 2004). But this can fluctuate between 30 and 35 vertebrae, and the typical 7–12–5–5–4 even occurs in only 38% of people (Williams & Pilbeam, 2021; Williams & Russo, 2015). In 24% of cases, people even have an extra tail vertebra. A spine of about 33 vertebrae has apparently proven to be evolutionarily very stable. Gorillas, chimpanzees ánd australopithecines also had 33. Exceptions are gibbons (31 or 32 vertebrae), bonobos (34) and orangutan (31 or 32) (Haeusler et al., 2002; Schultz & Straus, 1945; Williams, 2012; Williams et al., 2013). The number of cervical vertebrae has remained the most stable, all great apes and anthropoids have seven cervical vertebrae, as do virtually all mammals (Williams & Russo, 2015). Even a giraffe with its long neck or a whale that seemingly has no neck have 7 cervical vertebrae. In primates, only 2% have a different number of cervical vertebrae (Galis et al., 2022). The explanation is not that this provides the best-adapted neck, but one cervical vertebra more or less is associated with developing fetal cancer (Galis, 1999). A study of the construction of the cervical and thoracic vertebrae in human fetuses, in which fetal cancer was identified as the cause of death, found that the seventh vertebra showed a rib (Galis et al., 2021). That is, the seventh vertebra which is normally a cervical vertebra without a rib, has acquired the anatomy of a thoracic vertebra. A cervical rib was thus identified in no less than 55% of the fetuses studied (Galis et al., 2006). This all has to do with the genetic control of how different types of vertebrae develop at different locations along the body axis is regulated. This is in fact under the control of so-called *Hox* genes, some of which may or may not be expressed in each body segment, giving rise to a particular type of vertebra (Williams & Pilbeam, 2021). For example, the last cervical vertebra is determined by a combined expression of all *Hox4* and *Hox5* genes, while the first thoracic vertebra is determined by all *Hox5* and *Hox6* genes. Thus, a shift in what *Hox6* genes do determines whether or not a vertebra will form a rib, and thus become a cervical or thoracic vertebra (Bohmer & Werneburg, 2017; Wellik, 2007). The same genetic control mechanism explains the difference between the number of vertebrae in the other regions. For example, lumbar vertebrae are formed when *Hox9* and *Hox10* genes are active and sacral vertebrae are formed when *Hox10* and *Hox11* genes are expressed.

Australopithecines have, like us, 12 thoracic vertebrae, so one less than great apes (except for orangutans who also have 12). The number of lumbar vertebrae also shows a lot of variation, which, as we will see further on, has everything to do with walking upright: chimpanzees only have 4, australopithecines had 5 and all *Homo* species had 5. It is not always obvious to find clarity in the fossil material about how many vertebrae of each type are present. For example, one usually looks for the so-called transitional vertebra, which typifies the transition from the thoracic region to the lumbar region. This vertebra combines typical features of both verte-bral types, especially regarding the orientation of the joint surfaces (Williams et al., 2013). Sometimes you do find in literature that Lucy had six lumbar vertebrae and only four sacral vertebrae (Lovejoy & McCollum, 2010). By the way, one of Lucy's tho-racic vertebrae later turned out to be that of a baboon (Meyer et al., 2015). Thus, although researchers did not always agree on the number of thoracic and lumbar vertebrae in our fossil ancestors, the most acceptable hypothesis seems to be that from the australopithecines onward, a spine with 7 cervical, 12 thoracic, 5 lumbar and 5 sacral vertebrae remained stable (Haeusler et al., 2011; Pilbeam, 2004). Compared to great apes, with 4 lumbar vertebrae, a longer lower back developed from then on. The major difference between thoracic and lumbar vertebrae is that the latter bear only very small protrusions, while there are large ribs attached to the thoracic ver-tebrae. As a typical mammalian characteristic, this has proven useful for respiration (because the diaphragm causing the expansion of our lungs lies on the boundary between the thoracic and lumbar regions) and locomotion. In fact, a slender lumbar region can bend more easily, thanks also to the shape of the joints between the verte-brae. The fact that no loose ribs are formed in the lumbar region lies under the control of the *Hox10 gene*. As a result, ribs do form, but they remain small and fuse with the vertebrae (Williams & Pilbeam, 2021).

I mentioned the sacrum a few times, the bone formed by five fused vertebrae that connects the spine to the pelvic girdle at the level of the sacroiliac joint. How many vertebrae eventually fuse, and the extent to which there is also fusion with caudal vertebrae, can vary quite a bit, even among us (Schultz & Straus, 1945; Tague, 2011). In 7% of people, the sacrum and tailbone are even completely fused with each other (Tague, 2009). In Lucy, as might be expected, five small vertebrae were identified in the sacrum (Johanson et al., 1982).

## Strolling in Laetoli

Quite some anatomical evidence strongly suggest that Lucy and other australopith-ecines walked upright. But there is also very convincing indirect evidence, such as fossilized footprints. The most spectacular and also the oldest footprints were left by australopithecines, presumably afarensis, in volcanic ash at Laetoli, Tanzania, about 3.66 million years ago. It was Mary Leakey who in 1978 described what she and her team had discovered several years earlier (new prints were discovered nearby in 2016, by the way) (Leakey, 1978; Masao et al., 2016). At two sites, they discovered footprints, one track containing 22 footprints of three individuals, over a distance of 23.5 m and in a layer about 15 cm thick of repeatedly deposited ash (Day & Wickens, 1980). Prints of other animals were also found, including birds, hares, predators, rhinos and elephants (Leakey & Hay, 1979). From these footprints it was possible to derive a great deal of new information about the locomotion of austra-lopithecines, which could not be obtained from fossil remains of skeletal material. It even allowed some reconstruction of the whole scene of those individuals who on a particular day, 3.7 million years ago were walking (not running) through the volcanic ash while shuffling at a high speed (Hatala et al., 2016). The footprints, deposited in moist ash were 18.5 cm long and 8.8 cm wide and showed that the stride length was short, about 39 cm (Day & Wickens, 1980). Based on this stride length, it could be estimated that those two individuals had a hip height of about 54 cm and were about 115 cm tall, with a body weight of about 45 kg and that they were stepping at a speed between 3.5 and 5.9 km/h (the speed varies depending on the reference used in the way they walked and the degree in which the legs were stretched or not) (Harcourt-Smith, 2005; McNutt et al., 2021; Raichlen et al., 2008). The makers of the footprints described in 2016 would have been about 165 cm tall (Masao et al., 2016). This also led to the conclusion that the tracks may have come from afarensis, as there were not that many hominins of that size living in this area 3.66 million years ago (Suwa, 1984). Yet the most remarkable thing about the footprints was that they showed conclusively that the big toe no longer diverged from the second toe and thus that these australopithecines no longer had a grasping foot! This was the first evidence that from then on, our ancestors had lost grasping feet and later research also showing that this must be an adaptation to walking upright on two feet. Moreover, the Laetoli prints show that their makers were already pushing off with the big toe and other toes, thus like what we do (Berge et al., 2006). Also, the prints show that the sole of the foot was no longer flat, as in chimpanzees, but was already slightly arched (Leakey & Hay, 1979). The latter also, as will be shown, an adaptation to upright walking, as well as the fact that the feet were fairly close together. In the Laetoli prints, the feet were much closer together than would be the case in chimpanzees.

At one site, two individuals walked along the same trail, one large and one small one, with one individual stopping at a certain point to turn around. The American Museum of Natural History in New York recreated this scene, with a male and female australopithecine walking arm in arm across the ash-covered savannah, with the erupting Sadiman volcano in the background (Zaitsev et al., 2019). Whether this was genuinely such a romantic evening out between the two remains to be doubted, in part because the footprints were rather too close together to have been comfortable for walking side by side like this. Also, the two footprints show that they were deposited at different speeds, also not convenient when walking together (Raichlen et al., 2010).

Several studies have tried to reconstruct the Laetoli impressions, by having great apes and humans walk under all kinds of conditions (stretched or folded legs, walking or running) in a layer of sand, and then analyzing the impressions left with high-tech equipment. This showed that the impression the toes and heel made in the sand while stepping with fully stretched legs, and thus how the weight is transferred from one foot to the other, was very similar to that in the Laetoli ashes (Raichlen et al., 2010). In turn, other analyses of human and chimpanzee footprints showed that the

Figure 5.3    Diorama at the American Museum of Natural History in New York, depicting the makers of Laetoli's footprints with the erupting Sadiman volcano in the background.

Laetoli footprints were not quite the same as ours after all but not quite the same as those of chimpanzees either (Hatala et al., 2016). The differences that the human prints showed were mainly due to the degree to which the knee was bent while putting the foot down. This is because with bent legs, the foot is put down flat, while with stretched legs, the heel hits the ground first. Interesting was the comparison of footprints from people used to walking barefoot and those from Laetoli. Prolonged barefoot walking has an impact on the footprint. For example, the toes will extend slightly wider and the arch will be less obvious. That kind of print is then already more similar to Laetoli's (Crompton et al., 2012). So, one study suggests that they were already walking with legs extended, the other rather that they were walking with hips and knees flexed (so-called *bent-hip-bent-knee*) like great apes (Berge et al., 2006). If they were still showing adaptations to climbing trees, it would also make more sense that the legs were not yet adapted to fully stretched walking because climbing is better done with properly pliable legs (Crompton et al., 2012; Crompton et al., 2010; Hatala et al., 2016). Stretched legs, however, would be very helpful when walking upright: it requires less muscle force to stabilize the bent joints when one leg is supported. In addition, less energy is also lost this way, thus producing less unnecessary body heat. However, the analysis of more recently found footprints in Laetoli seems to point more in the direction of walking with extended legs, but still in a way that was not completely identical to our way of walking (Raichlen & Gordon, 2017).

## On two supports

Walking upright on two legs, we do it daily without giving it a second thought. And yet, this is far from obvious. That's because we are adapted to doing it very well. The question is whether this was as good for australopithecines, who still combined walking upright with climbing trees. Was Lucy strolling through the African savannah like we take a long walk to relax? Or did this imply a great effort to cover a large distance by walking on two legs? Walking on two legs involves a maximum of two points of support. If we stand still, all the body weight is concentrated on the two feet only. If we walk, we even alternate between one foot, two feet, one foot, two feet, .... Each time we put one foot down, gravity is directed along the spine and all the body weight is directed onto that one foot. When alternating between putting one and two feet on the ground, our center of gravity must constantly switch sides because the center of gravity must always be above the point of support to keep balance. We can do it, but so could australopithecines.

The makers of the Laetoli footprints probably already had an arched foot (although recent research raises questions on whether footprints can provide conclusive evidence about this) (Hatala et al., 2016, 2023). When a foot is placed on the ground, the inside of the midfoot barely touches the ground. This is thanks to two curvatures caused by ligaments in the foot sole that through tension arch the bones. A combination of a transverse arch across the midfoot and a longitudinal arch along the inside of the foot (from the heel to the base of the toes) ensures that when we put our foot on the ground, the toes, heel and outer edge of the foot rest on the ground but the inner edge of the midfoot does not. When you step with wet feet, you leave a footprint where the midfoot is seemingly very narrow because it only halfway touches the ground. We are even born with those arches, only the thick baby foot pads make the entire foot to touch the ground (Tardieu, 2010). Chimpanzees, on the other hand, do not have those arches; they have completely flat feet. Besides, if we have flat feet, it is because those ligaments are weakened, so we lose the arches. Even if the footprints would not be conclusive proof for presence or absence of these arches, a well-preserved foot skeleton of afarensis shows that they did have longitudinal and transverse arches. Clearly different from the first hominins, but not yet as pronounced as ours (DeSilva et al., 2018; Lovejoy, Latimer, et al., 2009). This only came about later, in *H. erectus* (Lieberman, 2013). Why is this now an adaptation to upright walking, you may wonder? These arches have three main functions during walking and running: shock absorption, reinforcement of the midfoot and storing energy (Bramble & Lieberman, 2004). Each time a foot touches the ground, the body weight will have lower leg pushing the foot flat, thus tightening the ligaments. Hence, most of the shock is absorbed. Then when we turn the foot further to push off with the toes, this tightened ligament will further tighten the midfoot bones, keeping the foot stiff and firm. During this tensioning, energy is accumulated in those ligaments, which is afterward released when the foot is rolled off and the toes detach from the ground. We even manage to adjust the tension depending on the stride frequency, thus regulating shock absorption (Holowka et al., 2021). Through the ligaments, we recover up to 17% of the energy needed to push off (Lieberman, 2013). Our Achilles' tendon does the same (it recovers 35% of required energy), but this did not yet happen in australopithecines (Ker et al., 1987). Indeed, their heel bone suggests that this tendon was still short, as in great apes. And a short tendon cannot stretch much, and thus store only very little of that energy.

In many ways, the foot skeleton shows that the australopithecine foot was adapted to absorb the shock on touchdown and transfer forces on push off. Their heel bone was already more robust and similar to that of later humans (Arlegi et al., 2023; Harper et al., 2021; Lieberman, 2013). Absorption not only takes place in the leg with which the foot first touches the ground but also the leg to which the Achilles tendon pulls to push off with the foot. So, it seems very likely that australopithecines put their foot on the ground heel first, just like us (O'Neill et al., 2022). Also, the ankle joint was already firmer and the toes were shorter, making the foot less useful for grabbing branches but more efficient for bouncing off while walking upright, without spraining the ankle all the time (Holowka et al., 2017). The rolling off of the foot in australopithecines was already more like in humans because now the big toe was also involved in the push off and no longer served as a grasping toe. Australopithecines already had a stiffer midfoot, with solid bones on the inside of the foot (Arlegi et al., 2023; DeSilva, 2022; Komza et al., 2022). This is also true for us, where as it supports the big toe, that side of the foot is stressed more. On top of that, we put our weight less on the outside of the foot when we unroll the foot, not as in great apes, allowing us to push off more forcefully with the toes (Crompton et al., 2012).

We even have a second shock absorber, in our spine. Our vertebrae do not follow a straight line but a kind of S-shape: in the neck they lie curved with the concave side to the back, in the chest the concave side lies forward, and in the lumbar region the concave side lies backward again (Whitcome, 2012). The chest area is very stiff, thanks in part to the large ribs connected to the sternum. But the neck and lower back, where ribs are only small protrusions, are quite flexible. Such a curved spine has several advantages when walking upright. It ensures that the center of gravity is closer to the hip joint, making us less likely to fall forward or backward when standing still. It also ensures that when gravity kicks in, the spinal curvature can increase so that shocks during walking are absorbed there as well. As already mentioned, a lumbar region with five vertebrae has remained evolutionarily stable from australopithecines onward. In chimpanzees, the spine in that region is still fairly straight, as can be derived from the angle formed between the anterior plane of the first lumbar vertebra and the posterior plane of the last lumbar vertebra. Should those vertebrae lie perfectly in a straight line, that angle would be 0°, because the planes would be parallel. When chimpanzees stand upright, that angle is 22°, so very slightly curved (Been et al., 2012). In australopithecines this had already almost doubled to about 41°. So, the lower back was more arched, a feature called lumbar lordosis (lordosis refers to a hollowing in the spine, with the concave side facing backward). In early human species, such as *H. erectus*, this was even slightly higher, at 45°. This curvature increased even further in sapiens, to about 51° (the exception in this trend were Neanderthals, with an angle of only 29°). The strong arch in the lower back is made possible because the vertebrae are somewhat wedge-shaped: at the front they are not as high as at the back (the angle between the upper and lower surfaces of the same vertebra can be 6°) (Masharawi et al., 2010). In sapiens females, this wedge formation can slightly increase during pregnancy, related to the fact that the presence of a fetus shifts the center of gravity forward, putting more stress on the spine and increasing the likelihood of falling forward. A more flexed lower back shifts the center of gravity slightly further back, providing a better balance (Whitcome et al., 2007). This explains why women are more likely to face so-called degenerative spondylolisthesis, where a vertebra shifts too far forward relative to the adjacent vertebra (Hay et al., 2015). Wedge-shaped vertebrae were already found in australopithecines, where the wedging was also more pronounced in females (Williams et al., 2022). An additional feature, already present in early hominins, is that the lumbar vertebrae were no longer enclosed in the elongated iliac wings of the pelvic girdle. While this is true in great apes, in australopithecines the shortening of these wings caused additional loosening of the lower back (Pilbeam, 2004). But all indications are that the way the lower back was loaded in australopithecines was still different from ours.

The first hominins already clearly had a pelvic girdle distinct from that of great apes and this was no different for australopithecines. For example, australopithecines and *Homo* species have a protrusion at the front of the pelvis (forms the so-called anterior superior iliac spine), which is not found in any other primate (Zirkle & Lovejoy, 2019). Its presence is associated with prolonged upright walking. Like us, australopithecines had obliquely directed lateral wings of the iliac bone, as opposed to transversely directed ones of great apes (Hammond et al., 2020). The plane of movement between the spine and pelvic girdle (defined by the sacroiliac joint) is also different. In great apes, this lies rather oblique, making forward and backward tilting of the spine difficult. In hominins, this joint moves in a more vertically plane, facilitating pelvic tilt. Lucy also had a remarkably wide pelvis, with the femurs also further apart (Gruss et al., 2017). This had advantages and disadvantages. The disadvantage was that this caused gravity to pull harder on the pelvis each time a foot was lifted off the ground. Earlier we talked about that kind of stabilization of the pelvis by muscles running from the oblique iliac wings to the femur (the lateral gluteus muscles). The wider the pelvis, the greater that downward force while supporting on one leg, the more muscle force needed to prevent the body's center of gravity from constantly going up and down while walking. The action of those lateral thigh muscles thereby changed dramatically, making the femur rotate differently than in chimpanzees: they pull the leg laterally instead of backward (Hogervorst & Vereecke, 2015; Stern & Susman, 1983). Oblique iliac wings also allow those muscles to pull more forcefully onto the femur.

So, two birds with one stone. More so, because a wider pelvis also gave Lucy the advantage that combined with relatively short legs, a relatively large step could still be taken by tilting the pelvis forward with each step (Nagano et al., 2005). We see this when we walk as well: people with a slightly wider pelvis take larger steps without their center of gravity rocking up and down more (Gruss et al., 2017). The energy consumption to cover a certain distance depends, among other things, on the number of steps, the speed of stepping, but also on how much the body moves up and down. The less energy this requires, the better. Most energy is lost in putting down and bouncing off the foot. Australopithecines thus benefited greatly if a step could be slightly larger, and the center of gravity did not fluctuate too much. So, is this the only explanation why the pelvis widened during evolution? Probably not. Things like regulating body temperature and allowing babies to be born smoothly probably played a role as well.

But back to a wider pelvis and femurs that are further apart as a result. This is a disadvantage during walking upright as it requires the center of gravity to swing more and more from left to right as we put one foot after the other on the ground. In order to support the left foot, the center of gravity must be above that supporting foot. Only then can the right foot be lifted from the ground to swing the right leg forward. Once the right foot is on the ground, the center of gravity must shift to the right, to above the supporting surface of the right foot. And so this goes from left to right, right to left, over and over again. This is why a chimpanzee waddles from left to right when walking on two legs. Lucy was the first fossil to show that australopithecines were already adapted to waddle less and thus expend less energy. The reason is the so-called valgus angle. This is the angle between the femur and tibia at which the knee is tilted inward. If the knee is tilted outward, this is called the varus angle. Should the lower leg be nicely in line with the femur, this angle would be 0°. This is approximately so in chimpanzees (Shefelbine et al., 2002; Tardieu, 2010). When we are born, we start with a varus angle of about 15°. Under the influence of walking upright, the knee slowly tilts inward (people who are paralyzed do not have a valgus angle). By the time we start walking at an age of about 1.5–2 years, the angle is about 0°. Women reach a final valgus angle of about 6.5° at the age of 9, in men this is about 11° at the age of 13 (Pujol et al., 2014). The inward tilt of the femur also shows in the angle the femoral neck makes with respect to the shaft (decreases from 163° to 126°). What about Lucy? For Lucy herself, this was difficult to measure as the knee was too much shattered. But based on other fossils found together with Lucy, that angle could be estimated at about 14°, so even more inward bent knees. The larger angle can be explained to the small stature, also seen in *Homo floresiensis*, an extinct human species from Asia that was about the same size as Lucy, that also had a valgus angle of 14° (Blaszczyk & Vaughan, 2007). Now what is the advantage of having such an inward angling knee for walking upright? It causes the feet to be closer together. So when walking with a valgus angle, the center of gravity has to be moved much less from left to right because the feet are much less apart. Clear energy gain, in other words! So from all this we can conclude that australopithecines were already adapted to walk efficiently on two legs, and that their children presumably did so as well (given that the angle in our case is induced by walking upright).

## Careful with energy

Much to do about energy, that is clear. You can even summarize evolution as 'surviving and having enough energy left to reproduce'. The less energy you consume while walking, the more remains available to invest in offspring. A lot of researchers have therefore tried to estimate how much energy it required for Lucy to walk upright, and which way required the least energy. Did she do so with hips flexed and knees bent, or was this too costly? Or was walking upright with legs extended, as we do, already energetically the most interesting for Lucy and other australopithecines? Any adaptation that causes the leg muscles to exert less effort to propel the body at the same speed results in energy gains. Important in this story are two types of movement energy: kinetic energy and potential energy. Simply put, kinetic energy is what makes us move, and potential energy is the energy we can accumulate as we move against gravity. If our center of gravity goes down, it goes with the gravitational force and thus loses potential energy. That energy is then converted into kinetic energy, which translates into a downward motion. If we push our center of gravity back up, then we start using kinetic energy to convert it into potential energy. Stepping with stretched legs is really nothing more than constantly converting those energy forms into each other. Stretched legs then behave like two stilts, with a point of rotation at the hip joint and a point of support on the ground. If we stand nicely vertically, then the center of gravity is at its highest point and the potential energy is highest. If we, by supporting ourselves on the left hip joint, tilt forward, the potential energy will decrease because the center of gravity will move slightly downward, with the force of gravity. This is accompanied by both a forward and downward motion, so the kinetic energy increases. We obviously don't let ourselves fall to the ground and swing our right stilt forward and put it on the ground. From then on, our center of gravity does not drop anymore: the potential energy is at its lowest point. The center of gravity then shifts from the left leg to the right leg, going back up, against gravity. We gain back potential energy, thanks to that forward motion. So kinetic energy is now converted into potential energy, until the moment our right leg is completely vertical and our center of gravity has returned to its highest point. From then on, this cycle repeats itself, but tilting forward on the right leg, and so we are off to taking steps.

Through that conversion of potential to kinetic energy and back, we lose very little energy. If we did not experience friction in our joints and with the ground, we would be able to walk endlessly without consuming energy. Wouldn't that be convenient! But that advantage only exists when the legs are kept straight. If we step with a flexed hips and bent knees, this advantage disappears (Wang et al., 2003). Then the potential energy will decrease about the same time as the kinetic energy decreases. Ditto for the increase in energy. This is because as we tilt forward and downward,

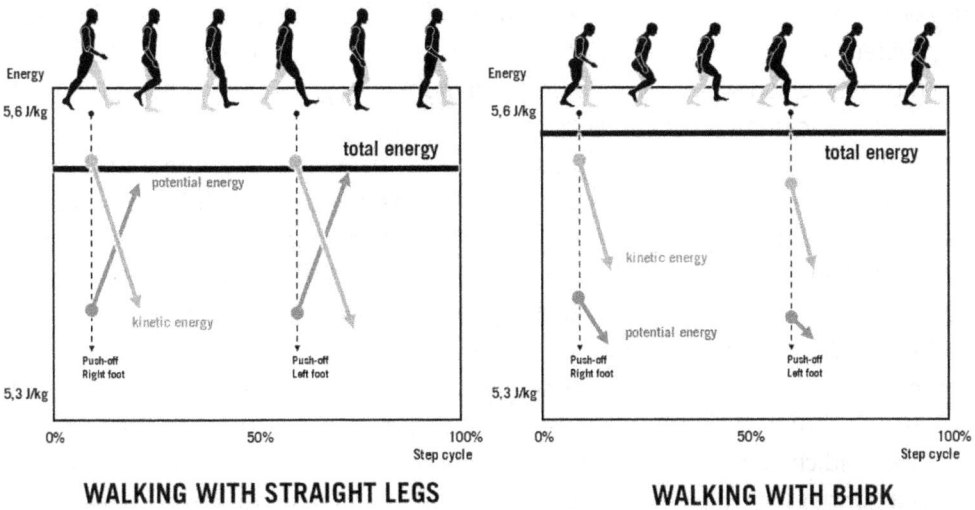

Figure 5.4   During walking with legs extended, potential energy is always converted into kinetic energy and vice versa (if one energy decreases, the other will
increase) (graph on the left). When walking with hips and knees bent (BHBK –
bent-hip-bent-knee), this is not the case, thus increasing total energy consumption (graph on the right). (Based on Wang et al. (2003).)

that motion must now be accommodated by folding the legs. So, we sink down a little further before we begin to push off. Not nicely converting potential into kinetic
energy (and vice versa) results in less energy being recovered between each step, so
the muscles must supply more energy. When people walk with their legs bent, they
also consume more energy overall. In fact, both slow and fast walking with stretched
legs still recovers more energy than slow walking with bent legs (between 50 and 70%
with stretched legs and only 27% with bent legs). When chimpanzees walk, they even
recover only 8% of that energy between steps (Crompton et al., 2010).

But how did Lucy do it? It was initially assumed that australopithecines were not yet
very efficient at walking upright and thus required a lot of energy (Kramer & Eck,
2000). The assumption was that they walked with hips and knees bent (so-called *bent-
hip-bent-knee*) and that efficient walking on two legs only became established in the
first large humans, such as *H. erectus* (Gruss et al., 2017). By running computer models
on them, scientists could make Lucy walk in a virtual environment, so to speak, under
a variety of different conditions. One simulation could make them walk and run like a
modern human, or like a chimpanzee, or something in between. Current models are
even so powerful and detailed that one can estimate how much force is supplied by
individual muscles while walking and running, or at what angles the muscles of bent
limbs optimally do their work (Sylvester et al., 2021; Wiseman, 2023). Each time, it
was calculated how much energy this would require. As it turned, models showed that

Lucy would have consumed less energy while walking than modern humans. However, the problem was that because of her short legs she would have had to switch to running much faster, so she still ended up using more energy (Kramer & Eck, 2000). This would also indicate that the distance Lucy could cover in a day was probably limited. The models also predict that even if Lucy walked with her legs extended, she still consumed more energy than if a modern human walked at an average speed. In Lucy's case, this was 3.8 J/kg of body mass per second for a speed of 1.8 km/h (Joule is a unit of energy). In modern humans, this is about 3.0 J for a walking speed of about 6 km/h (Nagano et al., 2005; O'Neill et al., 2018). Expressed in distance traveled, this would be 7 J/kg of body mass per meter traveled (Sellers et al., 2005). If you convert that for full body mass, Lucy is 27% energetically less efficient at walking upright than an adult modern human but would consume about the same amount of energy as a 7–10-year-old child, who is about the same size as Lucy (Steudel-Numbers, 2006). If you extend that for whole-day energy consumption, Lucy would consume even 50% more energy than early humans, such as erectus (Carey & Crompton, 2005; Steudel-Numbers, 2006). Should Lucy walk with hips and knees flexed, this energy consumption would have been significantly higher (Gordon et al., 2009; Schmitt, 2003). Also, the way the leg extensors in Lucy pulled on the lower leg suggests that they were walking with their legs extended (Wiseman, 2023). Chimpanzees who walk upright in this way use larger muscle volumes to walk and therefore use 32% more energy than if they would walk on four legs and 77% more energy than a human walking on two legs (Sockol et al., 2007). What computer models also showed is that Lucy was also not walking like chimpanzees do, their constitution was not equipped for that (Crompton et al., 1998). More recent models confirm that afarensis could run bipedally, albeit at a substantially lower speed than humans (Bates et al., 2025).

On another front, there were already energy gains for australopithecines walking upright, thanks to the forearms. Surprising perhaps, but when we walk, we save energy through our arms, by swinging them forward and backward in an inverted motion relative to that of the legs and at the same frequency as the legs. We compensate for the forward swing of our left leg by swinging our right arm forward. In this way we maintain the torque of our body, so it does not rotate all the way left and right every time we walk. It does require muscle power to move the arms but that still makes us save a total of 10% of energy while walking compared to if we kept our arms still (Yegian, Tucker, Gillinov, et al., 2021). Longer arms require more muscle power to move them, but we can adjust them by folding them or not. And that differs between walking and running. While walking, the arms hang down, almost stretched out. When we run, we pull the forearm up, to about 90° to the upper arm. By doing so, we make the overall arm a little shorter, so to speak, which makes it easier for the shoulder muscles to adjust the swing frequency as a function of step frequency. But keeping the arm like this requires extra muscle power and thus energy. It is probably no coincidence then that throughout our evolution the forearm has shortened. In chimpanzees the forearm is about 93% the length of the upper arm and in Ardi (first hominin) it was still 90%. In humans this is only 70–80%. In Lucy it was about 86%, so shorter than in our earliest ancestor but even longer than in humans. In modern humans it could be shown that the shorter forearms provide three to four times more energy gain while running compared to walking. A rough estimate suggests that the shorter forearms provide significant energy gains when running long distances. Even in australopithecines, this then provided a small energy gain compared to the first hominins.

# The sun is shining

The forest around the australopithecines gradually opened up further, leaving them less and less protected from solar radiation by the canopy. As their habitat is equatorial, radiation hits the Earth's surface straight down and thus exposes everything and everyone that walks and grows there to both a higher heat and radiation. From a natural selection perspective, it can be expected that australopithecines also developed adaptations that better protected them against both: a cooling system that prevented the body from overheating and skin that better shielded those harmful rays.

## Some like it hot

Being nice and warm is one thing, being hot is something else. All organisms have a healthy margin within which their system functions properly. Too much heat threatens to disrupt the body's normal metabolic processes, for which the organism must then invest energy to restore balance. Mammals and birds are able to keep their body temperature nearly constant, but it is at a cost (Glazier, 2008). Living in an environment with too low a temperature requires energy to keep warm. Living in an environment that is too hot also requires energy but to cool the body. Somewhere in between lies a balance that requires the least amount of energy. This balance can be maintained in three ways: allowing less solar heat to enter the body, getting more heat out of the body and generating less internal heat during activity.

### Less heat coming in

Living in a tropical forest is handy to reduce heat exposure, something that australopithecines over time could do less and less. In an open area like a savannah, body temperatures can rise quickly. Baboons living in day temperatures up to 41°C show a 5°C fluctuation in their body heat, even though they possess an ability to keep a stable body temperature (Brain & Mitchell, 1999). A body temperature of 39°C has been measured in humans in hot deserts, while it can drop to 24°C in a cold environment (Wheeler, 1994). Two strategies can avoid overheating without the need of an excessive cooling system: either be active when the sun is low or limit the amount of sunlight radiating onto the body. Many predators employ the first strategy, as they lie dormant in the shade during the day. Running after prey in such excessive temperatures would easily lead to overheating, not to mention that their muscles would not work optimally. Once the greatest heat subsides, they go out hunting. If this would have been a good strategy also for australopithecines implied that they would frequently encounter those large predators. Foraging for food at the same time as predators that can eat you may not have been such a good idea.

Limiting the amount of sunlight penetrating the skin is another option, by keeping the irradiated surface area small and the amount of heat penetrating into the skin low. The surface area being irradiated depends on the shape of the body and its orientation relative to the direction of the sun's rays. Quadrupeds, that hold their body horizontally, are being irradiated all over their back when the sun is completely overhead. This is even more so for quadrupeds walking in an open area around the equator at the warmest time of the day. However, as australopithecines walked upright, this changes the situation. Their exposed skin area would have been largely confined to the crown of the head and shoulders, reducing the irradiated area to only 40% of what it would have been if they walked quadrupedally (Wheeler, 1985). The scalp, which covers only 10% of the body surface area, would then shield up to 70% of the body surface that would otherwise be intensively irradiated. Using a doll equipped with heat sensors, scientists tested the effect of different types of wigs when putting the doll under a hot lamp, provided with some wind and some humidity. It confirmed that hair effectively blocks a lot of heat rays from a light source. On top of that, frizzy hair was found to do a better job at it than other types of hair (Lasisi et al., 2023). So, walking upright had more benefits than for locomotion alone!

Figure 5.5 Differences in body surface area exposed during solar radiation in a bipedal human versus a quadrupedal lion, with the dark zones indicating the exposed areas. (Lion modified from 3D model from Sketchfab, by Lary – https://sketchfab.com/3d-models/lilion-914d1dfc0d2b44c49103da81f230bfc5.)

The amount of radiation that then does penetrate also depends on several factors. And it is not just skin color that determines this. Saying that a darker skin necessarily absorbs more heat than a pale skin would be too simplistic, especially in warm-blooded animals such as mammals (Stuart-Fox et al., 2017). Pale skin reflects more light and absorbs less heat, but the structure of the fur and certain properties of the hairs themselves are also crucial. A fur with thicker hairs packed closer together will already allow less light to pass through. By erecting the hairs, mammals can even control the amount of transmitted light. Also, hair color determines how much light will pass through. Although white hair reflects more light, the light that is not reflected will penetrate the skin more easily than is the case with dark hair. To what extent australopithecines showed skin adaptations against irradiation, however, we do not know as fossilized hairs have not been found. We do not know whether they still had fur nor do we know if the skin was still pale, like the anthropoid ancestors. What we can expect is that they had as many hairs per square cm as we do, as even chimpanzees have as many. Only they have thicker and darker hairs (Kamberov et al., 2018).

Had australopithecines already lost the thick ancestral fur? We don't have fossils but evidence might come from a very unexpected source the evolution of pubic lice (Dunsworth, 2021; Light & Reed, 2009)! These lice only thrive in the hair at the pubic area in humans. One could say that the human pubic louse has made that tiny tuft of hair its natural habitat. Initially, it was thought that the species was evolutionarily derived from the hair louse (*Pediculus humanus humanus*), once our ancestors began to lose their thick fur. The ecological analogy is that the ancestral hair louse lived in a very large jungle of hair, initially spread all over the hominin body. However, as that jungle gradually started to disappear in several places, leaving only separated patches of hair, the descendants of that hair louse had to survive in the remaining patches. The head hair and the pubic hair can thus be seen as two dense remnants of that jungle, separated by unbridgeable expanses of areas with hair too small to survive in. As the lice population on the head thus became reproductively isolated from the population in the pubic area, they went their own evolutionary way and became separate species: the head louse (*Pediculus humanus capitis*) and the pubic louse (*Pthirus pubis*). If this scenario is correct, then the evolutionary separation of these two species from the common ancestor should roughly coincide with the loss of the hominin fur. Using DNA of the lice, it is possible to find out when these two species formed: a million years ago, and thus much later than when australopithecines thrived. A more exhaustive analysis of the lice genomes, however, suggests another scenario. In this scenario, the human pubic louse would have split off much earlier but not from the human hair louse. Rather, it would be derived from the pubic louse found in gorillas (*Pthirus gorilla*), where the split would have occurred as early as 4–3 million years ago. That's in the time window within which australopithecines lived! Pending more direct evidence, we can surmise that australopithecines may have already begun to lose their thick fur.

## More heat going out

If australopithecines no longer were covered by thick fur, this undoubtedly would have resulted in increased solar heat penetrating the skin. Even still with the fur, a body exposed in an equatorial forest opening up would increasingly require a mechanism to remove an excess of heat. Cooling can physically occur in four ways: conduction, radiation, convection and evaporation. Conduction causes heat to be transferred primarily in the skin and body, with one cell passing heat to the adjacent cell, just as the handle of a pan heats up when on a stove. Radiation causes heat to enter via electromagnetic radiation, as is the case for sunlight, but not to get out. Convection and evaporation are then the two ways that matter for how we and many other animals cool their bodies. All body parts that contain body heat and are in contact with a cooler outside air can release heat through the surrounding air that is moving against the body. This is convection, where heated air is constantly replaced by cooler air, drawing heat from the body. But this only works when there is displacement of air. For example, we lose a lot of heat (or absorb heat if the air is warmer than the body!) while breathing. The air moving along the mucous membranes of the oral and nasal cavities, trachea and lungs takes up heat from those mucous membranes. In cold air, this convection is even important for heating the air before it reaches the lungs (Mori et al., 2021). However, the efficiency with which heat is given off increases if the body expels warm fluids that can evaporate through convection, and thus extracting heat. Then we are dealing with cooling via evaporation. The mucous membranes in question increase heat loss due to the moisture they release along with it (they are covered with a layer of moist mucus, as the name suggests). So, convection and evaporation work closely together, as without air exchange the evaporation of moisture would be less efficient and so will be heat loss. A lot of mammals do this by panting; we do this by sweating (Carrier, 1984). Panting has the advantage that cooling efficiency does not depend on how much the air moves around the body: the panting itself causes the movement of inhaled air along the tongue. There is also no loss of body salts, which does occur during sweating. Panting also has disadvantages. For example, the cooling surface is limited to that of the mouth and tongue, whereas the surface area that can cool via sweating is much larger. In addition, cooling is not dependent on breathing at an increased rate, with the tongue dangling out.

The key question then is whether australopithecines already showed adaptations to sweating. As we shall see further on, we know that chimpanzees are much less able to sweat than modern humans are. Chimpanzees have on average about 10 sweat glands per square cm of skin, while this is 94 in humans (Kamberov et al., 2018). Chimpanzees also do not need to be able to sweat much, being shielded from solar radiation by the forest canopy. However, we have no hard evidence to know what the ability to sweating would have been in australopithecines. Much depends on whether they had lost their thick fur or not because sweating with thick fur does not go very well. A fur prevents the air from reaching the surface of the skin sufficiently, hence limiting efficient evaporation. Sweat is produced by sweat glands that lie embedded in the skin, that discharge the sweat onto the surface of the skin. The passing air then causes the warm sweat to evaporate, drawing heat away from the body. If australopithecines no longer had thick fur, they would have benefitted from being equipped with many sweat glands. This is because naked skin without sweat glands would have the opposite effect: the air convection heat would increase heat absorption by the skin. If they did still possess thick fur, then having many sweat glands would nevertheless have resulted in an inefficient cooling of the body. Maybe some new information will resolve this conundrum someday.

### Less heat being produced

We have already talked about the importance of efficient energy management. Making efforts to get rid of an excess of heat coming in ultimately is also about energy management. Everything is intertwined, also the body's heat management and the efficiency of upright walking. If Lucy required more muscle activity to walk via a *bent-hip-bent-knee*, this would have been associated with an increased body heat production (Raichlen et al., 2010). When modern people are asked to walk with their hips and knees bent, their body temperature increases by 1.7°C after only 15 minutes (Crompton et al., 1998). On top of that, they don't get rid of that heat very easily as even 20 minutes of rest does not suffice for body temperature to restore. The overall metabolism increases by 50% and the energy cost for walking with bent legs is double compared to walking with straight legs (Carey & Crompton, 2005). This triggers other effects, such as a doubling in lactic acid production, a byproduct of anaerobic muscle activity. Altogether, this shows that if a modern human were to walk for an hour in this manner, he or she would need an hour and a half to fully recover to avoid overheating. Not quite efficient. Of course, Lucy had a different constitution than modern humans have. If you ask a fish to climb a tree for 10 minutes, it will also need more energy than to swim in water for 10 minutes. But computer models seem to confirm that Lucy would likewise produce excessive body heat should she have walked that way. A switch to walking with straight legs was thus interesting not only for the walking itself but also to keep body temperature in balance while doing so.

## UV protection

If australopithecines still had a thick fur, it protected them well from the harmful component of sunlight, namely UV radiation. The light leaving the sun contains three types of UV rays. UV-A rays easily get through everything and reach the surface of the Earth, including the skin of animals walking around. Of the UV-B rays, most get absorbed by the atmosphere. Where sunlight enters almost perpendicular to the surface of the globe, only part of the UV-B rays will get through and reach the Earth. The rest are filtered out by the atmosphere. Virtually all UV-C rays are absorbed by the ozone in the atmosphere and do not even reach the Earth (Jablonski, 2012). So, animals living closer to the equator are exposed to greater amounts of UV radiation. And that's not so interesting because it causes damage to DNA, causing cells to die or no longer function properly. Throughout the animal kingdom, a solution to this problem emerged: melanin. Melanin is a dark pigment produced in the form of tiny granules in skin cells called melanophores. Those granules are also deposited in scales, feathers and hair. Cells rich in melanin can absorb 30–40 times more sunlight and protect 7–8 times better against UV damage. As early as 1833, Constantin Gloger observed in birds that occurred close to the equator that they had darker feathers than those living toward the poles (Nicolaï et al., 2020). A black scalp occurred 148 times independently in birds closer to the equator. This 'Gloger's law' applies equally to mammalian fur and, as we will see later, to human skin color. So, there is a relationship between skin color and UV radiation. Applied to australopithecines, this implies that they either possessed dark fur with pale skin, like chimpanzees still do, or that the fur was absent but their skin was highly pigmented with melanin. In this second scenario, again, there is a connection with sweat glands, as melanin also protects sweat glands from UV damage. Surprisingly, though, one would intuitively expect that a dark skin would cause an increased warming of the body. However, this does not seem to be the case, as warming is limited to the surface of the skin.

So, there are several arguments that may point toward australopithecines no longer possessing a thick fur, but having a dark skin equipped with sweat glands needed for cooling. A dark skin also brings benefits other than just stopping DNA damage. In fact, melanin also has properties as an antioxidant and as an anti-inflammatory component, providing additional protection against microbes penetrating through the skin. Even Darwin at the time saw a connection between darkening of the skin and protection against parasites (Darwin, 1871). Shielding from UV-B rays by melanin even has more advantages, such as not breaking down certain substances found in the blood in the skin, folate (also called folic acid) being one. This is a type of vitamin that is taken up through plant food and accumulated in the liver. However, it is an unstable molecule that is easily broken down by high temperatures and by UV-B rays. Folate is nevertheless important for the body, where it impacts the functioning of genes, regulating the amount of amino acids available (these are essential building blocks for the body), forming an insulating myelin sheath around nerve fibers (necessary for proper stimulus conduction), the production of neurotransmitters (molecules that transmit stimuli from one nerve fiber to another) and for the embryonic development of the nervous system. All are important for a properly functioning organism. As this is greatly affected in animals with little melanin being exposed to a lot of UV-B, it may threaten their chances of survival and reproduction. Australopithecines also subjected to natural selection would imply that individuals with slightly darker skin were favored and thus had a higher fitness. That's what evolution needs, to bring about a rapid shift in a trait in a population. Picture this: a pregnant Lucy with pale skin and without a thick fur walks in an environment where she is regularly exposed to high doses of UV-B. The folate in her body is massively broken down by the UV rays passing smoothly through her skin. As a result, she cannot pass on enough folate to her developing fetus, which nonetheless needs it to develop its nervous system. Chances are that that baby will eventually be born with an underdeveloped nervous system and therefore not be viable. Hence, Lucy does not pass on her genes to the next generation. It is no coincidence that even today doctors prescribe folate to pregnant women to reduce the risk of birth defects such as *spina bifida* (sufficient folate prevents 70% of neural tube defects) (Jablonski & Chaplin, 2000). Not only through the mother but also through the father, a folate deficiency can have a negative effect on reproduction, as has been found in mice (Lambrot et al., 2013). Putting all the puzzle pieces together, everything seems to indicate that the process toward a darkening of the skin had at least started in the australopithecines. How far this had progressed remains a currently unresolved question.

Let's throw in some genetic information into the story. There are multiple genes involved in the production of melanin, which makes it not always clear which genetic traits bring about exactly what skin color. One such gene is the *Melanocortin-1 receptor gene*, which is associated with hair and skin color. In fact, certain variants of this gene allow an increased production of melanin. By studying the evolution of that gene, it has been demonstrated that some variants became positively selected in the gene pool of our ancestors about 1.2 million years ago (Jablonski, 2012). This could indicate that the skin only then intensively darkened. But this is after when australopithecines roamed Africa, as except for *P. robustus*, all australopithecines had disappeared by then. A plausible scenario could be that the first australopithecines, which still underwent a limited exposure to elevated UV radiation, still possessed a pale skin covered by dark and dense fur. Over time this exposure increased, as the forested environment started to open up, and so the selective pressure for an improved cooling and a better protection against UV radiation began to increase as well. This process continued gradually but did not fully materialize until after australopithecines had already disappeared. This could have occurred as early as 2 million years ago, in early *Homo* species, shortly before an accelerated selection took place on genetic variants that allowed for a dark skin (Jablonski, 2012).

## Big brains not so important yet

Lucy opened up the way to a better understanding of how our ancestors were built. At the time of Darwin and beyond, the preconception prevailed that our ancestors must have been distinguished from other primates by a larger brain because surely our ancestors must have been more intelligent than apes. And then came Lucy, a small hominin who walked upright and had a brain about the size of a chimpanzee's. So, the hypothesis that we first got a bigger brain and then started walking upright had to be abandoned. But if the brain of australopithecines was not larger, was the way it was built already different? The neocortex was especially important because that's where all the processing of information, thinking and creativity happens. Fortunately, we have a few fossilized and other evidence that can tell us a little more about the australopithecine brain.

## Taung's child

Coincidentally, the first scientific discussion about how human-like the brain of the australopithecines already started with the first *Australopithecus* fossil: the child of Taung. Recall the struggle that the discoverer, Raymond Dart, had to face to get his view accepted on the importance of this fossil within our evolution history. Much had to do with his identification of brain structures in that fossil, for it contained a very well-preserved brain imprint. Now, well-preserved is relative. As a brain is surrounded by meninges, many superficial brain structures are masked by them and ultimately there is not that much to see on such an imprint, at least not to a non-trained eye (Hrvoj-Mihic et al., 2013). Taung's child showed some indentations and impressions of blood vessels of the meninges (Falk, 2009). One of those indentations Dart identified as the so-called lunate sulcus (literally moon-shaped groove) at the back of the brain. Indeed, in primates, this groove forms the anterior border of the neocortical area where all the information from the eyes is sent to. According to Dart, that groove lay far back, just as in humans. In monkeys and great apes, in fact, it lies much more forward. That was important evidence to Dart that Taung's child was a hominin, and not an ape or great ape. Soon critics of Dart's work made it clear that he was wrong because that indentation originated from a suture between two skull bones and therefore had nothing to do with a brain groove. Why did people attach so much importance to that groove, you would think? Because its position could show whether the neocortex was already changing, toward a complex organization as found in humans. In fact, Taung's child still had a small brain, about 382 cc. Based on the child's age, one suspects that the brain had already reached 92% of its adult volume (Holloway, 1970). This brings the adult brain volume to about 405–440 cc (Falk & Clarke, 2007; Holloway, 1970; Neubauer et al., 2012). At birth, australopithecines had a brain volume of about 180 cc (DeSilva & Lesnik, 2006). Knowing that the brain volume of a newborn chimpanzee is about 130 and 400 cc in an adult individual, it is clear that Taung's child had similarly large brains as a chimpanzee, perhaps slightly larger. Robust *Paranthropus* had a somehow larger brain (about 530 cc) (Falk, 1985; Holloway, 1972). Brain volume remained similarly small in australopithecines during the first 5 million years of human evolution, that is, after the split with the great apes (Lovejoy, 1981). So, the key question regarding Dart's research was as follows: was he correct that although brain volume had not yet increased in australopithecines, there had already been a change in the construction and complexity of the neocortex? Was the neocortex of australopithecines now anthropoid or hominin? A posteriorly positioned lunate sulcus might have been evidence of this at least. As it will become clear later, a lot of ink has been spilled on this and other brain cortices.

Looking at other australopithecines, there is not that much information based on endocasts. Some have been found from several species of *Australopithecus*, such as *afarensis*, *africanus* and *sediba*, and also from a robust *Paranthropus* (Holloway, 1972, 2015). The oldest go back just over 3 million years (the oldest endocasts of *Homo* species are about 2 million years old) (Hrvoj-Mihic et al., 2013). But not all of them are sufficiently well-preserved to yield much reliable information. In two cases, they are imprints of child brains, such as the africanus child of Taung and the afarensis child of Dikika (Alemseged et al., 2006). Ultimately, there are only about four endocasts from which some insights have been gained as to whether and how the neocortex was already more human-like in australopithecines. Here it quickly becomes clear that the evolution of our ancestors' brains involved much more than just an increase in volume.

## The 'Affenspalte'

The indentation pattern of brain imprints in australopithecines suggests that while it was still very similar to that of chimpanzees, changes were already beginning to occur in the neocortex (Falk, 2014; Gunz et al., 2020). How much or how little, that is fodder for debate among scientists (Falk, 1980; Gunz et al., 2020; Holloway, 1972). Changes manifested themselves through expansions of certain areas of the neocortex, as well as left-right asymmetries becoming established. The way different brain areas develop and the extent to which different types of cells proliferate in the brain also plays an important role (Holloway, 2015). For example, we have special neurons in our neocortex (called pyramidal neurons), which are very complex and receive information from lots of other neurons. The complexity of those neurons, such as the number of branches they exhibit, varies between different areas and between the left and right sides of the neocortex (for example, they are more branched in the left side in people who are right-handed and vice versa in left-handed people). Compared to other primates, these neurons in humans develop much longer (Hrvoj-Mihic et al., 2013). Whether this was already the case in australopithecines, however, remains unanswered because fossilized neurons have not yet been recovered. The presence and position of grooves on brain imprints must also be interpreted with caution. For example, research based on MRI images of brains of different chimpanzees showed that the position of grooves can vary quite a bit between individuals (Falk et al., 2018). This also showed that one should not simply conclude that certain patterns in brain imprints can be considered typical of specific hominins.

The biggest changes occurred in the prefrontal cortex (at the very front under the frontal bone) and the cortex at the back (occipital lobe and parietal lobe) (Beaudet,

Clarke, de Jager, et al., 2019; Holloway, 2015). This is not coincidental because these are the areas linked to intelligence and to where the main thought processes take place, where language processing occurs, where designing and making tools is controlled, where tasks related to relationships in space and time occur and where social behavior is controlled. In addition, we can expect that the evolution of the cerebellum has also been linked to the evolution of the neocortex because of its connection to numerous areas in the neocortex (Weaver, 2005). Such changes presumably occurred as early as about 3 million years ago (Falk, 2014). Although not every detail is clear, there are three levels at which it has been possible to gain some insight into the link between changes in the neocortex and important changes in the evolution of our ancestors' behavior: the lunate sulcus, Broca's center and the prefrontal cortex.

Scientists have been interested in primate neocortical organization for a long time. In 1907, Grafton Elliot Smith, an Australian anatomist, published the results of his comparison of the groove patterns in primate brains. One groove that received his special attention was the one separating the visual cortex (receiving information from the eyes) from the parietal lobe located in front of it (Falk, 2014). That slit was very noticeable in primates and he coined it the 'Affenspalte', literally 'monkey slit'. According to Smith, humans also had a slit like that, only it was more to the back and smaller. That cleft in humans he then called the lunate sulcus, or lunate groove. So, according to Smith, the Affenspalte and the lunate sulcus were one and the same cleft, only located differently, and showing that the neocortex in the occiput had become organized differently during our evolution. In humans, the visual cortex would have moved posteriorly and have become smaller due to an expansion of the parietal lobe.

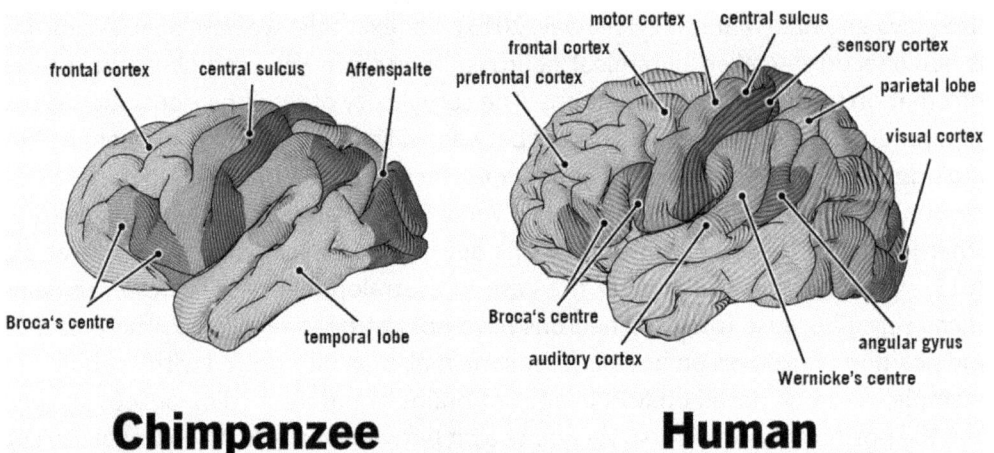

Figure 5.6   Neocortex in chimpanzee and modern humans. (Chimpanzee and human brain model from Sketchfab, by Gwuanthro – https://sketchfab.com/3d-models/comparative-brain-anatomy-chimpanzee-human-acf1799d-8f9a4242849579f2a782c8df.)

That idea may not be unfamiliar to you. It is no coincidence that Raymond Dart came to the same conclusion while studying his Taung child; after all, he was a disciple of Elliot Smith and had become head of the anatomy department in South Africa thanks to him. Meanwhile, however, we know that Dart was wrong and what he identified as the human lunate sulcus was an imprint made by skull bones. With what is now known about fossil brain imprints to gene action in the brains of modern humans, the conclusion is that australopithecines and we do not have a lunate sulcus (Allen et al., 2006). Thus, all indications are that the Affenspalte was lost before the origin of australopithecines. One explanation is the enlargement of that parietal lobe. Indeed, an enlargement or reduction of the white and gray matter in the brain affects the tension between cell layers, and so does the formation of grooves. So, the fact that some grooves 'suddenly' stop forming or are no longer superficially visible may be a consequence of this.

Elliot Smith had his Affenspalte, Paul Broca had his own neocortex area. This area is located in the lower fold of the frontal lobe, right next to the motor cortex. The Broca's center is also a motor center: it controls the muscles we use to speak. Now, the center does much more, such as processing information according to a certain hierarchy but also coordination when manipulating tools (Schenker et al., 2008). As in chimpanzees, the human Broca's center consists of two parts (a posterior Brodmann area 44 and an anterior area 45). During evolution, those areas have moved more posteriorly in humans, mainly due to a shift of area 45. However, Taung's child does not tell us much about Broca's center, but the brain prints from another South African species, *A. sediba*, does so (Carlson et al., 2011). This imprint suggests that the area where Broca's center is located was already different from that of great apes.

If we compare a chimpanzee's neocortex to ours, the biggest differences are mainly in the prefrontal cortex, that is, at the very front. Additional convolutions have developed, areas have become larger and neurons lie further apart (Schenker et al., 2008; Semendeferi et al., 2001; Semendeferi et al., 2011). One of the grooves from this prefrontal cortex is often found in brain imprints of australopithecines (i.e. the middle frontal sulcus). It is sometimes absent in chimpanzees but always present in humans. It has also been found in all australopithecine brain imprints. The consistent presence of that groove indicates that expansion may have already occurred in the prefrontal cortex, especially in that area where information from different senses is processed and organized to maintain attention, to manage our working memory and to coordinate goal-directed actions. Especially the most anterior part of the prefrontal cortex is extensively enlarged in sapiens, compared to other primates, and has been associated with crucial cognitive functions like enhanced working memory, improved performance to make links between different types of information and improved capacity to make abstractions of information (Levy, 2024). So perhaps australopithecines were already slightly better at coordinating and organizing more complex executive tasks, than is the case in great apes.

## Yet handy and smart(er)

Being able to perform more complex tasks, thanks to an expanded prefrontal cortex, certainly has an evolutionary advantage. We also know that a more specialized Broca's center is important not only for being able to produce language but also for controlling the hands with respect to tool-making. The question then is whether this was the same with australopithecines? They could use it by all means, for recall that they became increasingly dependent on hard fallback food. Underground roots had to be extracted from the hard and dry soil, not something you do smoothly without using some kind of tools. This can be done with a stick, bone or stone and requires only a little dexterity to dig with it (Bunn, 1999). Other animals also use tools to get access to food (Rutherford, 2019). These are usually objects taken from the immediate surroundings and so used unmodified (Barrett et al., 2019). In the broad sense of the word, this can already be considered a 'tool', namely the use of one object to alter another in its shape, position or structure. Thus, wrasses that use a stone to smash shells are using tools. Several primates do the same, such as capuchin monkeys and chimpanzees that use a stone to smash hard nuts that they place on yet another stone, or chimpanzees and orangutans that use a stick to fish for food with it (Brown, 2012; Kuhl et al., 2016; Van Schaik, 2006).

Usually, we associate another form of tools with humans, namely objects that are first modified to look differently, so that they become more efficient for working another object. A stone from which pieces have been chipped off to obtain a sharp cutting edge is more efficient for cutting flesh from a cadaver than a smooth and round stone. It was long assumed that these kinds of tools did not come into existence until the advent of *Homo*, and especially then with the discovery of the most primitive *H. habilis* … the handy man (Leakey et al., 1964). The remarkable thing here is that the discoverers of handy man, including the famous Louis Leakey, had previously found worked stone tools at the site where they had found the skull of the robust *P. boisei*. They named those tools as 'Oldowan technology', derived from their finding in the Olduvai Gorge in Tanzania. It was only with the discovery of *H. habilis* that they concluded that habilis must have been a more advanced toolmaker, and that the tools belonged to him. So, he was handier. But scientists today are moving away from the contention that *Homo* was the inventor of crafted tools, which is supported by several sources of evidence. In 2010, cutting marks were discovered on ungulate bones, clearly made with a sharp object. The surprise was that these were 3.4 million years old, and thus too old to have come from *Homo*. They were found in Dikika, in Ethiopia, where fossils of afarensis had previously been found from the same period (McPherron et al., 2010). Five years later, 3.3 million-year-old tools in the form of stone flakes were found at Lomekwi, near Lake Turkana in Kenya (Callaway, 2015). These tools were not only much older than the Oldowan ones (which are up to 3 million years old) but also showed other features (Mercader et al., 2021; Plummer et al., 2023). Scientists therefore refer to these tools as Lomekwi technology. By looking at the insertion sites of muscles on the hand skeleton of australopithecines, researchers concluded that they already showed some adaptations for an improved capacity to hold onto and manipulate objects, although not as sophisticated as in *Homo* (Kunze et al., 2024).

Recent research shows that it is not so simple that Lomekwi tools then belonged to australopithecines and Oldowan tools to *Homo*. For example, 3.8 million-year-old Oldowan tools have been found in Kenya associated with *Paranthropus*, i.e. a robust australopithecine, something previously suspected (Gibbons, 2023; Lovejoy, 2009; Plummer et al., 2023). So surely everything seems to point to a better ability to handle more complex executive tasks in australopithecines.

# So ...

Lucy and all her close relatives were thus much more than a brief flare-up in our evolution. They followed a roughly 3-million-year-long path to slowly transition from an existence as climbing forest dwellers to walking savannah dwellers on the African continent (White et al., 2015). They were adapted for this to handle hard fallback food, grasses and nuts, thanks to a large jaw apparatus. They had numerous adaptations to already walk well upright, but still retained features that aided climbing. Besides their small stature, relatively long arms, protruding snout and small brains in many ways they were already more similar to us than we might suspect. The wide pelvis, the inwardly placed knees and the foot vaults are testimony to this. Changes in the brain began to manifest themselves at the level of neocortex organization and explains why they were already able to make stone tools.

# 6

# Fully erect!

It was long believed that the transition from small australopithecines with small brains to large humans with large brains was accompanied by rapid and drastic changes in both anatomy and behavior (Kimbel & Villmoare, 2016). Meanwhile, we know that australopithecines nevertheless bore many similarities to *Homo* and even to *Homo sapiens*. Yet, it is a striking evolutionary trend break that we are now dealing with humans who are taller than 1.3 m and weigh more than 50 kg, who have a less robust jaw apparatus, possess a brain volume that begins to exceed that of australopithecines, who use and refine plenty of tools and ... who begin to leave Africa. We are talking about the third group of hominins, namely the first true humans of the genus *Homo* (Collard, 2003). In time, we situate ourselves at the start of the Pleistocene, about 2.5 million years ago, and focus here to about a million years ago.

## The first tall ones!

The search for the first true *Homo* is a lifetime quest of many anthropologists to keep looking in the sometimes most inhospitable places. If paleoanthropologists found any fossil that could qualify, the biggest challenge was determining where to draw the line as to whether something was *Homo* or not. As is true for many organisms, this was initially thought to be very easy, until more fossils were found that made the boundary between two species to become murky. This was no different from the boundary between *Homo* and non-*Homo*.

### May the first man rise

The most primitive *Homo* species is not the first one discovered. Linnaeus had already 'discovered' us as a species in 1758 (Linnaeus, 1758), Neanderthal is known from 1864 (King, 1864) and Java man (*Homo erectus*) since 1896 (Dubois, 1896). In 1931, Louis Leakey found fossils and stone tools during excavations, after which his wife, Mary Leakey, found the skull of the robust *Paranthropus* in 1959, and their son, Jonathan Leakey, discovered another skull in 1960 (Wood, 2014). It was only then that the Leakey's made the connection between that skull and the actual makers of the tools found earlier. Convinced that they had found the missing link that could prove that *Homo*'s origins were in Africa, and not Asia, the telegram they sent to South African anthropologist, Phillip Tobias, revealed, *"We've found the man. Come quickly"* (Gibbons, 2011). Later they discovered skull and jaw fragments, a foot skeleton and

DOI: 10.1201/9781003588795-7

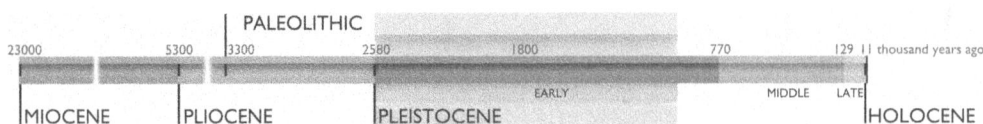

Figure 6.1    Time line showing the period dealt with in this chapter.

a hand skeleton with striking similarities to ours. By 1964, the new species was born: *Homo habilis*, the 'handy man' (Leakey et al., 1964). By then, no African fossils of erectus had been found, only Asian ones. This habilis had smaller brains than erectus (647 versus 840 cc in erectus), but larger ones than australopithecines, and had hands and feet that showed they could use tools and walk upright (Anton, 2003; Gibbons, 2011). Already three criteria that were considered 'typical *Homo*' at the time. Only there was a problem with brain volume. With no real biological explanation for it, an arbitrary limit was used for brain volume, the so-called cerebral rubicon (referring to the river, where once over it a 'point of no return' was reached). Fossil skulls with a brain volume greater than 700–800 cc were assigned to *Homo*, anything below that was not. To get around this, the limit was brought down to 600 cc, and so this habilis fossil now did meet the requirements to be called *Homo*.

It is obvious that currently, not everyone is convinced that this made any scientific sense. In addition, we now know that all considered, habilis bears quite a few similarities to australopithecines, more so than to erectus. Habilis has a more robust jaw apparatus and was also comparably small as an australopithecine (about 34 kg and 1 m tall) and thus significantly smaller than the typical erectus (Spoor et al., 2015). In addition to adaptations to walking upright, habilis also showed adaptations to climbing, like australopithecines (Wood & Collard, 1999). The arms were still longer than the legs and the femur showed more resemblance to that of a *Paranthropus* than of an erectus (Wood & Richmond, 2000). Also, the use of tools can no longer be considered unique to *Homo*, for australopithecines did so as well. It is therefore not surprising that for some, habilis cannot or should not be considered a *Homo* species, but rather *Australopithecus* or even something else (Gibbons, 2011; Wood & Collard, 1999).

This picture about the very first *Homo*, however, is even less clear than that. Indeed, among all the fossils considered, one can distinguish three forms: an East African with a small brain and small stature, an East African with a larger brain and medium stature and a South African with an as yet unknown brain volume and presumably medium size. According to some, these are all variations of the same species, *H. habilis*. Others treat that as three *Homo* species: habilis from East Africa with a small brain, rudolfensis from East Africa with larger brain and gautengensis from South Africa (Curnoe, 2010; Grine et al., 1996; Prat, 2022).

Taking all fossils combined, three species or not, we are dealing with a small human (about 51.6 kg for males and 31.5 kg for females) who lived in East Africa (and South

Africa) about 2.8–1.6 million years ago (Anton, 2003; Villmoare et al., 2015; Wood & Richmond, 2000). Equipped with a sizable jaw apparatus and small brain (somewhere between 500 and 800 cc), it was able to carve stones into simple fist axes, which one refers to as Oldowan technology (Wood & Richmond, 2000). Is this the great evolutionary change toward sapiens? Clearly not.

## The first true tall one

The big change was yet to come. Characteristics that largely typify us as sapiens came about 2 million years ago. Then, figuratively speaking, the new man emerged: *H. erectus*. Contrary to what the Leakeys may have thought, however, erectus '*was the man*', not habilis. Now, '*erectus was the woman*' is equally true, of course. Anyway, a human species arose and would create waves, figuratively speaking. In the end, it is erectus that has populated much of the world for nearly 2 million years. We have only been around for a sixth of that time.

### Boys are getting big

The story of erectus starts very early, in August 1891 when Dutch paleoanthropologist and geologist Eugene Dubois found a tooth during excavations at the Solo River in Java, and a month later, a skull roof. A year later, a femur, which looked very much like that of a modern man (Dubois, 1896), was added. Enough material to come to the conclusion that this had to be a new human species, which he called *Pithecanthropus erectus*: the large and erect ape-man with a brain volume of about 840 cc (Anton, 2003; Baab, 2015). Of course, this discovery was soon fodder for debate, and it took until the exposure of the Piltdown forgery that it was accepted that Dubois's fossils came from an elongated human equipped with a rather large brain (Anton, 2003). About 35 years later, Davidson Black would describe another human species from Asia, found in the Zhoukoudian cave not far from Beijing: *Sinanthropus pekinensis* (Black, 1929; Boaz et al., 2004). Several more discoveries of new species followed later, all with their own names, until Ernst Mayr had enough of the jumble of species and in 1950 argued that these were all the same species and were a *Homo* species (Mayr, 1950). From then on, *H. erectus* was born. By the 1980s erectus was widely known as a species that lived in Africa 1.9 million years ago and spread from South Africa to Java and China until about 400,000 years ago (Baab, 2015; Hammond et al., 2021). Now we know that this was even up to 100,000 years ago. But still, it causes some disagreement among scientists. For some, this is effectively one and the same species that lived for so long and ended up in many places. Others argue that the older and smaller forms, equipped with smaller brains, belong to a separate species, *Homo ergaster*. These are found in Africa and Georgia. The more recent and larger forms from Asia, with larger brains, then represent the true *H. erectus*. Still others think the Georgian fossils also represent a separate species, *Homo georgicus* (Gabounia et al., 2002). Although all these fossils show a lot of variation, research shows that this still fits within the range of what can be expected for one species that became widespread over such a large area (Baab, 2008b). So, we will speak of erectus only from now on.

After the discovery of this Java and Peking man, another spectacular find was made in 1984 that would teach us a great deal about the doings of erectus: the Turkana boy (Brown et al., 1985). This is a nearly complete skeleton of a boy who lived about 1.5 million years ago on the banks of the Nariakotome River in Kenya (Anton, 2003; Graves et al., 2010). Supposedly he was only eight to nine years old when he died and by then had a length of 1.47 cm. If one extrapolated that, taking into account the conviction that early humans reached adult size much earlier (as early as about 12 years of age), the Turkana boy would have ended up growing somewhere between 160 and 177 cm tall with a body weight of 60–83 kg (Cunningham et al., 2018; Graves et al., 2010). The brain volume was similar to that of the Java man, at 880 cc, slightly more than half that of ours (DeSilva & Lesnik, 2008). The spine and rib cage were on the other hand very modern (Anton, 2003).

Although the Turkana boy went on to become a large man, we now know that early erectus that lived in Africa from 1.9 to about 1 million years ago, had a slightly smaller stature than the later Asian ones (Potts, 2012). So the Java and Peking man was more recent (Java man is estimated at 1.3 million years and Peking man at 770,000 years (Matsu'ura et al., 2020; Shen et al., 2009)) and more specialized than the Turkana boy. The very oldest fossils are found only in Africa, the very youngest from Java (Baab, 2015; Hammond et al., 2021; Rizal et al., 2020). During all that time, they were making, using and transporting old and new tools. Early erectus used similar Oldowan tools as habilis did, but about 1.4 million years ago, innovation came about: teardrop-shaped and double-edged fist axes became the standard (Ayala & Cela-Conde, 2017). These Acheulean tools would also remain the standard for a long time and even be adopted by other species that would join later.

So erectus explored much of the world for nearly 2 million years. The question is why? Why did humans leave the familiar Africa about 2 million years ago? And how come erectus was able to dominate as a human species for so long, only to eventually disappear?

## Out of Africa

That erectus was to impress is also evidenced by the fact that not only was this the first *Homo* to leave Africa, but he even managed to get to East Asia very quickly. After the discovery of the Turkana boy, it was known that erectus had lived in East Africa, Java and China. A new surprise came in 1991 when a mandible of a hominin was found in Dmanisi, a village in Georgia, which represented the oldest fossil evidence of hominins in Europe: somewhere between 1.8 and 1.6 million years ago (Gabunia & Vekua, 1995). Meanwhile, skeletal evidence from several individuals shows that they were smaller than other erectus, being 40–50 kg and 145–166 cm (versus 32–70 kg and 122–185 cm in the others) (Lordkipanidze et al., 2007). Their brains were also smaller

(600–775 cc), more similar to habilis than to other erectus individuals, and they had tools with both Oldowan and Acheulean characteristics. These primitive features and their age (1.85–1.77 million years) suggest that this must have been one of the first migrations from Africa.

All the fossils combined show that erectus occurred from South Africa to the Far East and left Africa very soon after its emergence in East Africa. This resulted in a lot of regional variation, including the shape of the skull roof, the thickness of ridges on the skull (such as the eyebrow arch), the heftiness of the lower jaw and teeth, body size or brain volume (Anton, 2003). There was also regional variation within a continent, presumably linked to the prevailing climate. An erectus from tropical Indonesia was slightly larger than one from northern China (160 cm versus 150 cm) (Anton, 2003). Erectus not only explored the world, but he also thus became adapted to it. He probably even gave rise to other human species on both continents: *Homo rhodesiensis* in Africa (our ancestor) and possibly several dwarf species in Asia (*Homo floresiensis* on the island of Flores and *Homo luzonensis* in the Philippines) (Brown et al., 2004; Détroit et al., 2019; Westerman, 2018; Wood, 2020).

But what happened to make erectus leave the familiar continent of Africa, after 5 million years of human evolution there? Climate presumably had a lot to do with it. In the period before erectus, more than 2.6 million ago, Africa was warm and humid (Timmermann et al., 2022). Then followed a long period with a dry and cold climate, but also some periods of very many and rapid alternations between dry-cold and wet-warm (Carrión et al., 2011). All this had a lot to do with the cycles of Earth's movements relative to the sun. The strong fluctuations pushed humans into an environment that became unstable, resulting in a limited or unpredictable food supply.

The oldest evidence of migration is 2.4 million years old, based on tools found in the Middle East (Ayala & Cela-Conde, 2017). Three migrations occurred sometime between 1.8 and 0.7 million years ago, including the one to Dmanisi. The first still occurred with Oldowan tools, the other two with Acheulean tools. This time window is characterized by phases with striking climatic fluctuations in Africa and ice ages in the north (around 2.8, 1.7 and 1 million years ago) (DeMenocal, 2004; Potts, 2012). Thus, all indications are that it was not so much the hottest or coldest, driest or wettest conditions that triggered the emergence of and migration by *Homo*, but rather the fluctuations therein (Potts, 2013). So, one could argue that the Earth's orbit around the sun helped determine how we humans got there and that what characterizes us as *Homo* is more likely the result of adaptations to an unstable climate (Antón et al., 2014).

## Steak and fries ... almost

Surviving in an unstable climate is challenging; it makes the food supply unpredictable. This is especially true if you are an herbivore, whose food has a harder time dealing with such a climate. Plant-eating australopithecines may have had a problem, but a problem probably even greater for erectus who now had to provide a much larger body and larger brain with the necessary energy and nutrients. On top of that, everything points toward erectus having been extensively confronted with an environment consisting almost exclusively of grassland (Potts, 2012). And erectus was not a grass eater. So, did erectus leave Africa because there just wasn't enough to eat? The story behind erectus is a little more complicated than that, but meat and fire played a crucial role in this (Wrangham et al., 1999a).

## Grasslands full of steaks

We already know from australopithecines that they worked cadavers with sharp stones (McPherron et al., 2010). Why else than to eat the meat? Also, habilis had its Oldowan tools and the later erectus had sharp Acheulean tools. Do we have evidence that meat was important to erectus? By comparing 2-million-year-old cutting marks on antelope bones, left by experimentally working a cadaver in different ways, it could be deduced that the bones were first stripped of the large pieces of meat and then smashed into pieces using stones (Parkinson et al., 2022). Does this prove that erectus was also an active hunter and not a scavenger? In this case it does because the cutting marks were found in places where predators usually leave gnawing marks (O'Connell et al., 1988). As there were only cutting marks, this indicates the working of a self-killed animal and not one already processed by predators. Archaeological research confirms that eating meat was already on the rise among erectus, between 2 and 1.5 million years ago (Ben-Dor et al., 2021; Ferraro et al., 2013; Smil, 2002).

Switching from a plant-based diet with underground roots to an animal-based diet would not be that unusual. Sometimes herbivores supplement their diet with animal material, making up for any calcium deficits. There are known cases of elephants and deer eating small animals or gnawing on carcasses or seed-eating Darwin's finches feasting on blood from other birds (Grant & Grant, 2008; Meckel et al., 2018). Even fruit-eating primates occasionally like a piece of meat, such as capuchin monkeys, baboons or chimpanzees (Gilby et al., 2017). Were we totally dependent on fruit, we would even need 3–5 kg of fruit per day. Were we leaf eaters, we would need to play down more than 10 kg of leaves a day (Smil, 2002). And even then, we would only meet half of the protein requirement. You can accommodate this by constantly eating plant food and otherwise doing little exercise as gorillas do. Or you supplement this diet with more energetic and more easily digestible animal material, like chimpanzees do. Food is important for energy and building blocks. The energy we get from carbohydrates and fats, and building materials we get mainly from proteins. Plants contain mainly carbohydrates; meat contains more proteins and fats. So, combining both is very interesting, because then you are not completely dependent on what can be collected as plant material alone. Our intestinal tract is very similar to that of herbivorous great apes (for example, food passes through it slowly) but is 40% shorter than what you would expect from an herbivorous primate. In one aspect, our intestine is already different: in humans, the small intestine is the largest part; in great apes the large intestine is the largest. This indicates a dependence on high-quality food that can be digested in the small intestine, i.e., not so much plant material. That requires specialized stomachs (as in ruminants) or large intestine and caeca (as in some leaf-eating primates). If we look at the diet of current hunter-gatherers, we see that meat (or fish) can make up between 42 and 85% of the diet, depending on the supply of plant material in the environment (Crittenden & Schnorr, 2017).

If we extend this to erectus, everything seems to indicate that it strongly benefits from eating meat. Meat is easier to digest and provides the body with essential amino acids (protein building blocks), vitamins and minerals, and energy through fat. This does have some implications because getting access to animal material in an open grassland is not that obvious. This requires coordination and cooperation, as we see in chimpanzees and current hunter-gatherers, as well as it involves traveling greater distances (Lieberman, 2013). Whereas chimpanzees travel about 2–3 km/day and eat 10–40 g of meat, for hunter-gatherers it is more than 6 km and 270–1400 g (Kaplan et al., 2000). It would have been very unlikely for an erectus to get through the day with a body 1.7 m tall and 60 kg heavy if its diet consisted only of finding edible plants or digging out roots in a grass savannah. An antelope steak provides five times more energy than raw carrots of the same weight. Tools for cutting large pieces of meat into smaller pieces, in addition, improve digestion efficiency (Carmody et al., 2016). For example, a chimpanzee was once observed chewing on a piece of raw meat for about 11 hours.

Meat is an important source of essential building blocks, not only to grow a body but also to maintain and repair it. It contains large quantities of protein composed of essential amino acids, essential for building our bodies (Stahl, 1984). Plants not only have less protein but also less of those amino acids. According to the World Health Organization, a healthy amount of protein for humans is about 0.83 g/kg of body weight (i.e., 58 g for someone weighing 70 kg) (WHO/FAO/UNU Expert Consultation, 2007). There is also natural fat, which is broken down in our intestinal tract and transported to the liver, to then be deposited in adipose tissue under the skin and between the organs. When we come in need of energy, these fat reserves are drawn upon. One hundred grams of fat is good for 900 kilocalories of energy (Brant, 2019). A body constantly needs a minimum number of kilocalories, to provide minimal cellular function. Small wild animals contain little fat (less than 5% of their body weight). For large animals, such as large herbivores, it is double that amount (Smil, 2002). If you catch an herbivore, such as a 40-kg springbok, you suddenly have 20 times more energy available than if you were to catch a 4-kg hare. Antelope, buffalo, wildebeest, zebra … all large herbivores well adapted to survive in these grass savannahs. Large herbivores, full of meat and fat, present in large herds in erectus' new habitat. The meat just walked there in front of them, so to speak, and it was then only a matter of being able to grab it. And if they did so in groups, it provided 30–50 times more energy than it required energy to kill them.

## Steak well done with cooked potatoes

The steaks were up for grabs for erectus, so to speak. Using this to their advantage seems evident from several sources, such as cut marks, tooth wear, and growth patterns (Dominguez-Rodrigo et al., 2005). But erectus could do much more with the food, thanks to the discovery of fire. Now, it will not have been a true discovery because erectus and predecessors must have faced natural fires, such as those caused by lightning strikes. But erectus was most likely the first hominin to learn to understand, control and use the functions of fire. There are some other animals that also actively use fire (Rutherford, 2019). Chimpanzees and green vervet monkeys sometimes actively forage in burned areas in search of food, as they can better detect food and predators there. Some raptors even set fires to chase prey, picking up burning twigs at a wildfire and dropping them in another location.

The oldest direct evidence of fire use by hominins goes back about 1 million years, with burnt remains of bones being found in the Wonderwork Cave in South Africa (Berna et al., 2012). This shows that erectus did not start using fire only once he left Africa (Roebroeks & Villa, 2011). Indirect evidence strongly suggests that erectus used fire much earlier (Chazan, 2017). The rapid body enlargement and changes in the jaw apparatus, digestive system and brain also point toward erectus using fire even earlier to process food. Fossil evidence of possible controlled fires also shows this (Wrangham & Carmody, 2010). Burned pieces of soil in Koobi Fora (Kenya) about 1.6 million years old showed a microstructure of the soil material that indicated that these were not natural fires (controlled fires heated the soil four times more than natural fires) (Hlubik et al., 2017, 2019). Burned stone tools from that period also showed the effects of getting heated to high temperatures, temperatures one typically gets in campfires (Cutts et al., 2019; Milton, 1999a). Wood species known to burn for a long time were also used (Rowlett, 1999). Erectus will also have found it useful to sit around a campfire at night in a grass savannah, as it cools down to 15–16°C and provides a stage for large predators in search of prey. While the smaller habilis presumably still nested in the trees at night, as did great apes and presumably australopithecines, this was less evident for erectus in a grass savannah. Fire could then be an efficient resort to keep predators at bay (Sabater Pi et al., 1997).

Darwin already saw the use of fire as the most important cultural innovation in our evolution (Darwin, 1871). And cooking was not at least as important in that. The burnt bones from the Wonderwerk cave in South Africa and 800,000-year-old burnt plants and fish from a cave in Israel suggest that heating food quickly came about as soon as fire could be controlled and was used everywhere erectus went (Berna et al., 2012; Inchley et al., 2016; Zohar et al., 2022). Some even argue that the use of fire is as old as the erectus itself and that without cooking erectus would never have been the erectus it was (Wrangham et al., 1999a). No big body, no big brain, no sophisticated tools, no migration from Africa. Presumably, the benefits of heating food were discovered by coincidence. A piece of hard carrot that ended up in the fire became softer and thus more edible. A piece of meat exposed to heat tasted better. Soon it must have become clear that cooking food offered so many more benefits.

First, cooked plant foods become much more digestible. It breaks open hard seed casings and the cell walls made up of cellulose, releasing highly digestible and nutrient-rich contents (humans can break down very little cellulose themselves) (Cummings, 1984). Heat also breaks down other large molecules, such as proteins and starches, making it much easier for digestive enzymes to further digest the smaller molecules obtained. Cooked cassava provides up to almost 80% more energy than raw one. Cooking plant foods may well have provided a greater energy benefit than eating meat. Heat also neutralizes a lot of toxic and digestion inhibiting substances present in numerous plants, suddenly increasing the range of edible plants. Plants of the bean family, brassicas, asparagus, tomatoes, potatoes, and most cereals contain substances that actually prevent the action of digestive enzymes (Stahl, 1984). Also, young children may begin to eat plants sooner if they are heated. For meat, the situation is somewhat different. There, prolonged heating actually makes meat slightly tougher. But heating meat initiates a so-called Maillard reaction: amino acids react with carbohydrates and give a brown color to the meat, as well as a typical pleasant taste (along with the taste of fat broken down by heat) (Mottram, 1998). Even chimpanzees would prefer fried meat if given the opportunity (Herculano-Houzel, 2016). On top of that, fire also kills any potential bacteria and parasites that may be in and on meat, which also makes it last longer. That meat became important from erectus onward is also evident from a completely different point of view. For example, new species of tapeworms emerged during the erectus period, that is, parasites that are transmitted from an intermediate host to a final host through eating meat (Hoberg et al., 2001). Antelope and other bovine animals, the source of steaks for erectus, are the intermediate hosts for several tapeworm species that through the meat are transferred to felines, hyenas and canines as the final hosts. When erectus began to eat meat from those intermediate hosts, tapeworm species emerged that used humans as final hosts.

Cooking food also affected the way erectus had to handle food. Softer food implied that only 10% of the day had to be spent chewing, instead of 50% (Wrangham et al., 1999a). It also required less powerful chewing. Not surprisingly, the jaw apparatus of erectus became much less robust than that of habilis in a short period of time (Zanolli et al., 2022). This can be partly explained by the fact that the jaw muscles became smaller about 2.4 million years ago. One deduces this not from fossils, but from a mutation arising in a gene that makes certain essential proteins for muscle contraction (the *Myosin Heavy Chain 16* gene) (Stedman et al., 2004). Indeed, unlike macaques and chimpanzees, humans make little or none of these proteins in their jaw muscles, making them less bulky (Ciurana, Artells, Munoz, Arias-Martorell, Bello-Hellegouarch, Casado, et al., 2017; Ciurana, Artells, Munoz, Arias-Martorell, Bello-Hellegouarch, Perez-Perez, et al., 2017). This also makes them less powerful, which was not a problem if erectus had to chew softer food, but this also had an impact on the development of the skull (muscles pulling on developing skull bones affect the shape of those bones) (Currie, 2004). Nearly 5 million years of earlier hominin evolution with a large jaw apparatus gave way to a rapid change to smaller jaws and smaller teeth, as well as a shorter digestive tract (Organ et al., 2011). The latter is inferred from the shape of the chest (Aiello & Wheeler, 1995; Larbey et al., 2019; Wrangham & Carmody, 2010). A smaller jaw apparatus also manifested itself in a shorter snout (Bastir et al., 2021). Wear on the teeth suggests that incisors were used more for tearing and that the diet was more diverse and contained tougher foods, such as meat (Antón et al., 2014).

## Energy in an unstable environment

Eating differently and more efficiently is clearly the message we get about erectus. Being able to retrieve more energy from the same amount of food is evolutionary, extremely interesting. This was certainly a challenge for African erectus, which not only found itself in an increasingly drier grass savannah and unstable climate, but also had to maintain a larger body and larger brain (Zeller et al., 2023). Put another way, had erectus not found a solution to this energy problem, it probably would never have grown as large and its brain would have remained similar to that in an australopithecines. Cooked roots and meat offered a way out here. Hunting small animals involves less risk and large animals yield more meat and fat (Hawkes et al., 1991; Stahl, 1984). But hunting also requires more distance to travel and chewing raw meat costs about 9kcal/h for a human (van Casteren et al., 2022). So, nothing for free.

Each body consumes a minimum amount of energy, even when doing nothing. This is because of our basal metabolism. Even when we are at flat rest, all sorts of things happen that require energy. Cells need to be kept alive, organ functions like heartbeat and breathing need to happen, brain needs to coordinate all that and much more. How much energy is needed depends on how big the body is and how many energy-consuming organs there are. But also variation between mammalian groups may differ, where it was initially thought that primates showed an overall increased metabolic rate. However, when using proper scaling factors, including body composition, they show the same metabolic rate as other mammals. And yet, humans form an exception! We have both a higher resting metabolic rate, as well as an elevated activity metabolic rate (Yegian et al., 2024). For a small australopithecine this has been estimated to be between 900 and 1200 kcal/day but for an erectus 1.7 m tall and 60 kg heavy, it was a lot more and like for us: somewhere between 1400 and 1600 kcal/day. An erectus that had to search for food consumed up to 85% more energy than an australopithecine, not only because the body and brain, an energy-consuming organ, were larger but also because erectus had to travel greater distances in a grass savannah in search for that food (Ayala & Cela-Conde, 2017; Leonard & Robertson, 1997). Combining all the activities a body exerts in a day, a body can only consume as much energy as it comes in through food. This total energy consumption is about 1400 kcal in a chimpanzee, 1700 in an australopithecine and 2000–3000 in current hunter-gatherers (Lieberman, 2013). One hundred grams of meat provides about 100–200 kcal of energy (as opposed to only 10–20 kcal for 100 g of leaves and 50–100 kcal for 100 g of fruit) (Leonard & Robertson, 1997). If australopithecines ate similar amounts of meat as what chimpanzees do today, thSey would have consumed about 10 kg of meat per year. In current hunter-gatherers, this is about 80 kg of meat per year (Wrangham et al., 1999a).

But then, was meat really necessary, you may ask? In the end, people today survive perfectly well on a diet consisting purely of plants. The problem is that you cannot compare the accessibility to sufficient plant-based food in a supermarket with the situation in a grass savannah. Finding sufficient food requires time and distance, as well as time needed to ingest and digest sufficient food. The energy remaining, after the body has been searching for food using its energy-consuming brains and muscles, is energy available for basal metabolism, to chew and digest, to grow, to care for children, to play with each other and form social bonds, to look for mates, to sleep and much more. So, it is essential to gather enough food in a day to meet those energy demands. Turns out that the time primates spend looking for food is a maximum of 8 hours a day (the larger the primate the more hours) (Herculano-Houzel, 2016). The rest of the time they need to perform all those other necessary functions. Depending on how energy-rich the available food is, 8 hours may even be insufficient to gather the necessary energy, something orangutans face during periods when fruit is scarce. I had mentioned that orangutans and gorillas seem to have brains that are too small for their body size, or put differently, have a body that is too big for their brains. The explanation is that it is impossible for an herbivorous primate to gather enough energy within 8 hours to maintain a large body and a correspondingly large brain (Fonseca-Azevedo & Herculano-Houzel, 2012). The evolution of these great apes thus reflects an investment of available energy in body size at the expense of brain size.

# Runners are born

Erectus was large, had large brains and had to travel great distances to gather food and to hunt. If meat and cooking had to provide just that necessary energetic advantage, it seems logical that erectus would be able to avoid losing that advantage while pursuing wild animals. Remember that an open savannah now also meant that it became very hot during the day. Efficient walking and running thus became very important, even more so than for australopithecines.

## *Walk like an inverted pendulum*

Australopithecines already used energy sparingly while walking upright, thanks to adaptations in their skeleton and muscles. Walking with extended legs was already a possibility, according to some scientists. Others maintain that they still walked with bent hips and knees. In erectus, however, it is clear: they took large steps with their long legs and did so efficiently by keeping the legs extended, making their legs similar to ours but 50% longer than those of australopithecines (Bramble & Lieberman, 2004; Ruff et al., 2015).

Leg length and stride length are linked: the longer the legs, the greater the distance between two feet when placed on the ground. But longer legs also weigh more and have their center of gravity farther away from the center of rotation, so it takes more effort to swing them forward. Fortunately, the foot and lower leg weigh less than the upper leg, so the center of gravity is not too far away, and swinging is not slowed down too much (our foot is only 9% of the leg mass, while it is 14% in chimpanzees). If we want to adjust our walking or running speed, we adjust our stride length and stride frequency. Overall, stride frequency does tend to be lower in people with longer legs (Bramble & Lieberman, 2004). If we walk and run faster, we are mainly going to increase stride length and to a lesser extent stride frequency (Bramble & Lieberman, 2004). An erectus with long legs could thus increase stride speed significantly. We take steps of about 60–70 cm long. This increases to 2 m when we run, even up to 3.5 m in elite runners (Uchida & Delp, 2020). We walk at an average frequency of about two steps per second. Stride length multiplied by stride frequency gives us stride speed, so we walk at about 1.2–1.5 m/s (or 3.4–5.4 km/h). So having longer legs means having to take fewer steps to cover a given distance. For the same frequency of steps, this results in an energy gain, as experiments also show (Steudel-Numbers & Tilkens, 2004). Lucy would have used 50% more energy with her short legs. Leg length also played a significant role for a growing erectus. If we may assume that the legs grew similarly slowly as ours do (as opposed to the rapid growth in great apes and australopithecines), the femur grew up to five times in size from infant to adult (Tardieu, 1998, 2010). But, as we know, there was a lot of variation in erectus. The evolution from smaller African and Georgian forms (with legs relatively 10% longer than Lucy's) to larger Asian forms (legs relatively 27% longer than Lucy's) possibly indicates a selective push toward more efficient walking by lengthening the legs (Pontzer et al., 2010). This was accompanied by having to rotate the pelvis less, which is more efficient even at the expense of stride length. For example, our pelvis rotates about 8°, which contributes up to 5% to stride length. In chimpanzees, this goes up to 61°, increasing their step length five times (Knight, 2021). It may surprise you, but relative to body size, our stride length is 27% shorter than a chimpanzee's, even though our legs are 12% longer (Thompson et al., 2021).

When those long legs are extended, the large upper leg muscles (such as the *quadriceps*) have to produce less force to support the body as the upper leg is positioned almost straight above the lower leg. This is also useful for just standing still, as easily demonstrated by something we undoubtedly did as a kid. Haven't we all given someone just standing still a tap in the knee cavity, causing them to sink through their legs

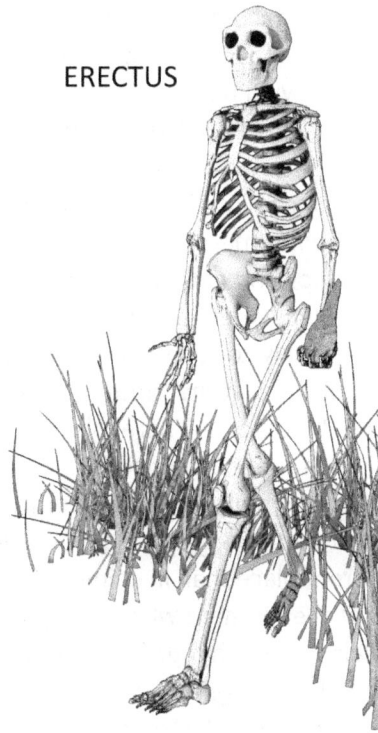

Figure 6.2 Erectus skeleton in a savannah environment, carrying an Acheulean tool. (3D model compilation of *Homo erectus* skull from Sketchfab, by the Société Géologique de France – https://sketchfab.com/3d-models/primates-homo-erectus-1afb3a6d39bd4f8b846705e6b11d774a; rest of the skeleton based on a modified 3d model of a sapiens skeleton.)

for a moment? Why? While standing straight, the thigh muscles are hardly active because the femur is in line with the tibia. When the angle between the two bones is changed, by slightly flexing the knee, that support falls away and the leg extensors must quickly correct it. Hence, we sag briefly but not completely. Standing upright but with hips and knees not fully extended thus requires a constant muscle activity. For example, a 74 kg human requires 22% less force to remain upright than an australopithecine weighing only 45 kg (Polk, 2004). Compared to Lucy, this would be as much as 37%. Now, the knee was not fully extended when at rest, only when viewed from the side. Given erectus had a femur and knee very similar to ours (Frelat et al., 2017), it probably had a similar valgus angle of about 9°. In habilis, it was possibly already 10°, lower than the 14° in australopithecines (Tardieu & Trinkaus, 1994).

Walking with fully extended hips and extended legs is likely to be unique to *Homo* (Hogervorst & Vereecke, 2015). Since muscles now need to generate less force, the volume of various leg muscles presumably decreased from erectus onward and small changes occurred in their attachments. Indeed, muscles that attach closer to the articulation of a leg bone can move that leg bone faster. With the same muscle strength, this does give less leg swing power, but the same amount of leg swing can be achieved with shorter muscles that have to shorten less. And it takes less energy to do so. To swing vigorously, more muscle fibers must then be activated, but that is out-weighed by the benefits of swinging with the legs extended. Such changes have hap-pened for the large *gluteus maximus* muscles and the hamstrings (back of the thigh), both muscles that extend the hips and can pull the leg to the back. Doing this while the foot rests on the ground will push the body forward, especially while running. If we start to walk, we are more likely to relax the calf muscles, making the body tilt forward from the ankle joint. Muscles at the front, necessary for stretching the knee, do need to be able to provide a lot of force. This explains why the leg extensors (*quadriceps*) in the upper leg are much more developed than the leg flexors (hamstrings). Here, the kneecap acts as a kind of pulley that more efficiently converts the force of these extensor muscles into straightening the leg. Some leg muscles also attach to the spine and pelvis in a different place than in chimpanzees. This allows the muscles to still have sufficient effect at the moment when the spine is fully erect and the hips are fully extended. This explains why a chimpanzee walks slightly hunched over when it walks upright: should it stretch fully, some muscles would be at such an unfavorable angle that they would not elicit movement. Other muscles would rather become tense during walking due to a shift in their position. For example, the *iliopsoas* muscle will be stretched as the leg rotates backward, then helping to swing that leg forward again. Whether the same was true in erectus remains a question. In any case, the pelvis was not yet fully modern yet (Ruff, 1995). Also, the neck on which the femoral head stands was long (relative to femoral length). As with australopithecines, this is advantageous for the lateral gluteus muscles to more efficiently stabilize the femur during walking. This may be an adaptation to the longer legs combined with the heavier body, as well as a larger distance between the articulations of both femurs with the pelvis, three factors that make it necessary to pull harder on the femur to keep the pelvis stable. That these muscles had to pull hard on the femur is demonstrated by the highly ossi-fied femurs in the African erectus, which were 12% firmer than those of australopith-ecines and even more than 40% stronger than our femurs (Ruff, 1995). Apparently, an evolution followed toward less robust femurs, because in Chinese erectus this is only 34%. Our skeletal-muscle configuration is presumably even better matched, so less substantial muscles have to pull less hard.

The foot of erectus still remains somehow elusive, even the Turkana boy did not yield much information. However, some foot bones were found in Dmanisi, showing that the foot was already very similar to ours and was less flexible than in australopithecines (Hatala & Boyle, 2022). The stiffer ankle joint helps to stabilize the foot when it hits the ground at every step and has to bear the full body weight (Lieberman, 2013). Just as australopithecines left fossilized footprints, so did erectus. The 1.5 million-year-old tracks from Ileret (Kenya) confirm that erectus already possessed a very modern foot (Bennett et al., 2009, 2016). For example, toes were short and the big toe was even more closely pressed against the second toe than was the case with the Laetoli's footprints, but not yet completely as much as in ours (although we do show a lot of variation) (Charles et al., 2021). These footprints are considered the oldest evidence of a foot being rolled during walking as we do (Dingwall et al., 2013; Hatala et al., 2023). The very recently discovered track ways in Koobi Fora (Kenya) of about the same age as those of Ileret confirm this (Hatala et al., 2024).

Walking with long and stretched legs, as in erectus, is often compared to the principle of an **inverted pendulum**. If you swing an object hanging from a rope, the weight keeps it going back and forth, moving down from the highest point each time, to reach the deepest point in the middle and go back up. A constant conversion of potential energy to kinetic energy (when the pendulum accelerates and moves down) and kinetic energy to potential energy (when the pendulum slows down and goes back up) ensures this. This is identical to what I mentioned earlier about the center of gravity of a hominin walking with outstretched legs, only that the pendulum point (corresponding to the contact point of the foot with the ground) is below the center of gravity and not above as would be in an ordinary pendulum. Hence, walking with fully stretched legs is compared with a pendulum hanging upside down (Saibene & Minetti, 2003). Every pendulum movement describes a step cycle consisting of a support phase (when one leg rests on the ground via the foot) and a swing phase (when that leg swings forward and there is support on the other leg) (Harcourt-Smith, 2013). We first touch the ground with the heel while the leg is extended. Then the foot tilts and is rolled off along the outer edge (because of the foot arching), until the center of gravity is now above the foot. Because of the inertia of that movement, our center of gravity continues to advance, causing us to tilt forward and repel with the ball of the foot. Meanwhile, our other leg swings forward in a slightly bent position, then straightens and touches the ground with the heel. This is a half-step cycle. Combine this with the contribution from elastic tendons in the hips and ankles, as well as legs being heavier at the top, then you know why it is that we need 47% less labor than a chimpanzee to walk on two legs. The long Achilles tendon, which first slowly stretches during the support phase and then shortens again like a rubber band as soon as the foot begins to repel, alone accounts for 39% of the power needed to repel the foot (Crompton et al., 2010; Ishikawa et al., 2005; Ker et al., 1987; O'Neill et al., 2022). Evolutionary changes in the morphology of the Achilles tendon probably played a key role in allowing a gradual shift to efficient endurance running, especially by increasing the speed range (Bates et al., 2025). In absence of contrary evidence, we can assume that this was similar in erectus, and thus erectus also wasted little energy when walking upright. In fact, we consume less energy while walking than an average mammal walking on four legs (Tucker, 1975).

So, the large erectus with extended hips and legs had numerous advantages that allowed very efficient walking. This might also explain another remarkable

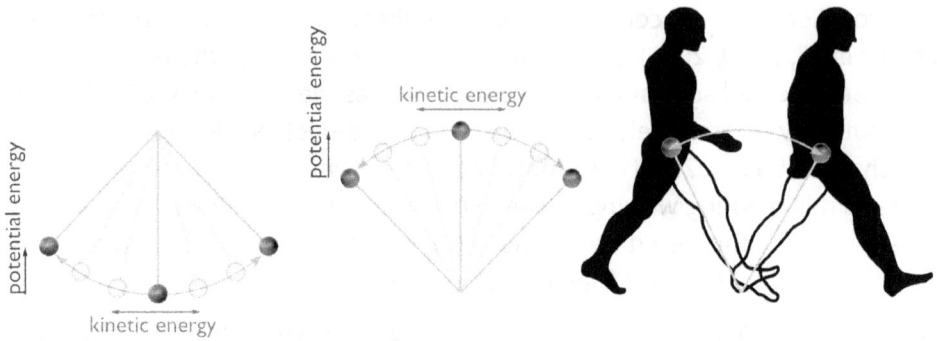

Figure 6.3   Walking as an inverted pendulum relies on efficient conversion of kinetic energy to potential energy. In an ordinary pendulum (left), the swinging occurs from a fulcrum below which a weight is suspended. In an inverted pendulum (middle), the swinging occurs from a fulcrum on which a weight is supported. When we take a step (right), our legs behave like the axes of an inverted pendulum, with the weight at our center of gravity at the pelvis and the tipping point where the foot is placed on the ground.

phenomenon in our evolution: from erectus, the sexual difference in body size was significantly smaller than in australopithecines. This could be inferred from the size of a female pelvis from Gona (Ethiopia) (Simpson et al., 2009). Male erectus were between 120 and 185 cm tall, the Gona woman was somewhere between 120 and 146 cm tall. A larger body also had other advantages, such as more energy that could be invested in a growing fetus and suckling a baby. Erectus also had a delayed growth and thus a longer phase of parental care (Anton, 2003; Hawkes et al., 1999; Smith, 1999). Erectus males were thus 1.15 times larger than females, compared to 2 times in great apes and australopithecines (Wrangham et al., 1999a). In addition, the pelvis also grew larger, which in turn had important advantages for giving birth, but more on that later (Brace, 1999).

## Running like a spring

We walk like an inverted pendulum, but does this remain energetically interesting as we increase speed? When we walk fast, we try to do that, but soon we bend our knees more when we place our feet on the ground. We also consume more energy when we try to walk like an inverted pendulum at speeds higher than 1.5 m/s (Uchida & Delp, 2020). We are clearly not built to walk like a pendulum at any speed, which explains why we switch to running from a certain speed (somewhere around 2.2–2.6 m/s) (Rakovac, 2021; Segers et al., 2007; Uchida & Delp, 2020). Unlike most quadrupeds, which exhibit multiple gait patterns, we are limited to two: walking and running. Exceptionally, we hop or run down hills very fast, doing so at a gallop. But unlike the gallop in quadrupeds, for us this requires more energy (Fiers et al., 2013).

Humans thus walk or run in natural conditions and are well adapted for it. Everything seems to indicate that this was the case from erectus onward: much of what typifies our way of running and how muscles and skeleton are involved, probably emerged somewhere between 2 and 1 million years ago.

If we don't run like an inverted pendulum, how do we do it? We can better compare running on two legs with a spring that compresses during the support phase and stretches during the repulsion. This pushes the body up and forward until the moment the foot comes off the ground. We then float for a moment, so to speak, until the next foot is placed on the ground (Saibene & Minetti, 2003). Where is this spring anatomically then, you might think? The main spring function is in the tendons and tendon plates in the leg muscles (but remember we also have shock absorbers in the feet and lower back that act like a spring). While running, we bend our legs when coming down. In fact, most joints in the legs and hips bend more during running than walking, something that increases with running speed (Uchida & Delp, 2020). When a leg bends, the leg extensors come into the picture (such as the *quadriceps* and *gastrocnemius* calf muscle): they ensure that the falling body is first slowed down by stretching tendons so that we don't end up falling on the ground, and then contract and stretch the tendons like a tensioned spring. The extensors of the hips then cause the body to be pushed forward. Our posterior gluteus muscle (*gluteus maximus*), which like the quadriceps is one of the largest muscles in our legs, is crucial during running. Surprisingly, however, this muscle is barely active during walking but very active during running and even more so during sprinting (Bartlett et al., 2014; Lieberman, 2013). Not only does it propel our body upward and forward, it stabilizes the pelvis so that we do not fall forward during this phase (Lieberman et al., 2006). This muscle has undergone a significant change during our evolution. In chimpanzees, in fact, it runs along the side of the hip joint, allowing the leg to swing out sideways. In humans, it has not only become much bulkier but also runs along the back of the hip joint where it attaches to the back of the femur. As a result, the leg no longer swings sideways but backward. When this occurs at the supporting leg, the pelvis and therefore the whole body will be pulled forward (Lieberman et al., 2006). You can actually think of running as a controlled, constant falling forward (we also tilt our torso 10° forward while running) (Lieberman, 2013). But the muscle will also be active on the swinging leg, as it decelerates the leg so that the foot can be placed down (Lieberman et al., 2005).

The lengthening of the legs in erectus was accompanied by many changes in muscle attachments and muscle functions. Natural selection will always have preserved those configurations of the muscle-skeleton that allowed for more economical use of energy. And this is not limited to the legs alone. How the rest of the body moves along has also changed in function of energy conservation. For example, we stabilize unnecessary movements of the body. When we run, the left and right legs are always alternatingly swung from front to back. As a result, our pelvis also rotates from left to right, which would make our trunk, arms and head also constantly swing along. We compensate for these hip and trunk movements by swinging the arms in the opposite direction: when the left leg swings forward, we swing the right arm forward and vice versa. By doing this, we save 3–9% of energy and reduce the rotation we have to do with our torso (Arellano & Kram, 2014). Energy consumption is further reduced by moving the center of gravity of the swinging arms closer to the shoulders. Ever wondered why we raise our arms when we run but leave them extended when we walk? Here you have the answer: raising the arms raises that center of gravity higher so that arm swinging requires less energy (Yegian, Tucker, Bramble, et al., 2021; Yegian, Tucker, Gillinov, et al., 2021). Also when our walking pattern is perturbated, we also very rapidly restore to the gait that consumes less energy. Within 1 minute, the left-right swinging pattern is restored to regain stability, followed after about 3–11 minutes with a reduction in the effort produced to walk (as measured from the work produced and the metabolic rate) (Brinkerhoff et al., 2024).

Swinging the arms also play a role in keeping the head stable. While running, this is important because that is where the organ of balance lies in the inner ear. A head that is constantly swinging back and forth is not conducive to maintaining balance and certainly not conducive to maintaining sharp vision (Lieberman, 2013). Try running between and over obstacles if the head is constantly swinging back and forth. Instead, our head makes compensatory movements. The inner ear registers the movements of the head and thus, through the brain, directs the neck muscles to make the head make counter movements. This keeps it fairly stable even though the rest of the body moves from left to right. Erectus, like us, had a larger inner ear than chimpanzees and australopithecines, it would have made it more sensitive to detect movements (but slower). In addition to the neck muscles, also a neck ligament is involved in stabilizing the head. The ligament starts at the 7th cervical vertebra and runs to the back of the skull and some neck muscles, such as the trapezius muscle. This muscle is also connected to the shoulder girdle, connecting movements of the arms to movements of the neck and head. That chimpanzees lack this ligament, then may not be a coincidence. Adaptations to efficient running on two legs are thus literally head-to-toe, and erectus was most likely already equipped with it.

## *Keep on running!*

We essentially have two gait patterns while most quadrupeds, like horses, have several. But why would a horse decide to switch gait patterns now and then? Again, everything has to do with energy. Not coincidentally, many mammals walk at a low speed and switch to another mode of locomotion once they go over 4–5 km/h. A horse does it, we do it. And just like a horse, we subconsciously know when that speed is reached and we need to run instead of walk, without the need to have a speedometer in our pocket (Rathkey & Wall-Scheffler, 2017). How do we measure it? Through our energy consumption. In fact, each of those gait patterns has an optimal speed: a speed at which the body uses the least amount of energy. For walking, that is approximately

4.7 km/h (D'Antona & Burtscher, 2022). If we walk slower or faster than that, we consume more energy to cover a given distance with a given body mass. Walking too slowly than what we spontaneously do requires more energy to cover a distance and requires more muscles to be engaged to slow down the speed. This is true for each of the gait patterns. Let's take the horse that has three main gait patterns: walk, trot and gallop. Once a horse reaches a speed above 5 km/h, it consumes more energy while walking than if it were going at a trot (Hoyt & Taylor, 1981). At a trot, a horse consumes minimal energy at a speed of about 12 km/h. As the speed further increases, consumption goes back up. Above a speed of 16 km/h, it will switch to galloping. Changing gears in a timely manner is an efficient use of energy, much like shifting gears in a car. When we switch to running, it is for the same reason. Once at a speed of 7–8 km/h, walking requires more energy than running. So race walkers who reach speeds up to 9 km/h don't have to do it to be efficient, they actually force their bodies to stay in an abnormal gait pattern when the body would rather be running.

If you compare the horse's energy consumption to ours, we do consume more energy relative to body mass and distance traveled. At the optimal walking speed, a horse will consume about 120 ml of oxygen (per kg of body mass and per km traveled), while ours will be 160 ml. Chimpanzees consume twice as much energy as a human

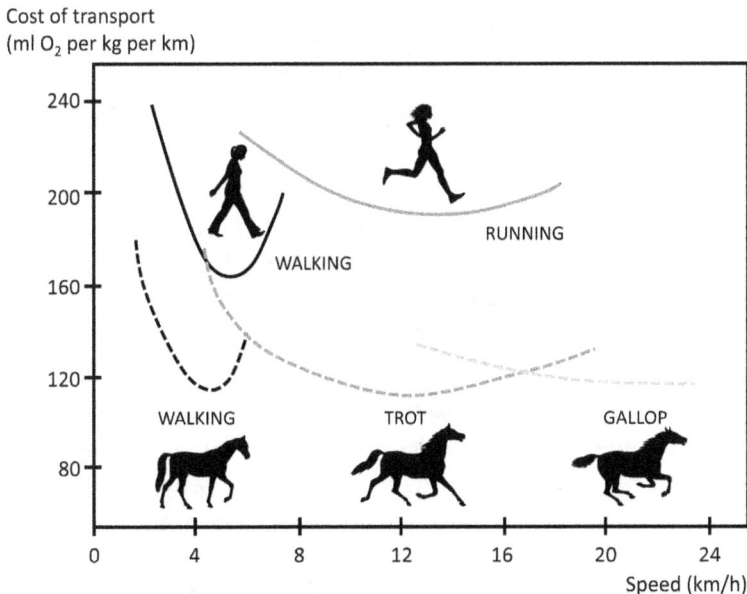

Figure 6.4   Energetic cost during different gait patterns in humans and horses, expressed as the amount of oxygen consumed per kg of body mass and per km traveled, and this as a function of locomotion speed. (Based on data from Carrier (1984); Hoyt and Taylor (1981); Rathkey and Wall-Scheffler (2017); Uchida and Delp (2020).)

(D'Antona & Burtscher, 2022; Sockol et al., 2007). At a speed of 12 km/h, a horse at a trot will still consume about 120 ml of oxygen, while we then run at a consumption rate of 190 ml of oxygen. We are not the only ones evolutionarily adapted to efficient walking and running; in fact, the entire evolution of the horse lineage reflects this. In general, running on four legs is also energetically better than running on two legs (Tucker, 1975). Hence, the fact that our ancestors started running on two legs cannot simply be explained as an adaptation to conserve energy. Just another confirmation that walking upright brought other benefits, large enough to offset disadvantages of increased energy consumption. Of course, this is all an average and there will be some variation. In fact, we all have our own preferred walking speed, determined primarily by body mass or leg length, but also by step frequency (Levine & Norenzayan, 1999; Pontzer, 2007). Research on walking behavior among people living in large cities in 31 countries surprisingly shows that factors such as socioeconomic context may also play a role: people from regions with higher purchasing power apparently walk faster. Time is money? Ambient temperature may also play a role: where it is colder, people walk faster. This research also shows that Irish people tend to walk the fastest and Brazilians the slowest.

A striking phenomenon in humans is that relative energy consumption (per kg/km) changes little during running. Sometimes it is even stated that it remains constant for speeds between 8 and 20 km/h (Carrier, 1984; Margaria et al., 1963). This would imply that it does not matter whether you run a marathon slow or fast, at the end of the marathon you would have used the same amount of energy in both cases. So, running slow and taking longer, or running fast in less time, makes little difference in total energy consumption. This has been called the 'running paradox' of our evolution, which would have come about from erectus (Carrier, 1984). The paradox would then explain why we can actually keep running without getting completely exhausted and overheated. In any case, this would have been very convenient for erectus who had to hunt during the day in hot grass savannahs. Even now, man is a very exceptional animal: it is practically the only one that can keep chasing its prey until it gives up from exhaustion (some predators do this too). This technique of persistence hunting is still being practiced by Bushmen from South Africa, Indians from North and South America and Aborigines from Australia (Lieberman, 2013). Simulations of energy consumption in both humans and prey animals of different sizes show that it pays energetically for humans to chase animals while running, driving them to exhaustion and overheating (Hora et al., 2022).

But then what makes us so unique? What do we have that allows us to keep going? Several factors are put forward. By walking upright, our chest is not compressed every time our legs swing. This is the case in quadrupeds. This allows us to breathe more freely while running, whereas in quadrupeds the breathing rhythm depends on the running rhythm, something that limits the efficiency in oxygen uptake (Carrier, 1984). It also affects panting in quadrupeds, something necessary to cool down. We do not depend on this. By the way, the Turkana boy shows that erectus possessed a rather large chest and thus had a large lung capacity (Bastir et al., 2020). Longer and elastic tendons in the leg muscles also ensure that more energy is recovered when we run like a spring and less muscle power needs to be generated (Bramble & Lieberman, 2004). Prolonged running is also favored by certain muscle fiber types, called slow muscle fibers. They consume a lot of oxygen and are relatively slow but have the advantage of being able to be active for long periods of time. Fast muscle fibers, on the other hand, are faster, more powerful but have limited endurance. Does it surprise you that we have more slow muscle fibers in our leg muscles than, say, chimpanzees (Marino et al., 2022)? This does make chimps more powerful (they produce 30% more force) but allows us to keep going longer at high speeds. Elite runners even manage to keep running a marathon at an average speed of just over 21 km/h (Bramble & Lieberman, 2004).

Does this explain why erectus was the first human to continue walking, so to speak, beyond Africa? It is estimated that if they lived in similarly large groups as current hunter-gatherers do (about 25 individuals), then the foraging range of such a group would have been 250–500 km$^2$ (Lieberman, 2013). To give you an idea, should this be the complete African continent, about 2.3 million erectus individuals would have been walking around (now there are more than 1.3 billion sapiens walking around there). Slow population growth combined with small groups breaking away would have allowed erectus to spread from East Africa to the Nile Delta in less than 50,000 years. To then disperse into the Middle East and so onward toward Eastern Europe and Asia might not have been such a big step for mankind.

# Keep it cool!

So erectus could efficiently run for prolonged periods of time. But how evident was this in a hot savannah? Once the body gets overheated, efficient running is quickly over. Could erectus then keep cool enough?

## *Skinny dipping*

We suspect that australopithecines started more than 5 million years ago as thick-haired, pale-skinned hominins that ended up being partially naked and darker-skinned about 2–1 million years ago. We are not sure. No fossilized skin or hair has been found from erectus to confirm this. Still, we know enough about erectus to have little doubt about that (Jablonski, 2021; Jablonski & Chaplin, 2000). Had erectus not been naked and dark-skinned, it would never have been such a successful species that could brave a hot savannah and, moreover, made it halfway around the world.

Survival in a hot savannah was and is a challenge. Erectus did walk completely upright which reduced the amount of radiation heat of the sun entering the body. For a diurnal hunter traveling long distances, however, that is not sufficient to keep cool. Fur would only be useful to keep warm at night, but erectus already had another solution for that: fire. Having to walk in the sun during the day with thick fur also had little advantages because cooling via sweating with thick fur is impossible. The selective pressure to lose the fur as much as possible was likely very high (Kushlan, 1980). How did the transition from thick fur with little to no sweat glands to naked skin with lots of sweat glands happen? How did we evolve from something like chimpanzees, thick fur and only sweating through the hand, foot palms and armpits, to a skin that appears naked and full of sweat glands? I deliberately say 'appears naked' because that's the way it is. We are not naked at all and in fact have as many hairs per square cm as a chimpanzee (only on the chest do chimpanzees have more hair) (Kamberov et al., 2018). That can't be right, you may think, because we are much less hairy than a chimpanzee, right? Everything has to do with different types of hair. When hair develops, vellus hair forms first: small (2–3 mm long) and thin hairs that contain little melanin (there is also a fetal phase with lanugo hair). Our body is full of vellus hair; you just have to look close at your skin to spot them. Only the palm of the hand, sole of the foot and the underside of the wrist are truly hairless. In some places, however, the vellus hair is going to become long, thick and highly pigmented. This is called terminal hair. Depending on the person, we find them on the head, around the mouth, under the armpits, around the genitals, ... you get the picture. There our fur does look more like that of a chimpanzee. But other than that, we are functionally naked: the hairs are still there but remain so small that the skin functions like naked skin: sunlight enters and sweat can evaporate as easily as if the skin were completely naked.

So, what about the evolution to a skin full of sweat glands? Actually, hair and sweat glands are not that different from each other, at least when you look at their development. Like teeth, scales, feathers, mammary glands or salivary glands, hair follicles (pieces of tissue in which hair will form) and sweat glands form as a thickening of a group of cells in the epidermis that grow inward. In hair follicles, these cells make the hair (by depositing keratin and melanin). In sweat glands, these become sweat-producing cells and cells that form the drainage canal for sweat that runs to the surface. Early in our development, our entire skin is dotted with such groups of cells. What happens to them is determined by several genes, such as the *Dkk2* and *Bmp* genes. When both are activated, no hair is formed, only sweat glands (Lu et al., 2016; Song et al., 2018). If the genes are blocked, then another gene is activated (the *Shh* gene) when only hair is formed then. So how come we still have both in the same piece of skin? That is because both processes happen but at different times. At an early stage, genes induce the development of hair follicles, in which vellus hair forms (and in some places eventually terminal hair). At a later stage, the other genes initiate the formation of sweat glands. The extent to which this occurs throughout the skin makes us unique among primates, a convenient novelty that probably originated from erectus. It can also be inferred from the genome that the skin has undergone extensive changes since the split from the chimpanzee lineage and that this occurred around the time when erectus was walking around (Jablonski, 2017).

But if it is so easy to form hairs and sweat glands, why did we lose fur while all other mammals in savannahs retained fur? Ultimately, fur shields some of the incident light and thus heat (not to mention harmful UV radiation). But those mammals are already seeking more shade or cooler periods of the day and walk on four legs instead of two. From the moment the sun is at a 40° angle, less solar heat enters when we walk upright than if we were walking on four legs (Wheeler, 1984). When the sun is at its highest, at the hottest point of the day, only 7% of the body surface is irradiated when erect. This is only 40% of the surface area that would heat up in a quadruped. Now, this obviously does not explain everything. Heat that an erectus needs to get rid of also comes from the inside because walking and running produces body heat (Chaplin et al., 1994). Let us not forget that erectus was well adapted to use as little energy as possible by walking like an inverted pendulum or running like a spring. Using less energy means producing less heat. That said, an efficient cooling system was nevertheless crucial for erectus.

## Sweating like a horse, cool as a cucumber

How did erectus still manage to keep body temperature under control in that hot grass savannah? A lot of evolutionary puzzle pieces seem to have fallen into place with erectus. It was erect (and received less heat) but also showed different body proportions. He was taller but so was she, the female erectus (slight sexual difference), and both were also tall in proportion to body weight (Wheeler, 1992b). Being tall and more slender is more interesting for walking upright, leaving more energy reserves for sustaining a pregnancy, for maintaining larger brains and much more. Being tall and slender also provides a favorable body surface area to body mass ratio to more efficiently dissipate heat, especially heat produced from the inside. How much heat is produced depends on how many cells that are active, hence on body mass. How much heat can be given off to the environment, via sweating, depends on how large the surface area of the skin is. Should erectus have simply been one and a half times an australopithecine, both in length, width and depth (i.e., $1.5 \times 1.5 \times 1.5$ times), its mass would have increased to the power of three (mass is related to volume expressed in cm³). Surface area, however, increases only to the power of two (expressed in cm²). So, a larger body produces proportionally more heat than it can get rid of because mass increases faster than surface area. Not so convenient for erectus in that hot savannah. In great apes and australopithecines, body length would increase exponentially with body weight by a factor of about 0.3, which is a normal factor for increase in length relative to mass. For erectus and other *Homo*, however, this is much higher, at 0.5. Erectus was thus much longer for the same body weight (Grabowski et al., 2015; Heymsfield et al., 2007; Jungers et al., 2016; Mchenry, 1992). Consequently, the body surface area is also relatively larger, about 1.5 m². In chimpanzee this is 1.3 m² and in australopithecines it was 1 m² (in sapiens it fluctuates between 1.5 and 2 m²) (Bryson, 2019). So, from erectus onward, humans were not only larger but also taller. Better to cool off via convection.

But a larger skin surface area also provides more space for sweat glands. Double gain to cool via convection and evaporation. We carry about 2–5 million sweat glands scattered throughout the body, around 100 per square mm (density is highest in the palms of the hands, soles of the feet, forehead and lower back) (Folk & Semken, 1991; Jurmain et al., 2018). About 90% of all sweat glands are eccrine sweat glands, the most important ones to regulate body temperature (horses and other ungulates do cool off via the other glands, the apocrine glands). Chimpanzees only have 10 glands per square mm, and only 65% of them are for temperature regulation (Best et al., 2023; Kamberov et al., 2018). We can easily produce 3–4 l of sweat per hour during heavy labor (Jurmain et al., 2018; Marino et al., 2022). Sweating like a horse, they say. Actually, that's not true because horses only have 3 sweat glands per square mm and thus do not produce as much sweat (Raghav et al., 2022). Sweat production also depends on the blood supply and as such has even been linked to skin darkening in erectus! Remember folic acid, vital but easily degraded by UV rays. In fact, folic acid causes blood vessels to open up more and allow more blood to flow (Stanhewicz & Kenney, 2017). A dark skin that blocks enough UV rays so that not all the folic acid is broken down, was also beneficial in allowing the body to cool down better via sweating. Sweat glands also accumulate glycogen, a substance that provides energy in the form of glucose to produce sweat (Best & Kamilar, 2018). A functionally naked skin that brings wind closer to the skin and is packed with highly active sweat glands thus gave erectus a lot of benefits for better cooling (Wheeler, 1992a).

And that's not all, because when all those properties are put together in a context of a warm grass savannah, they further reinforce each other. For this we need for a moment to fall back on the physics that take place when a gas (air) passes by an object and the exchange of heat between the two. Three factors are important here: the speed at which the air moves past the object (the faster, the more exchange of heat), the temperature difference between air and object (the greater the difference, the better the exchange) and the humidity (the drier the air, the better it can extract moisture and heat from an object). Moreover, the physics behind it manifests itself on two levels: the interaction between the wind and an erectus, as well as the interaction between the wind and the grass. The closer the air gets to a surface, the more friction and thus the slower it becomes (Wheeler, 1991). As a result, the temperature against the bottom of a grass savannah is higher than above the grass. The ground temperature in an open savannah can be up to 25°C higher than that in a forested area (Cerling, Wynn, et al., 2011). This is very hot for a body that would have to move within the grass. Quadrupeds are constantly in that hot zone with most of their body. A bipedal and long erectus, on the other hand, has much of its body sticking out above the grass, where it is much cooler. On top of that, the higher wind speed above the grass enhances the convection and evaporation of sweat. The naked erectus' upper body, covered with sweat glands and protruding above the grass where it is cooler and there is more wind, allowed very efficient evaporation of sweat. The dry air, typical of

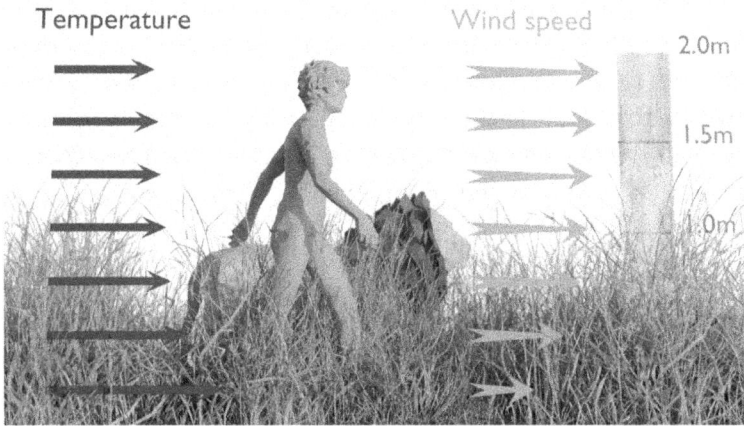

Figure 6.5    Differences in wind speed and heat of the air in a grass savannah, showing differences in exposure in a bipedal human versus a quadrupedal lion. (Lion modified from 3D model from Sketchfab, by Lary – https://sketchfab. com/3d-models/lilion-914d1dfc0d2b44c49103da81f230bfc5.)

an open savannah, made this extra conducive (Wheeler, 1991). And maybe, there was also a physiological efficiency gain, as it has been shown that sapiens seems to have the unique capacity among mammals, including other primates, that can consume high amounts of energy for their given body size and the environment in which they live (Yegian et al., 2024). This can only be achieved through an efficient removal of metabolic heat, maybe something that erectus would have benefited from as well.

Was erectus' entire body then tuned to limitless cooling? Surely not. The large brain already required additional cooling and a shorter snout limited cooling through the mucous membranes of the nasal cavity (Mori et al., 2021). Presumably another solution arose in erectus: a fleshy nose (Franciscus & Trinkaus, 1988). Great apes and australopithecines did have protruding jaws, visible as the snout, but had a flat nose. The skull of erectus suggests that the nostrils were surrounded by outgrown skin supported by cartilage. Possibly this allowed some cooling of the inhaled air in the nasal cavity anyway, even though the jaws and thus the snout had shortened as an adaptation to a softer diet. A nose also ensures that less moisture is lost during breathing. Not coincidentally, noses are longer on average in people from arid regions and in those with higher metabolisms (Kelly et al., 2023). Still, maintaining necessary body fluids was a challenge for a sweating erectus with a short snout. A 70 kg human would lose about 1.2 l of fluids as sweat between 6 a.m. and 6 p.m. in a warm savannah, where it is 35°C to 40°C (would be about 2 l for a whole day) (Wheeler, 1992b). This loss must be replenished daily, because we cannot afford to lose much. Already at a fluid loss of 5%, cooling through sweat no longer works very well. At 10% we lose consciousness and 20% is fatal. Based on the distance to travel between sites containing material for making tools and sites where tools were found, one could estimate how much fluid erectus would lose in a day. This was estimated at 4%, thus within the safety margin. Erectus would even have been able to keep running behind prey for 5 hours without having to drink (Hora et al., 2020). Drinking water to replenish that fluid loss was hence very important, because unlike great apes, erectus could not get the necessary moisture from juicy fruits (Pontzer et al., 2021).

# What big brains you have!

Erectus' big brain thus became a big challenge, because it demanded a lot of energy and needed to get cooled down. Was erectus able to provide that energy and provide additional cooling? Apparently so, as it survived for more than a million years, but how?

## *Bigger brains, thus more intelligent?*

We already know that primates have relatively large brains for their body size and also more neurons in the neocortex. From erectus onward, this culminated further with a brain volume between 600 and 1250 cc (Anton, 2003; Baab, 2008a; Bruner & Beaudet, 2023). Erectus brains increased in volume by an average of 160 cc per million years. Was this just because of a larger body or was there more to it? Erectus brains seem to have experienced an evolutionary tipping point, with the onset of an accelerated increase in brain size. Recent analysis even suggested that an important component of the evolutionary increase in brain size in hominins took place at an intraspecific level, each time leading to an increase as new species emerged. This would definitely have been important in erectus, considering its vast range of intraspecific variation in time and place (Püschel et al., 2024).

Before Darwin, brain volume was considered a reflection of intelligence, and thus intelligence could be expressed in numbers, which we still do today through IQ tests (Killgrove, 2021). In 1762, Albert von Haller had already noticed that larger animals had larger brains, so there had to be a connection (Herculano-Houzel, 2016). Ultimately, a larger animal needs more nerve pathways to efficiently connect all the body parts to the brain and more neurons to process all that information (Deaner et al., 2007). In 1973 Harry Jerrison introduced the 'encephalization quotient'. At a quotient value of 1, the measured brain volume perfectly corresponds to an expected size based on body size, using small insectivorous mammals as a reference. For sapiens this would be 7.5, meaning that so our brains would be 7.5 times larger than what you would expect for a mammal of about 70 kg. In chimpanzees this is only 2 and in great apes even less than 1. Thus, humans seem to have exceptionally large brains. Once taking into consideration other mammals as a reference, this quotient for humans already drops to about 3 (Holloway, 2015). Less extreme but still three times larger than normal and incidentally slightly larger in females than in males. Erectus had a quotient of 2 (Dmanisi erectus) to 2.6 (Chinese erectus).

Now, what does this number say about intelligence? If one takes into account the size of certain neurons (so-called Von Economo neurons), then absolute brain size does

BRAIN SIZE (g)

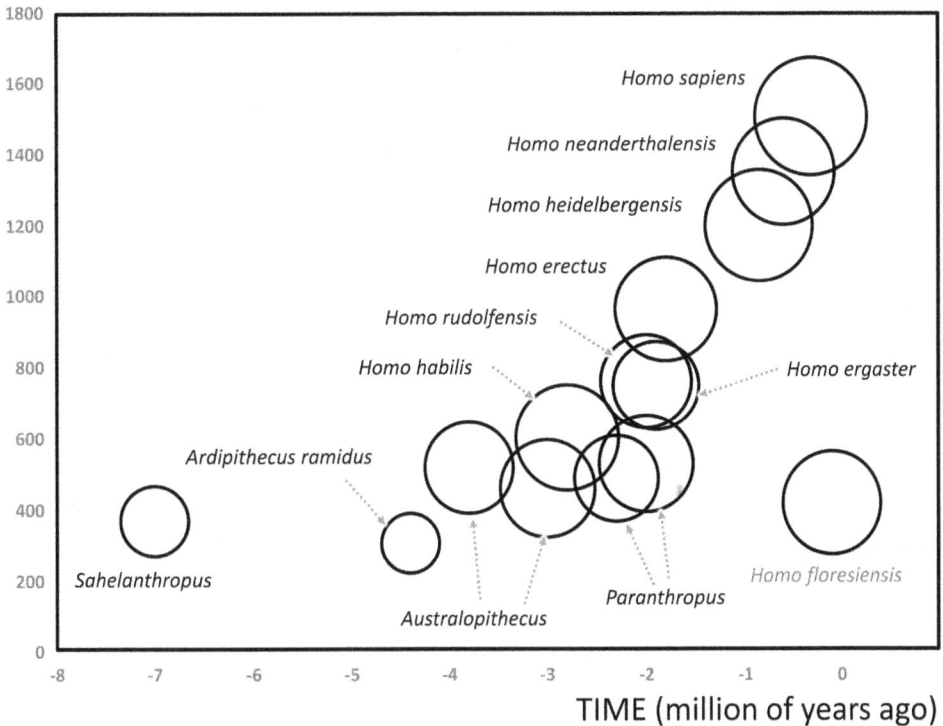

Figure 6.6   Increase in brain size (vertical axis) in hominins over time (horizontal axis), where two trends can be observed: A slower increase from *Sahelanthropus* (about 7 MYr ago) and an accelerated increase from *Homo habilis* (about 2.8 million years ago). The size of the circles represent the relative size of the brain, relative to body size (the data points used in this graph do not reflect that the brains of Neanderthals were actually larger on average than those of sapiens) (Herculano-Houzel et al., 2015; Herculano-Houzel & Kaas, 2011).

say something about intelligence (Deaner et al., 2007; Nimchinsky et al., 1999; Reader & Laland, 2002). But brain size is also influenced by diet, longevity or even population density (Walker et al., 2006). Also, not all brain parts contribute equally to intelligence. Just look at the neocortex, where more or less folds determine the number of neurons present. And here lies the core of intelligence: numbers of neurons. Actually, it is the connections between neurons that are the determining factor because new information is recorded in our memory through the formation of new synapses. Also important is the connectivity between different areas in the brain that helps determine how efficient and creative information can be handled.

So, did the brain and neocortex grow abnormally large from erectus onward? Although usually suggested as such, I'm afraid it is not. If you compare the relationship between brain size and body size in all *Homo species* with those in other primates, it is true that brains are relatively larger in *Homo*. Only here one assumes that all other primates are typical primates. And therein lies the error: great apes such as orangutan and gorilla do not follow that trend. Their brains are actually much too small for their body size. A gorilla with an encephalization quotient of 0.5 has a brain that is half too small. Looking at it statistically, the general relationship between brain size and body size predicts only 83% of what one actually measures in brain sizes in *Homo* and 88% in the other primates. Instead of considering *Homo* as abnormal primates, it is better to consider the large great apes as abnormal. If data of *Homo* species is taken together with that of other primates, but excluding the two great apes, then brain volume can be predicted with 95% reliability based on body size! Simply stated, it is statistically more reliable to say that erectus and sapiens have brains as large as you would expect for such a large primate (Herculano-Houzel, 2016).

We can extend this to the neocortex, because we have a typical primate neocortex, at least in volume (Holloway, 2015). But what about the number of neurons? Sapiens have about 86 billion neurons in the brain, 16 billion of which are in the neocortex. A

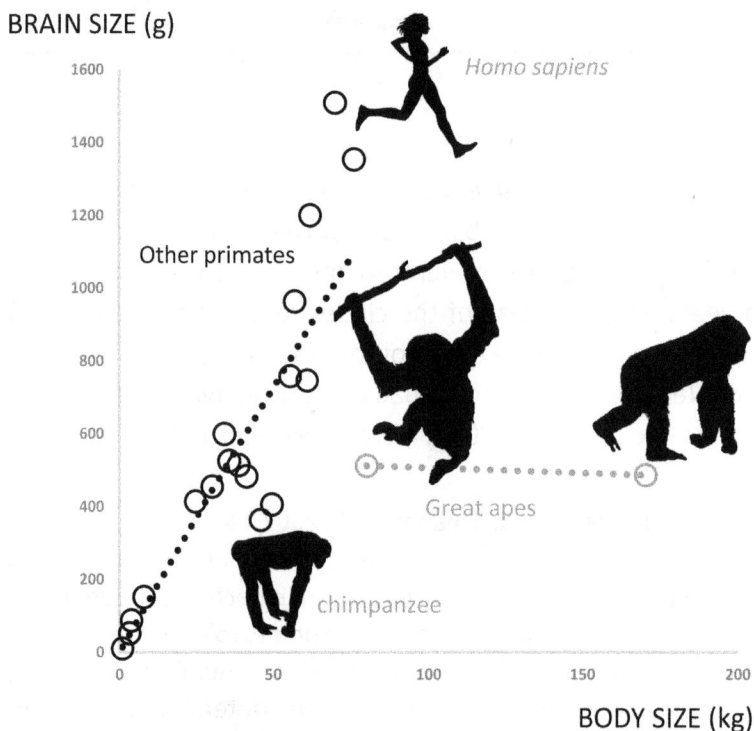

Figure 6.7   Relationship between brain size (mass in g) and body size (mass in kg): It is not humans that have abnormally large brains among primates, but rather the large great apes (orangutan and gorilla) that have abnormally small brains for their body size (Herculano-Houzel, 2016).

gorilla with a body mass of 170 kg and brain of about 500 g only has 33 billion neurons, of which about 9 billion are in the neocortex (Herculano-Houzel & Kaas, 2011). All primates, including *Homo* and great apes, do have something in common: the number of neurons is highly dependent on brain size. Then we can also predict how many neurons erectus must have had: about 41 billion for Dmanisi erectus to about 80 billion for Chinese erectus. Primates in general do have a higher number of neurons for a given brain volume than some other mammals (Herculano-Houzel, 2011, 2012). Neurons also take five times longer to develop and certain cells divide more frequently in primates, which could explain why so many are present in the neocortex (Kornack & Rakic, 2001). So, relatively speaking, we do not have a special primate brain but what is special is that we have nearly 90 billion neurons that can make many more connections, where just like in a computer data can be stored and edited. Now, just looking at the number of brain neurons does not say everything. For example, pilot whales (kind of toothed whales) have more than 37 billion neurons in the neocortex (twice as many as we do) and an elephant has three times as many neurons in the brain as we do (about 257 billion), but less in the neocortex (only 5.6 billion, most of it is in the cerebellum) (Herculano-Houzel, 2016; Mortensen et al., 2014).

## Changing in different ways

So bigger brains don't tell us everything. We also need to consider other kinds of changes in the brain to get a better picture of why, from erectus on, we became a very different human being. Chimpanzee and human genomes show that in human brains more than 100 unique variants of genes are expressed (Bradbury, 2005; van 't Hoog, 2005). Some are associated with enlargement or rearrangement of certain parts of the brain (McLean et al., 2011; van Straalen & Roelofs, 2017). Genes are also more active in the neocortex in humans (Hrvoj-Mihic et al., 2013). In turn, other mutations new to humans affect the rate at which cells divide and certain parts of the neocortex develop (such as the *NOTCH2NL* and *ASPM* genes) or the extent and timing at which neurons form connections among themselves (this occurs later in humans than in other primates) (An et al., 2023; Bradbury, 2005; Feng & Walsh, 2004; Rutherford, 2019; Vallender, 2019). Some of these typical sapiens genes evolved very rapidly, indicating a positive selection (Kun & Narasimhan, 2023; Vallender, 2019). Thus, they brought an advantage, which in some cases can explain why *Homo* brains started to look differently (Kriegstein et al., 2006; Vallender, 2019). When genes unique to humans (such as the *NOTCH2NL* and *ARHGAP11B* genes) are being expressed in monkeys, their neocortex becomes larger, more folds are formed and the number of neurons increases (Heide et al., 2020). This, however, is still no solid proof that they were positively selected because of its impact on brain enlargement (Allen, 2009). Of many genetic changes, we don't know why, when and from which species they came about, but thanks to the genomes of Neanderthals and Denisovan humans, we know of some whether they originated before or after the split with sapiens (Vallender, 2019). But we don't know that about erectus.

If we include the fossils in this story, especially those of children, we can derive a surprising amount of information about how brains changed evolutionarily. Considering a newborn chimpanzee with brains that are less than half as large as when they reach adulthood (130 g versus 380 g), their brains thus undergo about 34–40% of their growth while the fetus is in the womb, cut off from the outside world (DeSilva & Lesnik, 2006). A newborn sapiens brain weighs about 340 g, which is only 25–29% of the adult brain (about 1350 g) (Dobbing & Sands, 1973; Peterson et al., 2021; Vrba, 1998). Their brains experience more of their growth after birth and thus outside the womb. Sapiens do have longer gestations (8.8 months on average versus 7.4 months in a chimpanzee) and have slightly larger babies (about 3 kg versus 1.7 kg in a chimpanzee (Harvey & Cluttonbrock, 1985; Lieberman, 2013; Vrba, 1998). Incidentally, of all primates, sapiens have the longest gestation and largest babies, even larger than those of a gorilla (a baby gorilla weighs about 2.2 kg). A sapiens brain experiences a robust growth spurt during the first two years and reaches 80% of its final size at only four years of age (Hrvoj-Mihic et al., 2013). Surprisingly, a chimpanzee also reaches 80% of its brain volume after four years, but the spurt is much greater in humans. Moreover, our brains develop longer than in the chimpanzee, with neurons taking double the time to develop than in for example macaques. Human brains reach their final weight after ten years (already after eight years in chimpanzees) but the neural pathways are not fully developed until the age of 17 years (after 11 years in chimpanzees), and the myelinization process of the neurons is only completed at the

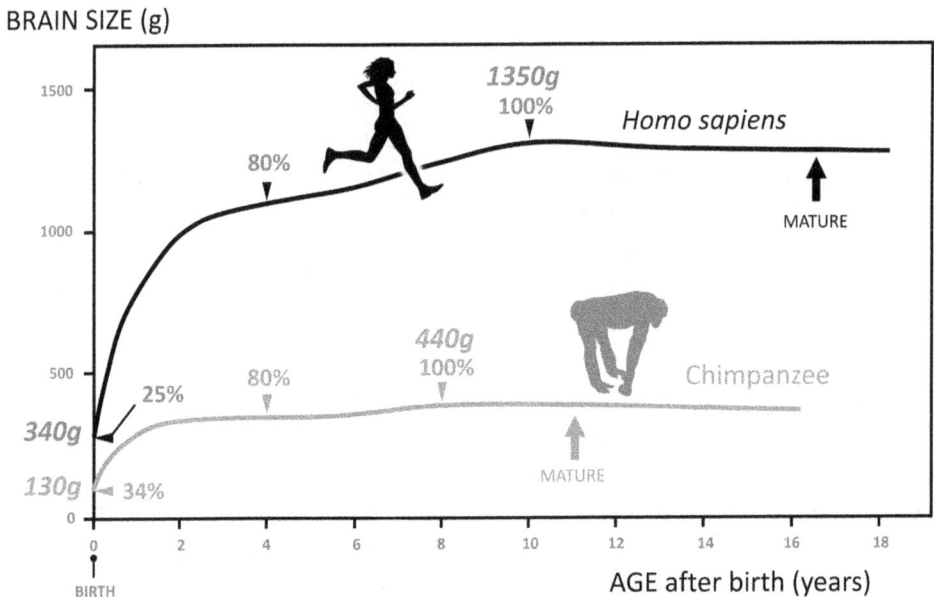

Figure 6.8   Brain growth after birth shows that human brains become much larger than chimpanzee brains by growing both faster and longer. (Data from Dobbing and Sands (1973); Peterson et al. (2021); Vrba (1998).)

age of 25–30 years of age (Lindhout et al., 2024). That is about halfway a human life, whereas this is already at one third of the lifespan in chimps and one 20th in mice.

But what does this say about erectus? Back in 1936, the famous Mojokerto child was found in Java, named after the site where it was found. It was a 1.8-million-year-old fossil of an erectus child that was one to six years old when it died. Based on the skull, they estimated the brain size to be about 660 g (Hrvoj-Mihic et al., 2013). Computer models based on 38 erectus skulls indicate that their brains weighed about 300–400 g at birth, which is about 35–40% of adult brain size (DeSilva & Lesnik, 2006). This is more in line with that of a chimpanzee than with sapiens (other models suggest it was more likely to be between 200 and 240 g, which would be more like sapiens). Larger baby brains may explain why erectus women are remarkably large compared to men, because larger babies with larger brains require not only more energy during pregnancy, but also a larger pelvis to get that baby head through (more on that later) (Brace, 1999; Martin, 1981).

Does it matter when the brain grows slower or faster? It's the final size that matters, right? It doesn't. On the one hand, larger brains take longer to establish the networks between all the parts (Barrickman et al., 2008). On the other hand, the moment the synapses form is crucial. Turns out that sapiens shows a slowed down development of neocortex neurons, resulting in neurons that carry more and more branched dendrites that also carry a higher number of synapses and thus an increased communication capacity between neurons, but also interactions between neurons and non-neuronal cells (Fang et al., 2022; Lindhout et al., 2024). The formation of these synaptic connections between neurons is stimulated by information coming into the brain, information from stimuli picked up from the outside world. Surprising perhaps, but synapses do not form homogenously. For example, the number of synapses in the neocortex increases sharply shortly before birth until 1 year after birth, only to decline again (Huttenlocher & Dabholkar, 1997; Stiles & Jernigan, 2010). In the prefrontal cortex, that is, where the main cognitive processes happen, this increase is even spread over the first five years after birth. After that, half the number of synapses disappear until adulthood, after which it remains roughly stable. A shift toward a slower growth rate during early years may already have emerged from early erectus, about 1.8 million years ago, as tooth growth in a 11.5 year old Dmanisi erectus showed that teeth were not fully developed (Zollikofer et al., 2024). However, considering the small brain size in these early erectus, it also shows that a slower early life phase did not immediately come with a notable increase in brain size (Guatelli-Steinberg, 2024).

Let's link that to the growth pattern of the brain before and after birth. In sapiens, 75% of growth occurs after birth, that is, outside the womb, at a time when only 40% of all synapses have been formed in the prefrontal cortex. After the first year of birth, the peak in numbers of synapses is reached at a time when the brain has already reached 60% of its final size. A human baby of one year old is thus equipped with a huge arsenal of connections between neurons, where information can be stored and

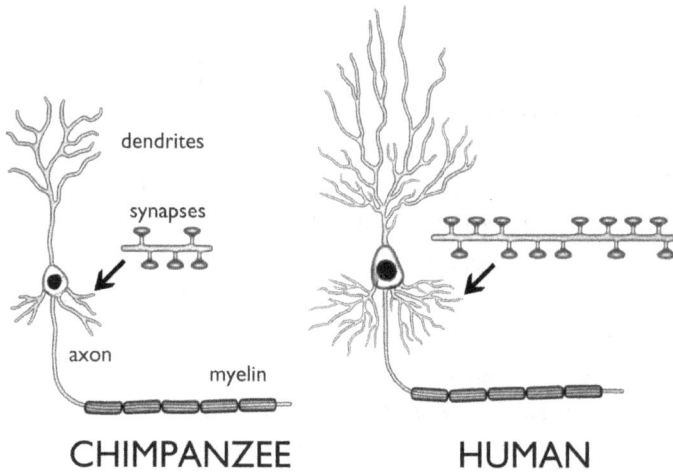

Figure 6.9    Difference in shape in fully matured neurons between sapiens and chimpanzee, showing the increased number of dendrites and the number of synapses they carry. (Adapted from Lindhout et al. (2024).)

processed, at the moment when it is fully stimulated by the environment outside the womb. Connections that are not reinforced by those stimuli disappear. Chimpanzee brains, on the other hand, spend up to 40% of their growth in the womb, which is a very low-stimulus environment. How their synapses develop is not known, but it does indicate that brains that still have a lot of growth potential outside the womb have more time to develop stimulated and powerful brains. It is thus interesting to allow brains to grow as much as possible after birth as more and the right connections can be established and thus intelligence built. The brain changes in erectus seem to have been accompanied by a shift from a more chimpanzee-like pattern to a more sapiens-like pattern in brain development. Whereas the arrangement of grooves in the neo-cortex in Dmanisi brains is even more similar to that of chimpanzees, in later erectus (younger than 1.5 million years) it is a typical sapiens pattern (Beaudet, Du, et al., 2019; Ponce de León et al., 2021). The important role of improved cognitive functions may also explain why the largest expansions during the development of the human neocortex take place at higher-order cognitive regions (Vickery et al., 2024).

## Energy guzzler

Larger and more complex brains with more connections between neurons can handle information about the environment better, faster, and more creatively and efficiently. With a better memory, one can remember more locations of food or material for tools. Being able to plan and cooperate better enables more efficient hunting. A more creative mind also translates into better tools. Thus, larger and more complex brains increase the chances of survival. Hence, evolution toward bigger brains is obvious, right? Actually no, at least if you also consider the cost involved as an organ can only

Numer of synapses (per 100 μm³)

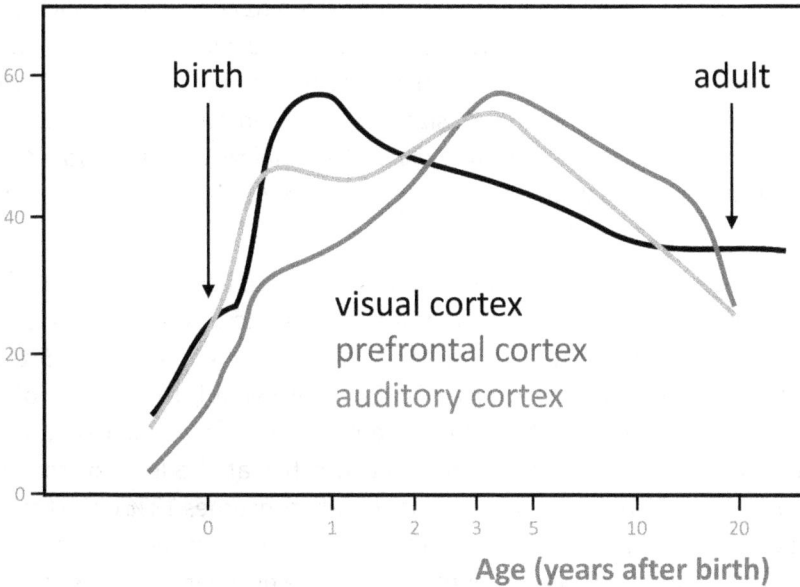

Figure 6.10   Ontogenetic changes in the number of synapses in sapiens in three neocortical areas. (Adapted from Huttenlocher and Dabholkar (1997).)

function if it has enough energy available, energy that must come through food. A minimum amount is already needed simply to maintain the body (the basal metabolism), not to mention the energy needed to perform all kinds of actions. Survival is therefore ensuring that the daily energy supply is at least as large as the total energy consumption, where digesting the food equally consumes energy! How large that supply should be depends on many factors: how active one is and how much distance one travels, the age, gender, whether one has to take care of children or not, the environment and so on. Also, not all organs consume energy equally: there are energy-devouring organs (such as the liver, heart muscle or kidneys) and low-consuming organs (such as fat, bones and skin) (McGrosky & Pontzer, 2023).

Brains are also true energy guzzlers, consuming the most energy after the liver. Although they only make up 2% of our body mass, they consume 20–30% of the energy needed for the basal metabolism (Herculano-Houzel, 2011, 2016). Even when we lie relaxed in a couch, our brain consumes 500 kcal/day at a constant power comparable to half that needed to run a 40-watt light bulb. To compare: a mouse's brain makes up 1% of its body mass but consumes only 8% of the basal metabolism. In most mammals this is between 2 and 8%, but in primates it is higher (Mink et al., 1981). Our consumption per kilogram of brain is even nine times higher than the whole-body average (Aiello & Wheeler, 1995). Energy in the form of glucose is constantly delivered to the brain through a blood flow of 750 ml of blood per minute. And there is little slack in that. If the flow decreases by as little as 1%, we are already risk of fainting

(Herculano-Houzel, 2016). Compared to the first *Homo,* the blood flow to our brain has increased six times (Seymour et al., 2016). Our brains consume as much glucose in about 25 minutes as there is in a lump of sugar (1 g of glucose that we get from carbohydrates through digestion provides them with 12 minutes of energy) (DeSilva, 2021). Based on erectus' body size and brain size, we can estimate how much energy they needed. Their brains consumed about 17% of the basal metabolism, about 260 kcal/day, a lot higher than in australopithecines in whom it was about 11%, but still quite a bit lower than in sapiens (Anton, 2003).

Erectus was large and had a large brain, two factors that require additional energy. On top of that, the energy consumption of the brain increases faster as it gets bigger than it does for the body (by a factor of 0.86 for brain, versus 0.75 the body) (Fonseca-Azevedo & Herculano-Houzel, 2012). However, it is the number of neurons that determines this, because they consume glucose to create signaling molecules (neurotransmitters) and transmit them to each other at the level of the synapses (Herculano-Houzel, 2016). A neuron at rest already consumes 13% of its total energy consumption, discharging neurotransmitters at synapses costs 80% of its energy consumption. On top of the 86 billion neurons, the same number of glial cells need to be energized. If we may extend this to erectus, it had to provide energy for 41–80 billion neurons, as well as for the larger body that traveled great distances and had to be additionally cooled. It is obvious that compromises had to be made, because the energy supply that can be provided in a day is also limited (Fonseca-Azevedo & Herculano-Houzel, 2012). Indeed, primates cannot spend more than 8 hours a day gathering plant food, which puts a limit on the number of maintainable brain-body combinations. This is not a problem for a chimpanzee, which is small enough in combination with its relatively small brain. A large-bodied orangutan but with a relatively small brain is just on the edge. Gorillas are above the limit, but they also use a little less energy because they are less active. Lucy, with about 35 billion neurons and 39 kg of body weight was safely below that limit. But sapiens is a paradox: the body is just way too big to sustain along with that big brain. And yet it apparently succeeded, and erectus was the first to do so. If we look at early Dmanisi erectus, they still had a combination that did not required special effort to get enough energy. In late erectus, with a larger body and larger brain, this is no longer the case. We already know the evolutionary solution to this: more energetic food and heating of food. Plant material and meat gathered and warmed within the 8 hours provides much more energy than if it were only raw plant material. Humans can therefore take up less food mass than what you would expect for their body mass (Simmen et al., 2017). Moreover, the fact that this allowed the jaw apparatus and digestive tract to shrink in erectus (as derived from the shape of the rib cage and pelvic girdle) brought an additional energy gain: less energy had to be invested in expensive organs (Aiello & Wheeler, 1995; Allen & Kay, 2011; Anton, 2003). For example, primates with larger brains also appear to have relatively fewer muscle fibers that also use glucose for their energy production (Muchlinski et al., 2018).

NUMBER OF NEURONS (billion)

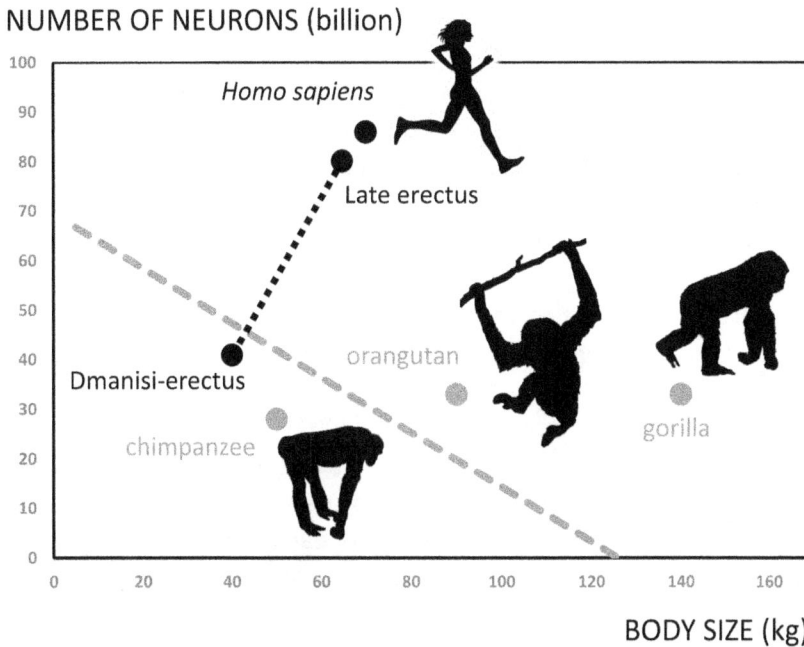

Figure 6.11 The amount of energy available to maintain a body and brain is determined by the amount of plant food that primates can gather. The sloping line represents the limit at which, for all combinations below the line, sufficient food (and thus energy) can be gathered in a day. For all combinations above the line, this is not feasible unless activity is restricted (as in the gorilla) or more energy is obtained from food (late erectus and sapiens).

Thus, the diet shift of erectus was a tipping point for the evolution of our brains, for without it our brains might never have grown as large. Moreover, other adaptations arose that improved the energy supply to the growing brain. For example, humans are very good at accumulating fat, more so than other primates (Lieberman, 2013). Fat is made from fat molecules and carbohydrates, absorbed through food or produced in the liver or adipose tissue. Fat is a very interesting material to accumulate energy: 1 g of fat provides 9.3 kcal, compared to 4.1 kcal for 1 g of carbohydrate or 4.3 kcal for 1 g of protein (Sherwood et al., 2005). Coincidence that a monkey baby accumulates only 3% of body fat at birth, compared to 15% in a sapiens baby? Probably not. In the last trimester of pregnancy, a baby accumulates a hundred times more fat (fat formation occurs only after birth in other primates) (Kuzawa, 1998). This fat helps explain how it is that our brains exhibit such a growth spurt during the first year of life. And even still, this limits the availability of energy for the rest of the body, because our body has a stunted growth shortly after birth, at the time when the brain grows the fastest (Kuzawa et al., 2014).

Conclusion is that the increased energy consumption makes it evolutionarily not so obvious that brains could have become so large. An additional problem is that this involves the production of additional metabolic heat and brains are very sensitive to an increase in body heat (an increase of 4°C to 5°C can already disrupt brain functions) (Falk, 1990). Not a good combination, is it? Unless the brain gets extra cooling! Artiodactyl mammals living in savannahs have a special network of blood vessels around the nasal cavity that can provide extra cooling to the blood going to the brain (Jessen, 1998; Wheeler, 1994). This allows them to keep the brain temperature 2.7°C below that of the body. Primates cannot. Cooling of the brain can only be done through the cooling of the body. So again, the specialized skin with its numerous sweat glands is also a crucial adaptation for this, which prevented the large brains from constantly overheating. Limiting the supply of warm blood to the brain was not an option because that would undermine the necessary supply of glucose. Yet we have a few perks that turn our head into a refrigerator (Irmak et al., 2004). Through special blood vessels running straight through the skull bones and cavities running around the brain, a fluid-filled space around the brain (the subarachnoid space) is cooled (Bruner et al., 2011; Falk, 1990). From erectus onward, more holes appear in the skull through which those blood vessels ran. The supplying blood is then cooled through the forehead (which contains a lot of sweat glands), nose and scalp. Even yawning would help cool the brain (Gallup, 2022). The ratio of brain mass to brain surface area also determines how easily they can be kept at a proper temperature, just as it did for the skin. This may explain why the brains of Inuit are on average larger (larger than 1450 cm³) than those of Aborigines (smaller than 1300 cm³), versus that of an average male European (about 1450 cm³). I don't think I need to stress here that this says nothing at all about intelligence?

## Why tools?

Larger and more complex brains are advantageous to more efficiently handle everyday challenges. An unstable and withering climate was already a major challenge for erectus (Carrión et al., 2011; Cueva-Temprana et al., 2022). That better tools became very important just during this period is therefore no coincidence (Hrvoj-Mihic et al., 2013). The relatively simple Oldowan technology made its appearance in Africa 2.6 million years ago and spread to Eurasia (Braun et al., 2019; Carrión et al., 2011; de la Torre, 2011; Semaw et al., 1997). Indeed, 1.6-million-year-old Oldowan tools are even known from Italy (Arzarello & Peretto, 2010; Arzarello et al., 2016). A particular tool technology is thus not necessarily unique to a single human species. At least three and possibly four species of hominins used Oldowan tools: an *Australopithecus*, a *Paranthropus* and two *Homo* (habilis and erectus) (de Heinzelin et al., 1999; Plummer et al., 2023; Semaw et al., 2020). But 1.7 million years ago, African erectus introduced a new Acheulean technology, with the typical teardrop-shaped fist axes that were sharp along two sides (hence called 'bifacies') (de la Torre, 2016). As early as 1.5 million years ago they were used in Israel and about 500,000 years ago they reached Europe (as in Happisburgh, present-day United Kingdom) (Carrión et al., 2011; Martinon-Torres et al., 2011). The early Dmanisi-erectus still combined Oldowan and Acheulean technology, but a later and more sophisticated Acheulean methodology still proved very successful. It is the technology that lasted longest, until about 100,000 years ago, the one that was used by all *Homo* species (except sapiens) and that was present in Africa, Europe and Asia (it survived longest in Europe) (de la Torre, 2016; Key et al., 2021). It took until the discovery of a whole new technique that allowed for the creation of long and thin stone blades, the Levallois technique of Neanderthals and sapiens (Key et al., 2021). What was first a functional hand axe later became the core to knock off finer cutting tools (Rosenberg-Yefet et al., 2021).

Making and using sophisticated tools requires fine motor skills and great dexterity. Knocking off stones and cracking nuts or bones also involves increased strain in the hands, especially at the level of the phalanges (Williams-Hatala et al., 2018). We already possess unique adaptations that have promoted precision grasping of objects (Marzke, 1997). Even the fact that 90% of us are right-handed is associated with performing complex hand movements. Bonobos also more frequently use one hand, the right hand, when they need to perform complex actions (Bardo et al., 2015). Erectus was presumably right-handed as well, as evidenced by the direction of scratches on their teeth (Xing et al., 2017). There is even a link with language: the right hand is controlled by the left side of the brain, with the left temporal lobe that has become highly specialized in the evolution of language. There is also a link with walking upright. Several primates are more likely to use the left hand to grasp something when walking on all fours but the right hand when upright (Westergaard et al., 1998). Some monkeys, moreover, use tools, such as stones to crack nuts, more when moving on the ground than when swinging in the trees (Falotico & Ottoni, 2023).

Getting access to food in an environment that became unstable and dry required increasingly complex actions (Antón et al., 2014; Potts, 2012). Other primates and even birds also use tools, such as sticks, to obtain hard-to-reach food (Byrne, 2001; Chappell & Kacelnik, 2002; Parker, 2015; Tanaka et al., 2007; Whiten et al., 1999). Roughly worked stones, like Oldowan's, are good for cracking hard structures like nuts or bones, but the sharp Acheulean stones were better for removing meat from a cadaver (Bunn, 1999; Kuhl et al., 2016; Roura, 1999; Toth & Schick, 2019). Being able to cut, slice or grind hard and hard-to-chew pieces thanks to tools saves time and energy because less chewing is required (Gowlett, 2016). Stones and sticks are also useful for keeping an attacker at bay or killing prey from a distance, although the first tools were not weapons (Lovejoy, 1981). Erectus was not limited to stone materials to make tools. A 1.4-million-year-old hand axe from Ethiopia was made from the femur of a hippopotamus (Sano et al., 2020). Erectus also knew that bone was easier to work than stone, as evidenced by an 800,000-year-old serrated tip (Pante et al., 2020). But these are rather exceptions, bone and other materials were only fully exploited from Neanderthals and sapiens onward. So, does the erectus relied on a little innovative technology for a very long time, perhaps a sign of limited creativity and intelligence? Is this perhaps partly underlying their eventual extinction (Shipton et al., 2018)?

# So ...

Erectus was clearly a game changer in our evolution. A large human, fully erect and able to walk very efficiently on two legs, as well as run for long periods of time. The erect and functionally naked body, full of sweat glands, and for the first time also a protruding nose, provided the life-saving solution to being able to do all this in a hot and open grass savannah. Running behind large grazers and using better cutting stone tools gave access to a new source of energy: meat. It provided energy much needed to maintain the large body, as well as that large and energy-consuming brain. Meat was an energy supply that provided stability during a time of instability and desiccation, but also during a time when the world was being explored. Erectus was the first to leave Africa and made it from Eastern Europe to the Far East, where it left its mark in several places, albeit in the form of stone tools, fossils but also small human species, such as in Flores and the Philippines. An important imprint was also left in Africa: our direct ancestor and our cousins.

# 7

# Conquer the world

The scene was set with erectus, a large and big-brained hominin that very efficiently walked and ran upright and had already explored half of the world. And yet this was not the end of our evolutionary story. We are different from erectus, as are some other new human species that appear on the scene, human species with whom we even shared Eurasia for a long time, with whom we faced tough climatic challenges, with whom love was sometimes shared, but of whom we were the only ones to survive. We are dealing here with a fourth group of hominins, with even larger brains than erectus and with a delayed growth pattern characterized by a longer childhood (Collard, 2003). This group shows a remarkable acceleration in cultural evolution, which has taken over the pace of anatomical evolution: from hunter-gatherers to astronauts, from a few wandering individuals to cities with more than 30 million inhabitants. The trail to these hominins starts about a million years ago.

## The sapiens track

We have been given the species name 'sapiens' or 'wise', but it is a fallacy to think that we were the only ones with the necessary intelligence to find creative answers to everyday challenges. Of course, what we accomplished in the last few centuries should not be compared with what human species were capable of several hundred thousand years ago. About 300,000 years ago, several human species discovered new methods of making better tools than the Acheulean ones of erectus. It is also a fallacy to think that evolution leading to such a complex intelligence and culture proceeded in a straight line from erectus to sapiens. Rather, it was a line that split somewhere between a million and 600,000 years ago in Africa. One line's scenario started early in Eurasia, resulting in Neanderthals and Denisova humans. The other scenario lingered in Africa for a while and produced the species that went on to explore the entire world ánd change it, to remain as the evolutionary autocrat among the hominins: sapiens. This story starts with our common 'antecessor'.

### The track splits

For a long time, the key question was about who the last common ancestor was of all sapiens currently inhabiting the world. Erectus was a candidate, because scattered in Africa and Eurasia, it could give rise to the sapiens there. This view is called the multiregional model, implying that the last common ancestor must have lived

DOI: 10.1201/9781003588795-8

| | | PALEOLITHIC | | | | | |
|---|---|---|---|---|---|---|---|
| 23000 | 5300 | 3300 | 2580 | 1800 | | 770 | 129 11 thousand years ago |
| MIOCENE | PLIOCENE | PLEISTOCENE | | EARLY | | MIDDLE LATE | HOLOCENE |

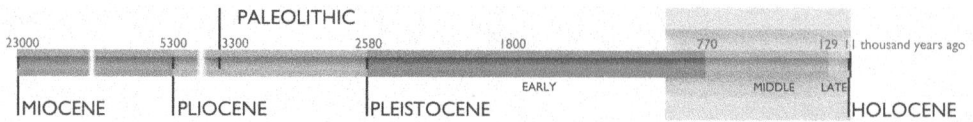

Figure 7.1   Time line showing the period dealt with in this chapter.

more than a million years ago (Thorne & Wolpoff, 1992). That anatomical differences between Chinese sapiens and erectus are different from those between Southeast Asian sapiens and erectus lies at the base of this idea (Chinese representatives have a less robust skull, smaller teeth, a more rounded forehead and a flatter nose). Should this be correct, then all human species since erectus should be called *Homo sapiens* (Stringer, 1996). In the 1980s, the first genetic analyses seemed to support another view, the 'Out of Africa' model, which posits that all modern sapiens originated from a common ancestor who still lived in Africa, long after erectus had moved into Eurasia. This new ancestor would then have eventually replaced all other species in Eurasia. They deduced this from mitochondrial DNA (the DNA outside the cell nucleus, present in the energy factories found in almost all our cells), which is genetic information passed from mother to daughter (Cann et al., 1987). Two important conclusions followed from this: the last common sapiens ancestor was significantly more recent than erectus, and the ancestor of all sapiens worldwide had an African origin. This was the so-called Eve hypothesis. Which species qualified, however, remained a big question mark. Because mitochondrial DNA was inherited only through the maternal line, it can cause that inheritance of genes is particularly sensitive to accidental changes in that DNA. This cast some doubt on this study (Thorne & Wolpoff, 1992). Yet the same scenario was later confirmed by genetic information inherited only through the paternal lineage (genes located on the Y chromosome) and also through other genetic information (Hammer et al., 1998; Maddison, 1991; Templeton, 2002). The vast amount of genetic information now available, as well as fossil finds from the earliest sapiens and early inhabitants of Europe, all confirm the 'Out of Africa' scenario (Stringer, 2003; Templeton, 2002). It is also clear by now that the last common ancestor of all modern sapiens was already a sapiens, originating in Africa more than 300,000 years ago.

But then who was the last common ancestor of all those other human species that roamed Eurasia before sapiens, such as the Neanderthals? And what was the relationship to the owners of some older fossils, who were clearly not yet sapiens or Neanderthals but also no longer erectus and which researchers lumped together for convenience: the 'archaic sapiens' (Ahern, 2015). Numerous books and articles have put forward multiple hypotheses. Here, in a very small nutshell, I try to summarize the scenario that seems to be gaining popularity among paleoanthropologists these days.

Everything started about a million and a half years ago with an African species derived from the African erectus: *Homo antecessor*. A species that made it to Spain via North Africa and Gibraltar, as evidenced by more than 800,000-year-old fossils found in the Gran Dolina and Sima del Elefante caves in Atapuerca (Bermúdez de Castro & Martinon-Torres, 2019; Hertler et al., 2013; Ribot Trafi et al., 2020). That area, south of the Pyrenees, was a climatically favorable and stable region during the period when ice ages manifested themselves in the rest of Europe, where human species could not penetrate until 600,000–500,000 years ago (periods between two ice ages then lasted longer) (Hertler et al., 2013; MacDonald et al., 2012; Roebroeks, 2006). Some recent findings of 1.4 million old stone tools found in Ukraine, however, also propose a scenario that this antecessor may simply be erectus that migrated from east to west prior to when this European cold barrier got established (Garba et al., 2024). Could the east to west series of fossil evidence, with 1.8 million years old Dmanisi fossils in Eastern Europe, 1.4 million years old Ukraine fossils, 1.2 million years old Italian fossils (from Vallonet) to the 1.1 million years old West European Atapuerca fossils reflect a West European dispersal of erectus?

This initially African species might have been the ancestor of two major evolutionary tracks: an African track from which (via *Homo rhodesiensis*) the African sapiens ancestor arose and a Eurasian track from which (via *Homo heidelbergensis*) the Neanderthals and Denisova humans arose (the recently described 'dragon skull', which some believe represents a distinct human species, *Homo longi*, is therefore located in that Eurasian lineage) (Athreya & Hopkins, 2021; Bermúdez de Castro & Martinon-Torres, 2020; Mounier et al., 2009; Ni et al., 2021; Stringer, 2021). Early Neanderthals from Atapuerca, found in the cave of Sima de los Huesos, incidentally show similarities to this antecessor (Arsuaga et al., 1997; Meyer et al., 2016). The rhodesiensis (known since 1921 from the 300,000-year-old Kabwe fossil from Zambia) and heidelbergensis (known since 1908 based on the 609,000-year-old Mauer jaw and footprints discovered in 2023 from Germany) are, by the way, two representatives of those archaic sapiens (Altamura et al., 2023; Athreya & Hopkins, 2021; Balzeau et al., 2017; Grün et al., 2020; Schoetensack, 1908; Wagner et al., 2010).

## From pinky finger to a human species

The unraveling of the Eurasian trail came about mainly thanks to genetic information that provided rather surprising insights into the affinities between species and even the existence of new species. The proverbial jaws dropped when in 2008 Svante Pääbo's team managed to reconstruct the genome of a Neanderthal using newly developed techniques, and two years later applied the same technique to a phalanx of a little finger found in Denisova, in Russia (Green et al., 2008; Krause et al., 2010). Svante, by the way, was awarded the Nobel Prize for his genetic research on human species in 2022. Why the jaws dropped at seeing the genome of Neanderthals, that's for later, first to that pinky bone.

Discovered in 2008 in a cave in Denisova, in the Altai Mountains of southern Siberia, Svante Pääbo and his team had the opportunity to remove 30 mg of bone tissue and apply their special technique to it (Pääbo, 2014). Genetic information from the DNA

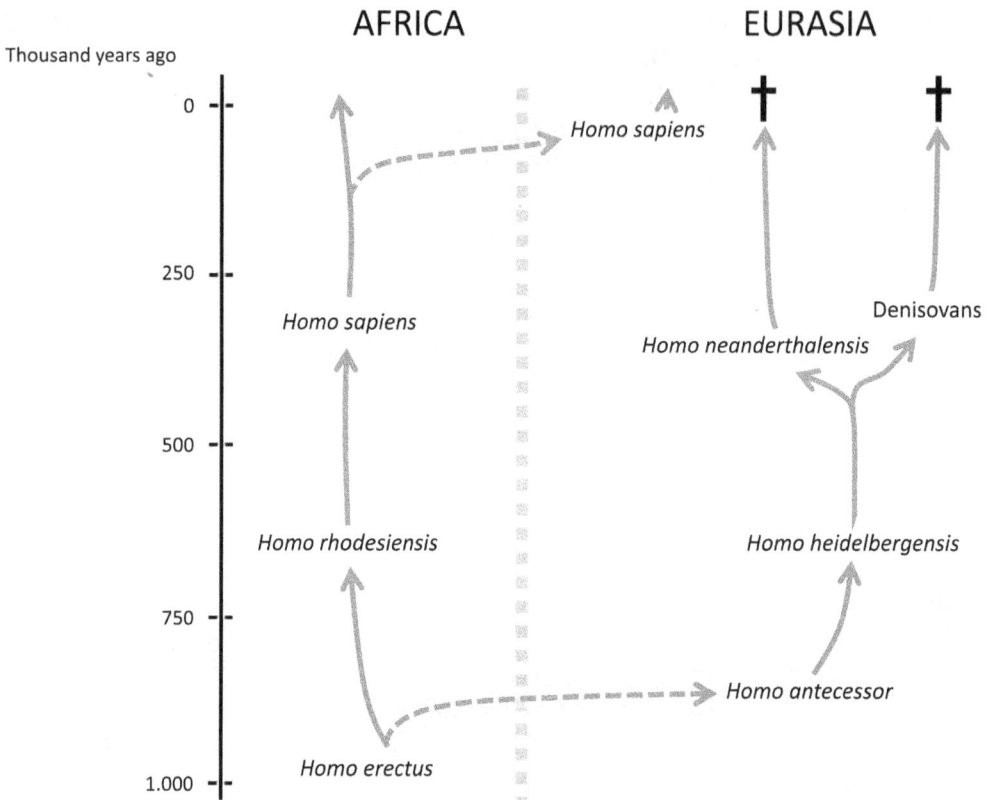

Figure 7.2   Two evolutionary tracks that emerged of the erectus ancestor, a hypothesis that reflects how Neanderthals, Denisova humans and sapiens arose, dispersed and some of which disappeared. Full arrows reflect evolution of species, dashed arrows show dispersal of a species from Africa to Eurasia.

(both mitochondrial DNA and DNA from the cell nucleus) resulted in two seminal publications in 2010 (Krause et al., 2010; Reich et al., 2010). By comparing this DNA with that of a Neanderthal and sapiens, the little finger bone was found to be genetically more different from sapiens than Neanderthals was. DNA from teeth confirmed this. The conclusion was: the little finger bone came from a woman, from a hitherto unknown human species, which had a common ancestor with Neanderthals and sapiens about 1 million years ago. So, for the first time a new human species was discovered based on what the DNA said. The woman lived about 50,000–30,000 years ago in the Altai region, at a time when both Neanderthals and sapiens were also present in that area (although Neanderthals may have been gone by then) (Douka et al., 2019). Moreover, the scientific jaws dropped once more when it was found that modern inhabitants of Melanesia (Pacific island group), aborigines from Australia and islanders from Southeast Asia have up to nearly 5% DNA derived from this new human species (Meyer et al., 2012; Reich et al., 2010). So Denisova humans must have crossed with sapiens, just as they crossed with Neanderthals (Bergström et al., 2021; Hubisz et al., 2020)! DNA from a bone fragment from the same Denisova cave showed that it came from a child, called 'Denny' whose mother was a Neanderthal and father a Denisova human (Slon et al., 2018; Wragg Sykes, 2020). It goes even further, as 1% of the DNA of Denisova humans appears to have come from yet another unknown and older human species (sometimes called 'super-archaic humans') (Hubisz et al., 2020; Prüfer et al., 2017; Rogers et al., 2020; Teixeira et al., 2021). And it gets even stronger: Denisova humans also occurred 200,000–100,000 years ago in Sulawesi and Laos (Southeast Asia), where they are believed to have also crossed with island-dwelling human species, such as the Flores and Luzon humans (Demeter et al., 2022; Teixeira et al., 2021). More on crossbreeding later, when talking about Neanderthals.

Everything indicates that we are dealing with a new human species, and yet people keep referring to it as 'Denisova humans' or 'Denisovans'. Initially, Svante Pääbo's team had planned to christen this new species *Homo altaiensis*, but eventually refrained from doing so (Pääbo, 2014). Meanwhile, names such as *Homo denisovensis* or *Homo denisensis* have already been suggested (Harvati & Reyes-Centeno, 2022). Because of all the crossing with multiple species and the limited fossil material (a pinky bone and a few teeth), it was not so clear to whom this species would be most closely related to and in which it would differ anatomically from them. Sometimes sapiens was considered the closest relative, sometimes the Neanderthals (Meyer et al., 2012). Hence, people mostly simply refer to them as 'Denisova humans'. The latter scenario is now widely accepted: Denisova humans are most closely related to Neanderthals, who split off from each other about 400,000 years ago (Brown et al., 2021; Prüfer et al., 2017). Their common ancestor then split off from the sapiens lineage about 800,000–600,000 years ago (Prüfer et al., 2017; Reich et al., 2010).

But what did Denisova people look like? The fossils provide little useful information. Other than a lower jaw from Tibet and the dragon skull from China (which would have come from a Denisova man rather than a separate species *H. longi*), little can be deduced from the available skeletal material (Chen et al., 2019; Gibbons, 2021b; Ji et al., 2021; Ni et al., 2021; Wragg Sykes, 2020). Genetic traits suggested they had a dark skin, brown hair and brown eyes (Meyer et al., 2012). In 2019, David Gokham and collaborators came up with a new technique that allowed to highly reliably predict traits of the skeletal anatomy based on DNA methylation. In fact, the expression of genes is constantly adjusted by this type of methylation (in which small molecules, methyl groups, are temporarily attached to the DNA, allowing genes to be expressed or not). In modern humans, many methylation patterns are known to be associated with specific skeletal features. For example, using this technique, they were able to predict with 85% certainty how a chimpanzee's skull would differ from that of a Neanderthal, based on the differences in DNA methylation (Gokhman et al., 2019). The same technique was then applied to the DNA of Denisova humans, which showed, for example, that they had a long face and wide pelvis, just like Neanderthals.

All in all, Denisova humans are known to have split off from Neanderthals 400,000 years ago, with fossil evidence from between 200,000 and 45,000 years ago (Brown et al., 2021; Dennell, 2019). The genetic information may explain why they disappeared, about the same time as when Neanderthals disappeared. The Denisova population showed remarkably limited genetic variation, which made them vulnerable (Meyer et al., 2012). Less genetic variation generally means there is less buffer to adapt as a population to a changing environment. It also showed that their population experienced a drastic decline between 250,000 and 125,000 years ago, at a time when the sapiens population was just increasing.

## Our closest friend

We do not know much about Denisova humans, which cannot be said about Neanderthals. Before the discovery of the Denisovans, it was believed that Neanderthals were our closest relative, even long thought to be our direct ancestor. New fossil finds and speculations have passed the review, some very exotic even. Everything starts with the mining of the chalk cliffs at Neanderthal in Germany, an area named after the poet and composer, Joachim Neander (Jordan, 1999; Wragg Sykes, 2020). The chalk rock was accessible for mining because it had been carved out over the years by the Düsselbach River, leading miners to find fossils in the Feldhof cave in 1856: a skull roof, a piece of pelvis and some long bones. Speculation about the origins could begin: from bears to humans from the period of the Biblical Flood. Even as recently as 1998, a book appeared in which Neanderthals were considered flood survivors (Cameron & Groves, 2004). I am going to skip this view, if you allow me. Already more acceptable misconceptions were the idea that they were skulls of deformed humans, as the eyebrow arches were indeed remarkably larger than those in modern human skulls. Diseases, such as rickettsia, were also put forward as explanations. The skull could also have come from a Russian cavalry soldier who had fallen into the swamp during a battle against Napoleon's army. The odd skull would then have been the result of some hard blows to the head during the fight, and the crooked bones could have been due to having sat on a horse for too long. Still, others were convinced that this was a prehistoric man, a species other than sapiens. Hermann Schaafhausen was already convinced of this in 1856, having studied the fossils carefully (Huxley, 1863). It was not until 1864, after similar skulls were also found in Gibraltar (in the Forbes mine) in 1848, that a certain William King decided to give it a new species name: *Homo neanderthalensis* (King, 1864). In retrospect, the Feldhof fossils turned out not to be the first Neanderthal fossils discovered: in 1829, Philippe-Charles Schmerling had found a child's skull at Engis, on the Meuse Riverbanks in Belgium, and by 1673, Neanderthal tools had already been found in London (which were first thought to have been left behind by Celtic warriors from Roman times). Iconic fossils include the Neanderthals from Spy (Belgium, discovered in 1887), 'Le Moustier 1', a Neanderthal child from Moustier (France, discovered in 1907), nearly complete Neanderthal skeletons from La Ferrassie and La Chapelle-aux-Saints

(France, discovered in 1908 and 1909), and the severely injured and ritually buried 'Shanidar 1' from Iraq (found in 1956) (Condemi & Savatier, 2019; Frayer et al., 2020; Gomez-Olivencia, Quam, et al., 2018; Wragg Sykes, 2020).

Since then, we must significantly adjust our image of Neanderthal prehistoric man. Don't just say 'Neanderthal' to any Neanderthal. There are the 450,000-year-old 'pre-Neanderthals' from the cave at Sima de los Huesos in Atapuerca, which have anatomical and genetic similarities to the older antecessor (Arsuaga et al., 1997; MacDonald et al., 2012; Meyer et al., 2016; Pagano et al., 2022). Neanderthals from 250 till about 100,000 years ago are then considered as 'early Neanderthals', followed by the 'late Neanderthals' (Richards et al., 2024). The Neanderthal evolutionary line, having survived four ice ages, split about 170,000 years ago into the Altai Neanderthals and the other, which split about 85,000 years ago into Croatian Vindija and Spanish El Sidron Neanderthals (Condemi & Savatier, 2019; Skov et al., 2020). The last survivors are also known from Western Europe, such as the approximately 28,000-year-old Neanderthals from the Gorham Cave in Gibraltar (Finlayson et al., 2006; Rey-Rodriguez et al., 2016). Neanderthals experienced multiple waves of migration between west and east and had spread over an area from Wales to the Chinese border with the Arabian deserts (Stringer, 2011; Wragg Sykes, 2020). The Altai Mountains were populated by Altai Neanderthals for about 190,000 years and by Vindija Neanderthals 60,000 years ago (Douka et al., 2019; Kolobova et al., 2020). Although anatomically quite convergent across these lineages, their genetic diversity turns out to be larger than initially thought (Fuchs et al., 2024). Ice ages mostly controlled their migrations, with caves in the Mediterranean being a refuge during the coldest periods (Fu et al., 2015). There were contacts between different Neanderthal lineages, but equally with Denisova humans and sapiens during that period. In Europe, the disappearance of Neanderthals lags about 3000–5000 years behind the appearance of sapiens (Bard et al., 2020; Prüfer et al., 2021).

That Neanderthals occasionally ran into sapiens is therefore not surprising. Several caves contain remains of both species, from the Goyet Cave in Belgium to the Denisova Cave in Russia. Dating of the remains left behind shows that they inhabited the caves at other times, such as in the Mandrin Cave in France where the earliest occupation of sapiens in Western Europe is evidenced (57,000–52,000 years ago) (Slimak et al., 2022). In 1997 they also found genetic evidence that Neanderthals could not be our direct ancestors (Krings et al., 1997; Scholz et al., 2000). At that time there was no mention at all of possible amorous encounters between the two, because the mitochondrial DNA of sapiens showed no traces of Neanderthal DNA. That changed when they were able to analyze nuclear DNA from well-preserved skeletal material of Neanderthals from Vindija and Altai (Pääbo, 2014). With $6 million in research funding, everything gained momentum, with the unraveling of the complete mitochondrial genome in 2008, a piece of the nuclear DNA in 2010 and the complete genome of several individuals as of 2014 (Green et al., 2010; Green et al., 2008; Prüfer et al.,

2017, 2014). Although already suspected, in 2014 came the earth-shattering news that there was now incontrovertible evidence that there were gene flows between Neanderthals, sapiens and Denisova humans and even from that super-archaic human species. Not surprising as Neanderthals and sapiens are 99.84% genetically identical to each other (by comparison, sapiens is 98.63% genetically similar to chimpanzees) (Rolian, 2014). Modern sapiens thus carries genes from multiple human species (just as late Neanderthals have sapiens genes) and that is not the same for every sapiens (I call it 'genes' here for now, but actually it is about alleles, i.e. variants of specific genes) (Harvati & Ackermann, 2022). Multiple hybridization events may have taken place over different time periods: a first one 250,000–200,000 years ago, a second one between 120,000 and 100,000 years ago and a third one between 60,000 and 50,000 years ago. The place of interbreeding during this second wave has been pin-pointed to have taken place in the Zagros mountain region in what is currently Iran (Guran et al., 2024).

Current African sapiens only carry 0.5% Neanderthal DNA, in Eurasian sapiens it is 1–4% (some limit it to 1–2%) (Iasi et al., 2024) (Tagore & Akey, 2024). In addition, inhabitants of the Far East (Melanesia to Papua New Guinea) also carry up to 5% Denisova DNA (Chen et al., 2020). Thus, their genome encloses DNA from three human species! Importantly, we are not talking about complete crossbreeding here, because ultimately, we only carry a very small amount of Neanderthal DNA. On the one hand, the probability of crossing was very low (one estimates an average of one cross within a couple once every 12 generations per 10,000 individuals), on the other hand, there is a minimum of three introgressions (Currat & Excoffier, 2004; Fu et al., 2014; Kuhlwilm et al., 2016; Vernot et al., 2016). This implies that a child born of a Neanderthal and a sapiens later reproduced with one of the parent species and not with another hybrid. That the sapiens Y chromosome, present only in males, shows no Neanderthal genes suggests that successful crosses occurred primarily between a sapiens male and Neanderthal female (although it cannot be ruled out that the genes simply disappeared) (Mendez et al., 2016; Neves & Serva, 2012; Petr et al., 2020). The reverse seems to have been genetic at least less evident (Mendez et al., 2016; Rogers, 2015; Sankararaman et al., 2014). Children were then integrated into the sapiens community, and once adults, they reproduced with a sapiens partner. In this way, Neanderthal DNA got further diluted generation after generation. And yet we incorporated more of that DNA than what one would expect according to this scenario. All indications are that several Neanderthal genes conferred an advantage to sapiens and as a result were positively selected. Some of those genes still offer us an advantage today, others are rather harmful. About 20% of sapiens today carries a Neanderthal variant of a progesterone receptor gene that has been associated with having more children, less miscarriages and less bleeding during early pregnancy (Zeberg et al., 2020). There are even Neanderthal genes that better protect us from COVID-19, while others rather make us more vulnerable (Luo, 2020; Zeberg & Paabo, 2020; Zhou et al.,

2021). Put all the Neanderthal DNA of all modern sapiens together, and you arrive at 30%–70% of the entire Neanderthal genome (Lowery et al., 2013; Vernot & Akey, 2014). So, our closest friend is still somewhat amongst us.

The Neanderthal genome also brought us many other insights, including why they eventually became extinct. This is because Neanderthals lived in very small groups, called clans, with about 35 individuals per clan (Condemi & Savatier, 2019). This is a typical group size for hunter-gatherers, who additionally lived at very long distances from each other. Neanderthals, like Denisova humans, were also genetically very little diverse: their clan consisted mainly of related males, their female partners and their children (Orlando et al., 2006). If there was genetic exchange, it was mainly through women joining other clans (Vermassen, 2020). By 100,000 years ago, genetic diversity was very low, and the total population size was estimated at 70,000 individuals (Bocquet-Appel & Degioanni, 2013). Over their 5000 square kilometer range, that equates to about five Neanderthals in a 70 km by 70 km area. And that was during the favorable interglacial times. During the tough ice ages this was much lower, about one clan on an area like Belgium and Luxembourg combined (Briggs et al., 2009). All this, combined with early sexual maturation, lowered fertility, increased inbreeding, a diet heavily dependent on meat and food competition with sapiens, a rapidly changing climate or exposure to diseases brought by African sapiens, may have caused the Neanderthal population to reach a critical level where population size and genetic diversity simply became insufficient for continued survival (McGrath et al., 2021; Prüfer et al., 2014; Ramirez Rozzi & Bermudez De Castro, 2004; Timmermann, 2020; Weaver & Hublin, 2009; Wißing et al., 2019). From then on, sapiens could become autocratic among hominins.

## The wise one

About 800,000 years back in time, to the start of our African track, when it split off from the Eurasian track. There are no fossils of an African antecessor (sometimes a fossil from Mauritania is associated with it), but there are fossils of rhodesiensis that gave rise to African sapiens (Mounier et al., 2009). Fossil evidence of the start of sapiens comes from Morocco's Jebel Irhoud, with 315,000-year-old fragments of skulls, mandibles, an upper arm and a pelvis (Hublin et al., 2017). Although the first fossils were already known from 1961, it took until 2017 to prove that these were very primitive sapiens and over 300,000 years old. Already known where the 230,000–160,000-year-old fossils from Herto and Omo Kibish in Ethiopia and a 260,000-year-old skull from Florisbad in South Africa (Grün et al., 1996; McDougall et al., 2005; Scerri et al., 2018; Vidal et al., 2022; White et al., 2003). The fossils from Morocco were surprising, not just because of their age. The skull did resemble that of early sapiens, yet some features were missing that were always argued to be unique to *H. sapiens* (Gosling et al., 2022; Hublin et al., 2017). For example, a bony chin was not yet present, the bony eyebrow arch still largely extended above the eye, the face still protruded slightly forward and the brain case was still elongated (rather than spherical). Some think this was another species (*Homo helmei*), others consider this to be an early sapiens (Endicott et al., 2010). Either way, about 300,000 years ago multiple *Homo* species lived together in Africa (sapiens, naledi and rhodesiensis). This is also the period when Africa experienced drastic changes in climate and thus vegetation, with East Africa becoming more humid again after a long period of drought and slowly transitioning back from a grass savannah to a forest savannah (Gosling et al., 2022).

Sapiens was the only one able to cope with these changes, which may explain why they were managed to spread very quickly throughout Africa, but also that they could easily leave Africa to explore the rest of the world (Timmermann et al., 2022). Meanwhile, we know that the oldest fossil sites do not reflect the original distribution of sapiens in Africa. As early as 300,000 years ago, sapiens had a population structure with a southern, a western and an eastern branch (Ragsdale et al., 2023; Scerri et al., 2018). From a southern tribe, genetically strongly represented by today's Khoi San population, the other two branches split off from it about 135,000–120,000 years ago. From the eastern branch, 125,000–80,000 years ago, a group of sapiens left to explore the Middle East (Armitage et al., 2011; Gibbons, 2021c; Lawler, 2011). Soon after, this already led to a first amorous encounter with Neanderthals. From the Middle East gateway, the world became explored in several waves. An early dispersion reached Australia 75,000–62,000 years ago, giving rise to the Aborigines (Freidline et al., 2023; O'Connell et al., 2018; Rasmussen et al., 2011). In a later dispersal, about 38,000 years ago (possibly even earlier), sapiens reached China and gave rise to the Han population (Liu et al., 2015; Michel et al., 2016; Rasmussen et al., 2011). From this population, several groups of sapiens continued northward about 35,000 years ago, crossing the Bering Strait (a land bridge between Siberia and North America) before it was completely closed by an ice sheet that formed during the 'Last Glacial Maximum' (an ice age between about 33,000 and 14,000 years ago) (Bennett et al., 2021; Raff, 2021). Sapiens already made it to the current New Mexico by 23,000 years ago, as evidenced by footprints in White Sands National Park.

From the Middle Eastern and Chinese populations, sapiens experienced multiple waves of migration into Europe, with some of the oldest evidence (45,000 years old) left in the Bulgarian Bacho Kiro cave, remains of a sapiens woman from Zlatý kůň in the Czech Republic, remnants of a man from Ust'-Ishim in Western Siberia and evidence of human presence at the Ilsenhöle site in Ranis, Germany (Fu et al., 2014; Hajdinjak et al., 2021; Mylopotamitaki et al., 2024; Prüfer et al., 2021). Genetic evidence show their origin to be deeply rooted within an early split from African sapiens. Many of those early sapiens, by the way, carried a lot of Neanderthal DNA, especially from during the third wave of interbreeding (Fu et al., 2015; Sümer et al., 2024; Trinkaus et al., 2003). However, these were not the first sapiens in Europe, as to everyone's surprise, 210,000-year-old skulls have been found in Apidima, Greece (Harvati et al., 2019). Also, in Misliya, Israel, fossils were found that may be about 185,000 years old

(Hershkovitz et al., 2018). These point to another very early dispersal from Africa. For sapiens, the period between 43,000 and 41,000 years ago was accompanied by a favorable climate, which explains why they made it as far as Western Europe by then (and maybe even earlier as 57,000 years ago) (Mellars, 2006; Slimak et al., 2022). What may also be surprising is that these are not the ancestors of today's Europeans. There is little to none of their DNA that can be traced to Europeans today (an exception being the inhabitants of Sardinia). The ancestors came with later waves of dispersal, one 12,000 years ago with the Natufians from the fertile crescent in the Middle East and the Anatolian farmers from present-day Turkey, and one 8000 years ago with the sapiens from the Pontic steppe (north of the Caspian Sea) (Krause & Trappe, 2020). This is the phase in our evolution when European hunter-gatherers were replaced by farmers and cattle raisers. A huge switch in how our evolution continued to be driven, largely by ourselves.

# From hunter to farmer to health issues

The hunter-gatherer behavior of erectus continued for quite some time but was greatly influenced by prevailing climatic conditions. Neanderthals and Denisova humans were primarily reliant on meat at times when they faced ice ages. In between ice ages, there was a wider range of animal and plant foods to rely on. Sapiens, who stayed longer in Africa, were less affected. Africa did cool a bit and did get drier during the ice ages, but it was not covered by ice sheets. Long after sapiens had already left Africa, only about 10,000 years ago, it went through a remarkable shift in eating behavior: hunting and gathering had to give way to growing and cultivating. Sapiens became an agriculturist. Neanderthals and Denisova humans could not benefit from this, as they had disappeared by then. Whereas their diet was dependent on changes in the environment, for sapiens it depended more on new technological developments (Fiorenza et al., 2011; Wißing et al., 2019). The switch caused a complete reversal in diet, behavior, interaction with the environment and thus our own sapiens evolution.

## *Hunter-gatherers become farmers*

How Denisova humans used the environment to get food, we don't really know. Given the similar distribution in space and time as Neanderthals, we can surmise that this may have been similar. We do know a lot about Neanderthals and, of course, sapiens about how their diet changed over time. Neanderthals went through quite a trajectory from carnivore to omnivore, which we can deduce from several sources of information. First, there is the environment itself and thus the supply of food. A continent covered in ice is a continent with few plants that could survive it, let alone plants that produced energy-rich fruits or so. Grasses and shrubs on the tundra did survive and were the food for large herbivores, such as reindeer, horses, woolly rhinos and mammoths (Ben-Dor et al., 2016). It is estimated that Neanderthals had to get a minimum of 74% of their energy from animal protein and fat (in carnivores it is 70%) (Ben-Dor et al., 2016). Survival in the cold also requires more energy (Heyes & MacDonald, 2015). For example, Neanderthals are estimated to consume 3500–6000 kcal/day, which is more than a world class athlete (Condemi & Savatier, 2019; Sorensen & Leonard, 2001). That's about 11 Big Macs a day. For a nursing mother, this could even be 7000 kcal (Wragg Sykes, 2020). Stable isotopes from skeletal material from 100,000 to 36,000-year-old Neanderthals from Belgium (Spy, Scladina, Goyet, Engis), seem to confirm this: the diet of quite a few Neanderthals contained a proportion of meat similar to that of hyenas and wolves (Bocherens et al., 2001, 2005). Especially woolly rhinos and mammoths were part of the diet and to a lesser extent reindeer (Germonpré et al., 2014). Neanderthals were clearly big game hunters, hunting for large tundra grazers, although they also consumed smaller prey like fowl and cormorants (Bocherens, 2009; Bocherens et al., 1999; Goffette et al., 2024). A remarkably large, bell-shaped thorax and abdominal cavity, and a pelvis about as wide as that of erectus, also points in that direction (Garcia-Martinez et al., 2017; Gomez-Olivencia, Barash, et al., 2018; Lopez-Rey et al., 2024; Sawyer & Maley, 2005). It provides room for a large liver and kidneys, which would be necessary in a very meat- and thus protein-rich diet (Ben-Dor et al., 2016). This is because the liver converts animal proteins into glucose. The kidneys then dispose of nitrogenous wastes through the urine, which result from the digestion of the proteins. The diet was not always so constrained among Neanderthals: they also ate rabbits, turtles, dolphins, beavers, birds, seashells, fish, as well as intestines, eyes, tongues and brains of the larger animals (Condemi & Savatier, 2019). They probably even cooked mussels to open them and even ventured into cannibalism (Frayer et al., 2020; Giaobini, 2006; Trinkaus, 1985; Wragg Sykes, 2020). They cooked grains or ate dates and underground roots. The diet of Neanderthals inhabiting the Goyet Cave before the last Ice Age was almost as broad as that of early European sapiens (Wißing et al., 2019). Axes from fire pits, tooth wear patterns, sediments from caves and microfossils in dental plaque confirm this (Henry et al., 2011, 2014; Madella et al., 2002; Power et al., 2018).

As far as is known, Neanderthals never grew crops or raised animals. Sapiens did but not until long after it reached the Middle East and even Europe. Sapiens even reached North America before conditions in the Middle East became favorable enough to make this transition. Early European sapiens were still hunter-gatherers, equipped with the technical skills to tap into a wide range of foods. Migrations along coastal areas provided access to marine foods, such as fish but also shellfish, even while still in Africa (Niespolo et al., 2021). Neanderthals also used their teeth to process meat and hides, as evidenced by wear patterns and chipped pieces of enamel. The success of sapiens relied entirely on sophisticated tools (Condemi & Savatier, 2019; Towle et al., 2021, 2022). After a long period of Acheulean technology, the Levallois technology emerged 300,000 years ago, both in Africa and Europe (Doronichev, 2016; Garcea, 2022). Now, instead of using large stones chipped off along two sides, they used fine blades that could be extracted in large numbers from a single stone via a specialized technique. Both Neanderthals and sapiens used these but sapiens were more likely to succeed in refining them further, and the one that could eventually use the technical skills to make tools for agriculture.

## Sandwiches with porridge

Use of fire was native to Neanderthals and sapiens from the beginning (Goren-Inbar et al., 2004). It was essential for heating themselves, deterring predators and processing materials (such as bitumen and resins) and food. A cave could have multiple fires in different places, depending on their function (Wragg Sykes, 2020). The first fires were simply made on the ground, later people learned to control smoke production by making it in pits or trenches and working with suitable types of wood. Fires were made with manganese dioxide powder and sticks (Heyes et al., 2016). Neanderthals cooked bones to extract an oil-like paste, heated grains and seeds and smoked meat with them to longer preserve it (Henry et al., 2011; Wragg Sykes, 2020). The structure and chemical properties of charred remains show that seeds and grains were soaked and ground to make a porridge or broth after heating (Gibbons, 2015; Kabukcu et al., 2022; Wrangham & Carmody, 2010). Dental plaque from Shanidar Neanderthals contained remains of wheat (*Triticum*) and barley (*Hordeum*), while sapiens also used peas (*Pisum*) (Condemi & Savatier, 2019). Meat was an essential food source during the ice ages, cooking plant material undoubtedly formed a significant proportion of the diet between the ice ages. Neanderthals and sapiens also showed physiological adaptations to be able to eat a lot and store energy when food was plentiful, to be able to rely on it for long periods of time when food was scarce (Shubin, 2008). This becomes problematic when there is never food scarcity, as is today in an industrialized world (Lieberman, 2013). Early Neanderthals from Spain, who lived during a severe ice age, show a variety of skeletal disorders that are even associated with some form of hibernation, where general metabolism slows down at times of food scarcity and cold (Bartsiokas & Arsuaga, 2020).

About 20,000 years ago, sapiens gained momentum. In China, pots were treated with fire and the first grains were grown (Boaretto et al., 2009; Snir et al., 2015; Wu et al., 2012). Nine thousand years ago, they already appear among the inhabitants of Anatolia (Dolbunova et al., 2022; Krause & Trappe, 2020). Remains in these jars show that they were used to prepare all kinds of food, including animal material from the sea. The oldest breads are even as old as 14,000 years old (Arranz-Otaegui et al., 2018). The transition from hunter-gatherer to farmer happened at that time among the Middle Eastern Natufians, between 14.5 and 11.6 thousand years ago (Stock et al., 2023). That switch to more sedentary crop growers may have been driven by a disruption in the climate, called the 'Young Dryas', which caused a cold dip during the warm, post-ice age climate. This could have forced Natufians to settle down and domesticate plants and animals, though still combined with hunting. But agriculture could just as easily have started after this, when the warmer climate got restored (Balter, 2010). Either way, around 12,000 years ago, sapiens made a drastic shift in how to use the environment and how to behave as a community, a shift that would strongly guide sapiens' further evolution. The Neolithic or New Stone Age was ushered in, the era when we largely exchanged animal protein for carbohydrates, when we concentrated our effort on providing as much food as possible in as small an area as possible, when we switched from a highly diverse to a restricted diet, when we went from wandering groups to sedentary large communities. We used grinding stones to grind grains, we organized the first barbecues for large groups, we fermented grains to make beer, we selected varieties of plant and animal species to ensure higher yields and less aggressive animals, we traded farmed goods with hunted and gathered goods (Curry, 2021a; Dietrich & Haibt, 2020; Dietrich et al., 2019; Valamoti et al., 2021; Wang et al., 2021). Food through farmed animals, such as goats and sheep and later cattle, also provided a buffer if harvests were bad However, it brought us into contact with all kinds of new diseases but also provided a new food source: milk (Hagen et al., 2023; Krause & Trappe, 2020; Smil, 2002). Eight thousand years ago, the Anatolian farmers emerged from this, some of whom moved west and north into Europe, causing the original European hunter-gatherers to retreat (Richerson et al., 2021). Genomic evidence shows that this turnover was fast and it came with limited mixing between these hunter-gatherers and the new farmers (Allentoft et al., 2024; Rowley-Conwy, 2024; Simoes et al., 2024). By 6000 years ago, agriculture predominated in Europe, and we were set off for a vicious cycle of more labor for greater yield, which produced more children and larger communities, which in turn demanded more land and livestock, eventually arriving at a socioeconomic system revolving around agriculture and trade (Harari, 2018). In Africa, agriculture made its appearance a little later, about 5000 years ago (Jablonski, 2021).

## Milk ... the white engine

'Milk ... the white engine'. In 1983, this slogan was supposed to encourage Dutch people to drink more milk but is also appropriate for a particular aspect of our recent evolution. Indeed, the availability of milk was the engine that powered important genetic changes in part of the sapiens population, i.e. those who had discovered that milk from cattle also offered benefits to humans. It is a fine example of how cultural evolution can bring about genetic evolution and do so quickly and several times independently.

The oldest traces of sheep and goat milk consumption go back about 7000 years, with remains of milk proteins in dental plaque and milk fats in ceramic pots in Africa (Bleasdale et al., 2021). In Europe, traces have been found in dental plaque going back 5000 years (Stock et al., 2023). Even today, milk is very important because it contains a lot of useful nutrients, such as proteins (casein, beta-lactoglobulin), fats, minerals (such as calcium), vitamins and sugars. The main sugar is lactose or milk sugar, which makes up 4–5% of milk. And yet milk is not suitable for everyone and was neither suitable for consumption by early sapiens. Lactose is the culprit here. Most people cannot digest it or find it very difficult: they are lactose intolerant. This is so from a certain age because as a baby, basically everyone can digest lactose. This makes sense because breast milk also contains lactose, even more than cow's milk (about 7%) (Waelkens, 2009). For this we must go back for a moment to the origin of mammals, the group of animals that had developed specialized mammary glands that provided nutritious milk for the young. Lactose is the main source of energy for a newborn mammal and thus humans too (Deng et al., 2015). Babies are therefore equipped to digest lactose without difficulty, producing the enzyme lactase that breaks down lactose into glucose and galactose (two smaller sugars that the gut can absorb). Lactase is produced in the small intestine as early as the eighth week of pregnancy, with production reaching a peak around birth. As long as there is lactase, people can digest lactose without a problem. But here comes the problem: several months after birth, lactase production declines and lactose can no longer be digested without causing problems. From then on, bacteria break it down in the colon, releasing gases. This is accompanied by all kinds of discomfort, such as diarrhea and abdominal pain. This is why most sapiens worldwide stop drinking milk from a certain age.

Maybe you drink milk without experiencing any problems? If so, chances are you have a mutation in the *MCM6* gene, which spread rapidly in some European and African sapiens populations sometime between seven and 3000 years ago (Bleasdale et al., 2021; Burger et al., 2007; Burger et al., 2020; Krause & Trappe, 2020). Normally, the *MCM6* gene is activated several months after birth, preventing another gene, the *LCT* gene, from producing lactase any longer. In sapiens carrying a mutant form of this *MCM6* gene (various mutations are known), this gene is turned off and lactase production continues. They are then what is called 'lactase persistent' or 'lactose tolerant' (Segurel & Bon, 2017). This is true in 90% to even 100% of today's Central and Northern Europeans but only in 50% of people from the Mediterranean region and even only 10% of Chinese (Condemi & Savatier, 2019). Pastoralist African and Arab populations independently of European people have acquired mutations that also cause lactase persistence (Campbell & Ranciaro, 2021; Stock et al., 2023). These mutations spread in those areas along with people in whom livestock production was a source of meat and especially milk (Leonard et al., 2010). Milk production has also increased dramatically since then, precisely because of a domestication process in which humans selected animals for their capacity to do so. Whereas a Neolithic cow produced about two liters of milk 8000 years ago (needed to suckle calves), this was already 15–20 l in cows from the Middle Ages and is now about 50 l (Krause & Trappe, 2020). Annually, cows now produce about 500 million tons of milk (Smil, 2002). An average North European consumes 489 l of fresh milk per year (Segurel & Bon, 2017). Being able to drink milk for long periods clearly had a selective advantage for sapiens, as a source of sugar, calcium, vitamin D, as well as pathogen-free fluids. This may all explain why these mutations were able to spread so quickly (Burger et al., 2020).

# Walking with big brains comes with consequences

For more than 60,000 years, both Neanderthals and sapiens resided in the Middle East and Europe and even shared joys and sorrows. Were they similar then? Neanderthals survived four ice ages, hard times which in the northern hemisphere translated into ice sheets that covered almost all of northern America and Eurasia with thick layers of ice (Batchelor et al., 2019). The Western European ice sheet shrank and grew repeatedly, alternating during interglacials and ice ages, forcing Neanderthals to continually adjust their distribution range. Sapiens entered Europe about 60,000–45,000 years ago, when the continent was almost completely ice-free. About 26,000 years ago, when Neanderthals had already disappeared, sapiens faced the last major ice age. In neither of those ice ages was the Middle East or southern Europe covered by ice (although the ice age did come close between 190,000 and 130,000 years ago). The Iberian Peninsula became a refuge for sapiens, from where they were able to re-colonize Central Europe 18,000 years ago (at the end of the ice age). This eventually brought these hunter-gatherers into contact with Middle Eastern-born Anatolian farmers (Krause & Trappe, 2020). That Neanderthals were strongly challenged by climate is clear. This is evidenced in part by unique skeletal adaptations, as walking and running in the cold was not so evident.

## *Walking and running in the cold*

Ever since erectus, walking and running upright was further refined but in different climatic contexts for Neanderthals and sapiens. All indications are that Neanderthals were adapted to living in cold and rough terrain and moved about in a slightly different way than erectus and sapiens. They had a shorter neck with large neck muscles, a stiffer lower back and a more vertical sacrum (which connects the spine to the pelvis) (Gomez-Olivencia, Barash, et al., 2018; Spoor et al., 2003). Their pelvis was wider than in sapiens, their limbs were shorter (especially the forearm and lower leg) and they had a robust ankle and stiffer foot (Adegboyega et al., 2021; Pablos, 2015; Spoor et al., 2003; Wragg Sykes, 2020). The construction of their inner ear, where the balance organ is located, suggests that they were less agile. All this indicates that they were not as good runners as sapiens. Where sapiens was adapted to an endurance running, Neanderthal was adapted to an endurance walking (Spoor et al., 2003). This is not to say that Neanderthals were incapable of prolonged running. There is evidence that they could chase a horse to exhaustion at a jogging speed (Hora et al., 2019).

Were these adaptations to walking in ice and snow? Surely adjustments in the foot point in that direction (Sorrentino et al., 2021). Or was this because they stayed mostly in mountainous areas and thus had to climb slopes a lot? For a long time, it was also thought that this was why they had heftier knees than sapiens, necessary to allow the leg extensors to pull harder on the lower leg. But that turns out not to be the case (Trinkaus & Rhoads, 1999). What does stand out is the more curved femur (De Groote, 2011). This used to be assigned to a disease, but everything points toward

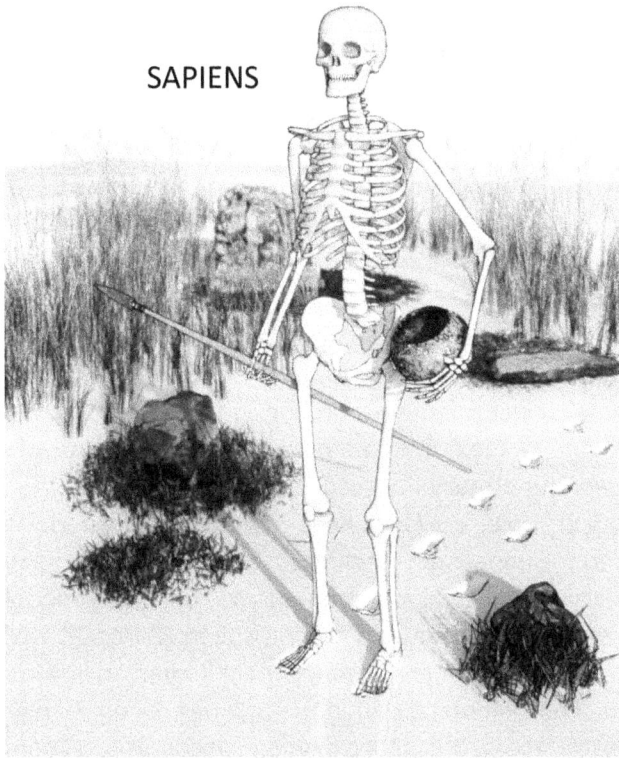

SAPIENS

Figure 7.3    Sapiens skeleton in a tundra environment during the last ice age.

adaptation: greater curvature is especially common in people who exhibit increased activity, such as hunter-gatherers who travel numerous miles daily, or people who live in mountainous areas (De Groote, 2008). In fact, a curved femur allows more leg muscles to become better packed around the femur and forces are better absorbed in the direction that the leg is curved (Bertram & Biewener, 1988). The more firmly ossified tibia also points in that direction (Shaw & Stock, 2013).

Most of the typical Neanderthal traits are more likely to be associated with adaptations to life in the cold (Spoor et al., 2003). For example, Neanderthals were generally heavier and more sturdily built. A sapiens of the same period weighed about 60–75 kg, but a Neanderthal weighed 66–80 kg (Heyes & MacDonald, 2015). The large nose and larger brain are also associated with living in the cold. The bell-shaped thorax, as found in the Shanidar Neanderthal, reflects the shape found in modern sapiens known to be adapted for living in cold environments (Lopez-Rey et al., 2024). In general, sapiens living in colder regions today on average have a wider body and shorter limbs (especially the lower leg and forearm) (Weaver, 2009). The opposite is also true, as longer lower limbs have been shown to allow a more efficient heat loss during endurance running (Struska et al., 2024). This is no coincidence either, it is a trend fairly common in mammals and known as Bergmann's Law (1847) (states that larger and wider bodies occur in colder regions, slimmer bodies in tropical regions) and Allen's Law (1877) (states that short extremities occur in cold climates, long extremities in warm climates). The explanation? A wider body with shorter arms and legs loses less heat to the environment (Ruff, 1994). A Neanderthal with broad shoulders, wider pelvis, short forearms and legs then seems to be better adapted to survive for long periods in cold areas (Ruff, 1991).

## Running economy

Sapiens faced an ice age late in their evolution, at a time when their technological evolution was already much further along. Bipedal locomotion and being able to run for long periods of time is deeply ingrained in our bodies and brains. From a distance of 13 m, we even recognize the gait of a familiar person (without seeing the face) and activate a brain region normally involved in the recognition of movements such as in the face (Hahn & O'Toole, 2017). Sapiens is the only human species that has managed to understand the underlying mechanism of efficient running and adjust its own running behavior accordingly. Sapiens manages its running economy, i.e. consuming as little energy as possible to run at a certain speed (Barnes & Kilding, 2015). It is expressed as the amount of oxygen consumed per kilogram of body mass for 1 minute (or over a distance of one kilometer). A higher running economy indicates the engagement of more muscle fibers, thus more oxygen consumption (Thomas et al., 1999). An elite runner manages to run fast with fewer activated muscle fibers. Over 10 years, the running economy of marathon runners goes down 15% on average (Jones, 2006). A lot of factors influence this, such as training (and way of training), body weight, running style, joint flexibility, shock absorption in the sole of the foot, body temperature, heart rate, respiratory rate and so on. That we can improve this by training shows that people have developed a mechanism whereby by running, they also learn how to further optimize energy consumption.

## Born to be wild

Neanderthals had a wider pelvis. How wide the pelvis was in hominins has been subject of debate among scientists since 1960 as the so-called 'obstetrics dilemma' (Gomez-Olivencia, Barash, et al., 2018; Haeusler et al., 2021; Ruff, 1991). This states that the pelvis in *Homo* is an evolutionary compromise between large brains and walking upright: a wider pelvis with a wider birth canal allows the birth of babies with larger brains but makes walking and running on two legs less efficient. Hence, the evolution of bigger brains conflicted with the evolution of better running. Even now, this causes births to be accompanied by complications. Among hunter-gatherers, maternal mortality during birth can be as high as 1 in 130 births. Up to 17% of all maternal deaths are directly associated with complications due to the fetus not making it through the birth canal, determined by the size of the head and width of the shoulders, and less by the baby's body weight. In addition, this results in increased infant mortality if the caesarean section cannot be performed in time (annual mortality of 2 million babies worldwide). This was probably also a problem among Neanderthals, as no fossils have been found of women over 40, only men (Haeusler et al., 2021).

Is this a typical sapiens problem or did this come about during the transition from australopithecines with their small brains to *Homo* with large brains? In primates, the size of a baby relative to the mother can vary greatly (Fischer et al., 2021; Moffett, 2021; Tague, 2011). In *Homo,* this is fairly stable, about 5–6% of the body mass (MacLean et al., 2009; Martin, 2007; Ponce de Leon et al., 2008). In prosimians and monkeys it fluctuates between 4 and 16%, with the outlier being squirrel monkeys (genus *Saimiri*) where it is 34% (Harvey & Cluttonbrock, 1985; Stephan et al., 1981; Trevathan, 2015). They have special elastic ligaments between the pelvic bones that can relax during birth, doubling the diameter of the birth canal (Laudicina & Cartmill, 2022). In great apes, birth weight is relatively small (about 3–4%), except for gibbons where it is 8% (DeSilva & Lesnik, 2006; Geissmann & Orgeldinger, 1995; Lynch et al., 1983). It is perhaps not surprising that this is that low in large great apes, for recall that they have abnormally small brains relative to their body size. Looking at the first hominins, that ratio is comparably low in Ardi (3%), but higher and fairly constant in the australopithecines (about 5–6%) (Laudicina et al., 2019; Leutenegger, 1972, 1987). A similar pattern is seen in the ratio of brain mass in the infant to maternal body mass, which fluctuates between 0.2% in the gorilla and 2.2% in the squirrel monkey. In *Homo,* this is higher on average than in australopithecines, at 0.7 versus 0.6% (see Figure 7.4). Taking infant brain size relative to adult brain size, the brain in *Homo* is smallest at birth (28–35%, versus 38–39% in australopithecines, 37–57% in great apes, and 48–60% in prosimians and monkeys) (DeSilva & Lesnik, 2008). This is remarkable because compared to australopithecines, *Homo* mothers have larger babies relative to their body size but do have smaller brains at birth relative to adult brain size. This is admittedly relative, because in absolute terms the brain of a *Homo* baby is larger: about 151 g in chimpanzees and 180 g in australopithecines, but 230 g in habilis, 310 in erectus, 405 g in Neanderthals and 400 g in sapiens.

Let's look a little deeper into the birth canal and the path a baby takes during birth. The birth canal is not just a simple cylinder through which the head and shoulders must pass. It is a complex 3D structure whose shape at the entrance is different from that in the middle or at the exit (Laudicina & Cartmill, 2022). In no primate is the size of the head and shoulders of a baby so tight to fit through the birth canal, as in humans (Dunsworth et al., 2012). Even the skull bones shift relative to each other because the passage is so narrow (Ami et al., 2019; Falk et al., 2012; Tague, 2012). Primates with a snout make their skull diameter smaller by sliding through the canal with the snout forward (Laudicina & Cartmill, 2022). Except for gibbons, this problem poses less of a problem in great apes because of the small baby brains (Rosenberg & Trevathan, 2002). In hominins with short snouts and larger baby brains, a forward snout yields little. Their birth canal also shows a different profile: in apes and great apes, the profile

## Brain mass baby
(in % relative to the mother's body mass)

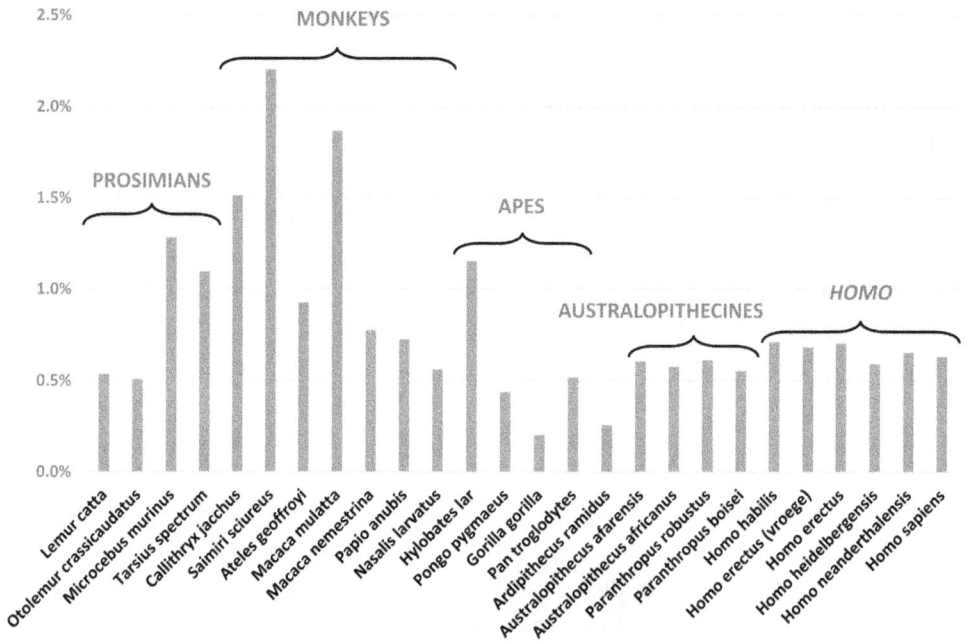

Figure 7.4    Ratio of brain mass in a primate baby to maternal size (expressed in body mass). (Data from DeSilva (2011); Harvey and Cluttonbrock (1985); Leutenegger (1973); Lynch et al. (1983).)

in cross-section is an ellipse that is longer front-to-back than left-to-right (Laudicina & Cartmill, 2022). In hominins it is different, there the ellipse at the level of the entrance is twisted 90°, so to speak. Ardi had a wider pelvis than great apes and their babies still had small brains (Lovejoy, Suwa, Spurlock, et al., 2009). Lucy had a birth canal entrance that was wider than deeper, even similarly wide to sapiens (about 12–13 cm) (Gruss et al., 2017; Hausler & Schmid, 1995; Rak, 1991). This brought about a significant change in the way a hominin baby is born. Indeed, at the onset of birth, the elliptical infant head must descend into the opening of the birth canal while facing sideways. As it passes through the canal, the head and shoulders must then rotate 90°, eventually being born with the face turned backwards (toward the mother's back) (Desilva et al., 2017; Gruss & Schmitt, 2015). The uterine muscles even contract spirally to achieve this rotation. In other primates, the baby's head starts and ends facing forward, toward the mother's ventral side, without the need for rotation. Thus, the evolutionary change in the entry profile of the birth canal in hominins has caused their babies to lie in an atypical position at birth (Stansfield et al., 2021). Human babies born facing the mother's abdomen, referred to as 'stargazers', are babies born in a typical primate manner but atypical for hominins (happens in 8% of births) (Trevathan, 2015).

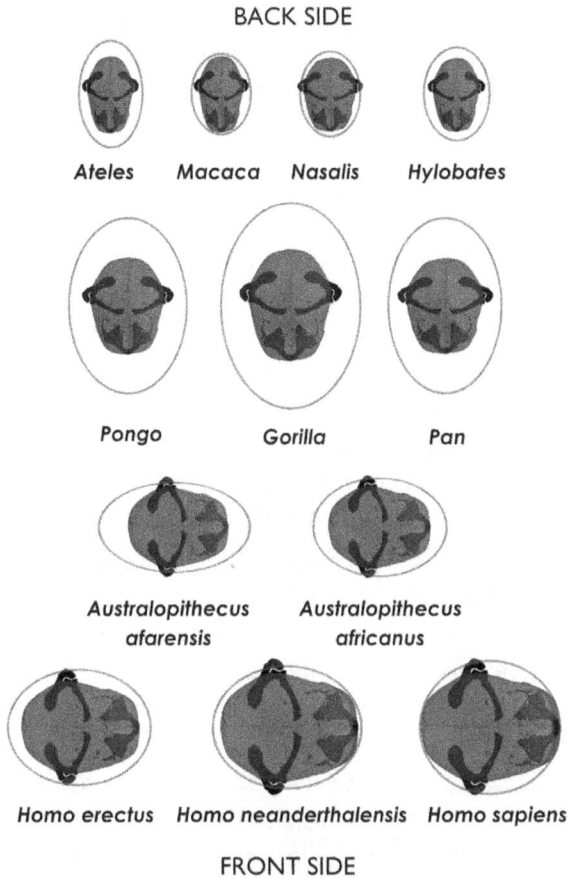

Figure 7.5   Relationship between the size of a baby's skull and shoulders and the shape of the birth canal entrance (ellipse) in primates: Monkeys and gibbons (top row), great apes (second row), australopithecines (third row) and humans (bottom row). (Based on Laudicina & Cartmill (2022); fetal skull 3D model modified from Sketchfab, by Eric Bauer – https://sketchfab.com/3d-models/human-fetal-skull-v2-7abc3f97d138424baf3b26a1df3b3bac.)

Now, turning is one thing, getting through the canal with difficulty is another. Although babies of australopithecines already had to turn their heads during birth, the canal was wide enough for the head to pass without difficulty and only have to undergo a partial rotation to pass the broad shoulders (Desilva et al., 2017; Hrvoj-Mihic et al., 2013; Trevathan & Rosenberg, 2000). In erectus, where women were larger and had wider pelvises, this problem may have been even more limited (Simpson et al., 2009). From Neanderthals, and especially sapiens with large-brained babies and a narrower pelvis, this was no longer the case, with the birth canal providing very little additional space. On top of that, a mother herself could less easily free the umbilical cord or pull the baby out of the birth canal if it got stuck, due to the head being turned backwards (Rosenberg & Trevathan, 2002; Trevathan, 2015). This suggests that they had to appeal to their social context, namely group peers who provided birth assistance (Mitteroecker & Fischer, 2024; Trevathan, 2015). The midwifery profession may have had its origins then.

We know that brain size is related to the number of neurons and that the number of neurons is a better suited measure of cognitive performance. We also know that the number of neurons increases only very slightly after birth. It follows that brain growth after birth is associated with an increase in the number of glial cells (the other brain cells), a growth of neurons and an increase in the number of connections between neurons (Lynch et al., 1983). To put it succinctly, the evolutionary strategy seems to have been to get as many neurons into the brain as possible in a way that it keeps the head small enough to get through the birth canal, and once born to invest as much as possible in the growth of neurons and the connections between them. Ultimately, it is those connections that are important. This is also interesting from an energetic standpoint, because the smaller the baby and its brain, the less energy the mother must provide. If we take a sapiens mother and chimpanzee mother of equal body size, a sapiens mother will still have to invest almost twice as much energy in the pregnancy because the baby will be almost twice the size of that of a chimpanzee (pregnancies last only a little longer in sapiens, i.e. 267 versus 228 days in chimpanzee) (DeSilva, 2011). As a result, a mother or both parents have to invest longer in postnatal care, because a human baby with a poorly developed brain remains cognitively and motorically helpless much longer. In other words, humans are altricial and not precocial, even though pregnancy in humans takes the longest of all primates (even gorillas have a shorter pregnancy of 8.5 months) (Dunsworth et al., 2012; Ernest, 2003). That phase when babies are still developing and dependent on their parents for survival is called the infantile period. An extended infantile period, in which infants must also be carried, is thought to have originated from *Homo,* but possibly only after erectus (Graves et al., 2010; Le Cabec et al., 2021). The reason? An australopithecine who had yet to climb up trees required a lot more energy to do so (carrying a child requires 16% more energy). The growth pattern of teeth also suggests that life stages in australopithecines were more similar to those of great apes. It also teaches us that an increased need for energy during gestation is thought to have been established about 6 million years ago, in australopithecines, but only translated into an accelerated increase in brain volume 1 million years ago, in erectus (Dean, 2006; Lieberman, 2013; Monson et al., 2022). So where did erectus get that energy? Very likely from meat and the use of fire (Lieberman, 2013).

But is this picture accurate? Ultimately, birth is triggered by a fetus and the placenta going into a kind of chronic stress because insufficient nutrients can be provided (Mclean et al., 1995; Valleau & Sullivan, 2014). A mother, based on her body size, is limited in the amount of nutrients that can be delivered per minute and by which molecules can be passed through the surface of the placenta (Ernest, 2003).

In fact, women consume twice as much energy during and after pregnancy then before pregnancy (Dunsworth et al., 2012). Nor is it so unequivocal that a wider pelvis would necessarily also require more energy while walking (Warrener et al., 2015). It would even help to walk energetically more efficiently while carrying a child, by adjusting stepping speed and step length (Gruss et al., 2017). And then there is the connection to regulating body temperature, as a wider pelvis in Neanderthals was advantageous in keeping the body warmer and implied in a wider birth canal (Whitcome et al., 2017). A wider birth canal is also associated with thinner pelvic bones, which in turn limits the strength of the pelvis in supporting a fetus and the intestines (Ekaterina Stansfield et al., 2021). You may guess it, the picture is again not so simple.

# Becoming pale, not yet an option

Of erectus, we now know that it must have been functionally naked and dark-skinned. We deduce this from the ecological context in which it dwelt, but also from genetic changes that indicate increased skin pigmentation. Now we are dealing with three human species that faced new, extreme challenges in Eurasia: the ice ages. Denisova humans, Neanderthals and sapiens, all three had to endure thousands of years of bitter cold while the northern part of the planet was largely covered in ice, interspersed with interglacials when a much more favorable climate was accompanied by a very different, more survivable environment. Denisova humans and Neanderthals had to endure more, as they survived four ice ages like this. With Sapiens arriving later in Europe, it only had to face the last major ice age. Survival during the ice ages was a matter of being able to cope with the cold, a more limited food supply but also a different light regime with more limited exposure to UV rays. And that determined much of what we look like as humans today and how sapiens came to explore the world.

## *Cooling down*

Ice ages are denoted by the abbreviation 'MIS' (Marine Isotope Stage) and a number. Even numbers refer to glacial periods, odd numbers to interglacials. Only in Eurasia were ice ages accompanied by ice sheets that covered vast areas of the northern part. People were thus pushed back to the tundra areas outside the ice caps and the southern and forested Mediterranean. It is no coincidence that the Levant, the eastern Mediterranean in the Middle East, was such a crowded area by human species (Groucutt et al., 2021). Caves also provided the necessary protection, heating them with fires (Condemi & Savatier, 2019). Neanderthals and Denisova humans had only just separated, about 400,000 years ago, or they were confronted with their first ice age (MIS 10, from 370,000 to 340,000 years ago) (Railsback et al., 2015). Forty thousand years later, the next one was there (MIS 8) which raged for about 60,000 years. A slightly longer warmer and more humid period was followed by MIS 6, a cold and dry period 190,000–130,000 years ago. About 70,000 years ago, sapiens also appeared on the scene, first in the Middle East, sharing love and suffering with the other two species during MIS 4, which raged briefly from 70,000 to 60,000 years ago. By the time the last ice age arrived (MIS 2), 35,000 years ago, sapiens was virtually alone in Europe because Denisova humans and Neanderthals had all disappeared. This ice age lasted until about 12,000 years ago and is marked by the Last Glacial Maximum, about 26,000 years ago. At that time, it was so cold that about 8% of the Earth's surface was under the ice sheets and the sea level was 120 m below the current level (Batchelor et al., 2019).

Surviving here thus implied adapting to the cold or making one's way to better places. Not surprisingly, Denisova humans and especially Neanderthals have multiple anatomical adaptations to surviving in cold regions, in addition to behavioral adaptations such as crafting skins to clothe themselves with (Wragg Sykes, 2020). Recall Bergmann and Allen's laws linking the shorter arms and wider body of Neanderthals to being better adapted to the cold (Ruff, 1991). The larger brain, on average slightly larger than sapiens, also retained heat more efficiently. And the large nose helped to better warm inhaled air before it entered the lungs. Cold and dry air is not so good for the lungs (Elad et al., 2008). Air passing through the nose and mouth warms up and absorbs moisture. Not coincidentally, we spontaneously breathe through the nose even though the air takes a more difficult path through the nasal cavity. The nasal hairs stop dust particles and contact with the nasal mucosa is greater. The blood in the nasal mucous membranes gives off heat and the mucous membranes themselves give off moisture. Cold and dry air during ice ages could thus be warmed and moistened before reaching the lungs. Thus, 10,000 l of air pass through our nose each day, reaching a temperature of 31–34°C and a humidity of 90–95% by the time it reaches the lungs. To do this, we consume 350 kcal and 400 ml of fluid per day. Thus, the larger nose of Neanderthals provided a larger contact surface between the inhaled air and the nasal mucosa. Also, the shape of the nasal cavity makes the passing air swirl, which further promotes the exchange of heat and moisture. The shape of the nasal cavity in Neanderthals and sapiens from cold regions shows that both efficiently cause the air to warm, but in a slightly different way and with more air movement in Neanderthals (de Azevedo et al., 2017; Wroe et al., 2018).

Surviving cold periods is possible if sufficient fat can be accumulated. This allows energy reserves to be built up during times when there is plenty of food, to be used during less favorable periods or when the future was uncertain about food availability (Zihlman & Bolter, 2015). It is surprising to note that sapiens living in polar regions have an equally thick or even thinner layer of subcutaneous fat than people from the tropics (Kuzawa, 1998). We also have very little brown fat (1–3% of our body weight), a type of adipose tissue that produces body heat through combustion and is mainly used in mammals during hibernation. The other type of adipose tissue, white adipose tissue, is an energy reserve and acts as a temperature insulator but does not produce heat. We have more of that: about 12% at birth. And yet several things point toward us, and presumably Neanderthals and Denisova humans, having an increased ability to store adipose tissue. Bonobos have an average of 4% (in females) to less than 0.1% (males) subcutaneous fat. In sapiens, this is much higher (36% in females studied and 20% in males) (Zihlman & Bolter, 2015). It is also higher in hunter-gatherers, who have much less access to energy-rich food, than in bonobos (20% in females, 11% in males). There is an essential minimum of body fat in sapiens, because once this falls below 5%, the body starts to break down muscle tissue to get energy.

Genes also point toward an increased ability to store fat. The surprise is that sapiens has gene variants involved in fat metabolism derived from Denisova humans and Neanderthals (Gittelman et al., 2016; Vernot et al., 2016). Their much longer evolution during ice ages had promoted these genes for an improved fat metabolism. When sapiens encountered them and crossed with them, it could immediately enjoy the benefits of those gene variants (such as of the *PLPP1* gene or the *KITLG* gene that is also associated with skin color) the moment sapiens itself faced the cold (Yang et al., 2018). But even when sapiens left Africa, variants have emerged that have made it easier to explore colder Eurasia (Tobler et al., 2023). These are variants of genes that promoted energy production during fat burning or fat deposition.

## All the colors of the rainbow

Australopithecines and erectus evolved a dark skin, but sapiens today reflects a wide range of skin colors. Was this because they were simply less exposed to harmful UV rays in the north and dark skin was not necessary? Or is there an adaptive story behind it, where natural selection caused them to fade? All indications are that there was effectively a positive selection for a paler skin, which, moreover, was not the same for all sapiens. The reason for this selection: vitamin $D_3$. Sexual selection could also explain variation in skin color, but this would only have had a limited effect at a late stage (Aoki, 2002; Jablonski & Chaplin, 2000). That women are on average paler than men has nothing to do with this but is all about vitamin $D_3$ (Dunsworth, 2021).

The epidermis mainly contains keratin-producing cells (keratinocytes), with cells at the base that produce the pigment melanin (Johnson et al., 1993). These melanocytes contain small vesicles or melanosomes within which the precursor of melanin pigment is produced and accumulated. The genes then control whether this precursor is converted into dark eumelanin (dark brown to yellow) or paler pheomelanin (brownish red) (Sturm et al., 1998). Eumelanin is very good at absorbing light at all kinds of wavelengths, including the highly energetic and harmful UV-B rays (Jablonski, 2004). The more eumelanin, the less likely the skin will get burned and damaged by UV-B (Johnson et al., 1993). You would think that it is the ratio of the two pigment types that would determine skin and hair color, but this is not so. Regardless of the darkness of the skin, the ratio is always about 74% eumelanin and 26% pheomelanin (Del Bino et al., 2015). Skin color is rather determined by how much melanin is present, with eumelanin being responsible for UV absorption (pheomelanin can do this much less) (Jablonski & Chaplin, 2000). The pigments are released to the surrounding kera-tinocytes, which themselves are constantly being renewed and advance toward the skin surface as they accumulate keratin. These cells die and surface as packages of dead cells, full of keratin and melanin, only to be constantly shed as skin flakes. We all have a genetically determined base color, which is adjusted to some extent by the amount of sunlight we are exposed to. The latter gives us a summer tan. The genetic machinery for the base color is evolutionarily fixed. A dark-skinned African may sit indoors as long as he or she wants, the skin will never look as pale as of someone from Scandinavia. Conversely, a Swede may spend whole days in the sun, he or she will never become as dark as some Africans. And yet they have about the same number of melanocytes in their skin. Only in dark-skinned people, the melanocytes produce many more and larger melanosomes (Johnson et al., 1993).

At the equator, UV radiation is strongest because it is less reflected and absorbed by the ozone layer. The closer to the poles, the more is blocked by the ozone layer. Good news you would say, all to the north! Better not as this comes at a cost. Too much UV is detrimental because it breaks down folic acid, but too little undermines the pro-duction of another vitamin, vitamin $D_3$. We get this vitamin into our bodies in two ways. Either we take it in through food or we make it ourselves. Via food occurs mainly through offal, meat from marine mammals and fish. We make it ourselves ... indeed, through UV rays, which convert a substance in the skin (7-dehydrocholesterol) to previtamin $D_3$, from which the functional vitamin $D_3$ is then formed (Chaplin, 2004; Jablonski, 2004; Jablonski & Chaplin, 2000). We need this to absorb calcium from food through the intestinal tract, but it also plays an important role in the immune system. A pregnant mother needs extra vitamin $D_3$ because she must also provide calcium for her fetus (Dunsworth, 2021). To get the daily dose through food, you would have to ingest very large amounts of meat and fish. So being able to make it yourself via UV rays is essential. Now let this have been a challenge for human species that lived in the northern regions. We can divide the world map into three zones. A first zone lies around the equator, where enough UV radiation exists on an annual basis for some-one with pale skin to make the necessary vitamin $D_3$. In the zone just beyond that, such as southern Eurasia, one faces insufficient UV rays to make enough vitamin $D_3$ for at least one month a year. Then in the far north (and south) lies the third zone where it becomes very difficult for someone with pale skin to still make enough vitamin $D_3$ on an annual basis. For dark-skinned people, zones 2 and 3 are much more extensive (they must spend five times longer in the sun to make the same amount of vitamin $D_3$) (Jablonski & Chaplin, 2000). Coincidence that dark-skinned people are mostly found in zone 1 and only very pale-skinned people can survive in zone 3? No. And yet it is not entirely correct, for the Inuit from Greenland have a remarkably dark skin for someone from zone 3. However, they manage to get enough vitamin $D_3$ through their diet consisting mainly of oil-rich marine mammals and fish. In Tibet people even have become darker due to recent genetic modifications as an adaptation to living at high altitudes (Yang et al., 2022).

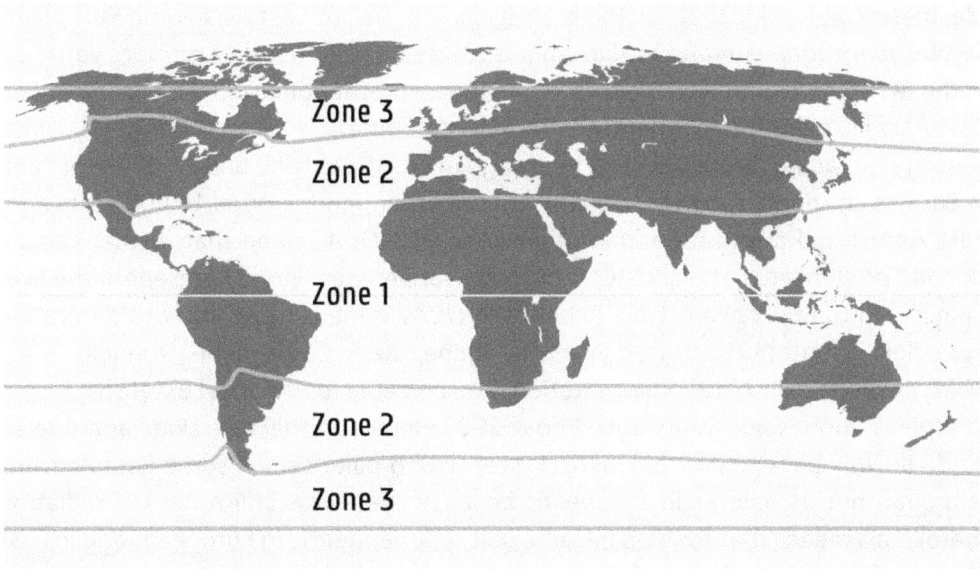

Figure 7.6   World map showing the three zones in which modern pale-skinned people receive sufficient or insufficient UV rays to make the necessary vitamin D3. Zone 1: sufficient UV on average each year to make enough vitamin D3; Zone 2: minimum one month per year people cannot make enough vitamin D3; Zone 3: insufficient UV on an annual average to make enough vitamin D3. (Based on Jablonski & Chaplin (2000).)

If the hypothesis is correct that this is all adaptive, then it should show in the genes (Izagirre et al., 2006; Jacobs et al., 2013). Let's go back a million and a half years, to erectus. Their skin darkened under the influence of genetic change in the *Melanocortin-1-Receptor* gene (*MC1R*). When the protein expressed by this gene binds to a hormone that stimulates melanocytes, their melanosomes produce mainly the dark eumelanin (Norton et al., 2007). There are many variants of this gene, which determine the variation in how much eumelanin or pheomelanin will be produced, and thus how variable skin color can be. A million and a half years ago, there was a selection pressure that greatly reduced the number of variants to only those that produce dark skin (Jablonski, 2021; Quillen et al., 2018). Virtually the whole erectus population thus became dark-skinned. About 300,000 years ago, sapiens appeared in Africa, when a selection for dark skin was further enhanced by mutations in two other genes (*DDB1* and *MSFD12*) (Jablonski, 2021). In South African humans, variation in the *MC1R* gene increased again shortly thereafter and so did variation in skin color. About 70,000 years ago, sapiens arrived in the Middle East, just before the MIS 4 ice age, which was accompanied by multiple mutations (Tobler et al., 2023). On the one hand, there was selection for gene variants that kept the skin dark (*DDB1* and *MSFD12*) in the branch that gave rise to Aborigines from Australia and peoples from

Melanesia (Jablonski, 2021). In the branch leading toward East Asians and Europeans 50,000 years ago, variation in skin color increased earlier because of more variation in the *MC1R* gene and later because of changes in other genes. Both European and East Asian sapiens independently experienced a mutation in the *KITLG* gene, which caused skin fading (around 32,000 years ago in European tribe and 21,000 years ago in East Asian tribe) (Yang et al., 2018). On the other hand, around 10,000 years ago, East Asians had a mutation in another gene, the *OCA2* gene that further caused skin fading (Edwards et al., 2010). This gene variant was already present in the Han Chinese. In the European strain (where the *OCA2* gene is more likely to determine eye color), a mutation occurred in another gene, the *SLC24A5* gene (Donnelly et al., 2012; Quillen et al., 2018). This mutation is responsible for the most extensive fading in sapiens about 6000 years ago. About 2000 years ago this pale skin variant even made its way back to Africa, where it gave rise to paler skin in some East Africans (although not as pale as in Europeans because there was still more UV radiation there). This means that the skin became pale independently in Europeans and in East Asians. Also, the very pale skin in Europeans came about late, long after they were already present in Europe (Adhikari et al., 2019). The genome of a 34,000-year-old sapiens woman from the Czech Republic shows that she still had a dark skin, brown hair and brown eyes (Callaway, 2021; Gibbons, 2021c; Hajdinjak et al., 2021; Prüfer et al., 2021). Even the so-called Cheddar man who lived in Somerset (United Kingdom) 10,000 years ago was still dark-skinned (Bryson, 2019; Jablonski, 2021). So why only 6000 years ago, you may ask? Well, that was the period when sapiens shifted from being a hunter-gatherer (with vitamin $D_3$-rich diet) to becoming an agriculturist with a diet consisting mostly of crops. And crops are a vitamin $D_3$-poor diet. Anatolian farmers already had the fading variant of the *SLC24A5* gene (Krause & Trappe, 2020; Quillen et al., 2018). Agriculture helped determine why Europeans very quickly became the palest among sapiens (Bryson, 2019).

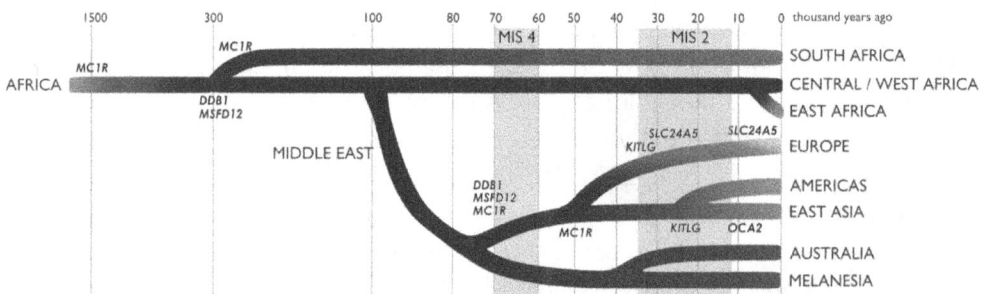

Figure 7.7   Genetic evolution of skin color in sapiens across geographic populations (abbreviations refer to genes that underwent mutations which affected skin color). The darkness of the skin is schematically represented by the gray values. (Based on Jablonski (2021); Murray et al. (2015); Quillen et al. (2018).)

But then what about Neanderthals and Denisova humans? Their genes too indicate a lot of variation in skin and hair color, such as the *MC1R* gene (Condemi & Savatier, 2019; Wragg Sykes, 2020). Eventually, they too thrived in UV-deficient areas, but with a diet largely composed of meat. As pale as the palest sapiens they presumably did not become. They did give sapiens a boost for rapid genetic adaptation, because sapiens obtained certain gene variants for paler skin through interbreeding with Neanderthals (such as from the *OCA2* gene) (Dannemann & Kelso, 2017; Norton et al., 2007).

# Survival through behavior

Erectus laid the foundation for larger brains in our evolution, but it didn't stop there. Colonization of the northern hemisphere required a whole palette of new adaptations. That precisely Neanderthals (and presumably Denisova humans) and sapiens were the human species with the largest brains, therefore was no coincidence. Adapting to an extremely changing environment is much faster through creativity and technology than when it requires anatomical adaptations. Brains of more than 1700 cc, the knowledge and dexterity to make simple stone tools to process meat and skins, fire to heat themselves and food, and the courage to explore new territories when things are not going well … these are all suitable ingredients to cope with a European context with its extreme conditions. We also see the most striking changes in behavior in Neanderthals and sapiens (we don't know much about Denisova humans). New materials to work with, new concepts to get more out of a stone, ways to make tools more efficient and even make shelters, it reflects increased creativity. Add to that a spoken language that made transmission of that new knowledge more efficient than ever before. All thanks to big, globular and healthy brains.

## *Big, globular and healthy brains*

### Big brains

Neanderthals, Denisova humans and sapiens share that large brain. Chinese skulls suggest that Denisova humans had brains of about 1700 to perhaps 1800 cc (Wu et al., 2022). In Neanderthals and sapiens, this went up to about 1740 cc (Jordan, 1999; VanSickle et al., 2020). This is larger than in their predecessors, such as antecessor from Spain, suggesting that they independently developed larger brains (Wong, 2015). It is often argued that on average Neanderthals had larger brains than sapiens, i.e. 1450 versus 1350 cc (Jordan, 1999). However, this is somewhat an oversimplification. Historical fluctuations in volume and statistical arguments suggest that this should be taken with a grain of salt (Pearce et al., 2013; VanSickle et al., 2020). Indeed, sapiens brains are thought to have shrunk about 12,000 years ago, due to a reduction in body size during a warmer period (DeSilva et al., 2021; Lieberman, 2013; Ruff et al., 1997). Whether brain shrinkage effectively occurred was recently questioned (Villmoare & Grabowski, 2022). Neanderthals also experienced some fluctuations because brains were larger in more recent forms (Pearce et al., 2013).

Yet there are some notable differences in the Neanderthals and sapiens brains. Neanderthals have a skull profile with a sloping forehead, while it is nearly vertical in sapiens. This indicates a difference in brain shape, especially than a larger frontal lobe with prefrontal cortex in sapiens (Wragg Sykes, 2020). The difference in brain shape is already present from birth, even though they are similar in size (about 400–430 cc) and grow out in a similar way (Gunz et al., 2011; Ponce de León et al., 2016). The cerebellum and parts of the neocortex (frontal and temporal lobes) have an early and rapid growth. In sapiens, the visual lobe grows out less strongly than in Neanderthals (Neanderthals also had larger eye sockets, by the way) (Alagoz et al., 2022; Pearce et al., 2013). After that, the brain grows slower and becomes slightly rounder. Hence, the Neanderthal brain did not grow like that of great apes, although previously thought (Gunz et al., 2010, 2012; Neubauer et al., 2010). This similar growth may help explain why cognitive and social skills were already advanced in Neanderthals. In fact, a huge dataset of more than 120,000 MRI scans of modern humans, with ages ranging from 16 weeks after conception to 100 years old, shows that our large brains continue to increase in volume up to an age of about 12 years, only to shrink again afterwards (Bethlehem et al., 2022; Kozlov, 2021). The neocortex reaches its maximum thickness as early as at the age of 2, while the white matter (which contains all the connections between brain centers) is at its largest only at the age of 29.

There was also variation in brain volume within both species, as between the sexes. Based on data from more than 2,500 British women and men each, it was possible to deduce that although absolute brain volume was slightly larger in men, there was also more variation in men (Ritchie et al., 2018). Also in Neanderthals, men, who are also slightly larger on average, had slightly larger brains (average 1600 cc in men and 1300 cc in women) (Jordan, 1999). Women have a thicker neocortex and a larger corpus callosum, which is the connection between left and right hemispheres of the brain. In the neocortex, some areas are larger in men, others larger in women. To a certain level, such anatomical trends can explain a difference in average (the word 'average' is very important here, because obviously there is a lot of overlap) behavior between men and women, with the way the brain anatomy does its job also being influenced by a variety of factors, such as hormones and memory (Nelissen, 2015; Pease & Pease, 1999).

## Globular brains

Neanderthals and sapiens do have similarly large brains, but sapiens is distinguished from other human species by its globular (spherical) brain. In others, such as Neanderthals, they are rather elliptical. Globular brains only came about during the evolution of sapiens and were not common until 100,000–35,000 years ago (Neubauer et al., 2018). Early sapiens had a general brain shape, somewhere between that of erectus and Neanderthals. During development, the globular shape appears when the rate of general brain growth is largest. It is manifested primarily by a bulging of the

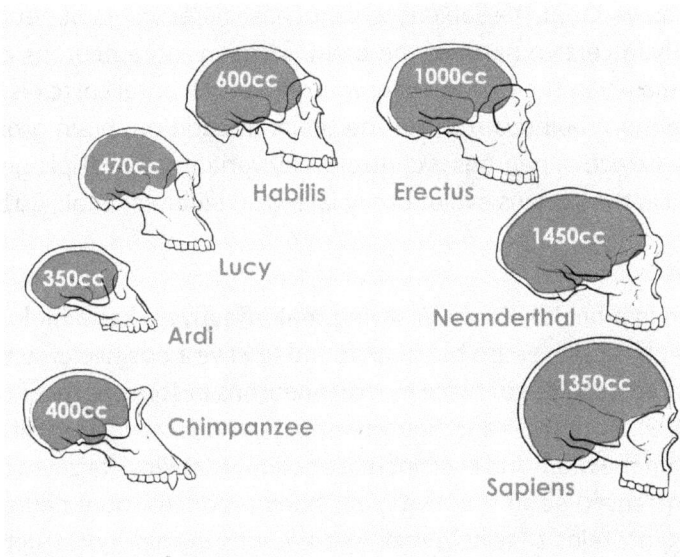

Figure 7.8   Brain size and shape in chimpanzee and hominins, illustrating the large and globular (spherical) brain in sapiens.

parietal lobe and of the cerebellum. The parietal lobe is involved in cognitive special-izations, such as forming a mental picture of the reality around us, or interpreting sym-bols and numbers (Lieberman, 2013). In addition to its role in controlling movements, the cerebellum is also important for memory, language and sociality (Neubauer et al., 2018). At a later stage, a change in the skull, especially with the shortening of the face, also induced further sphericity in the brain.

Just as certain genes are thought to be responsible for the evolutionary enlargement of brains, there are genes involved in bringing about the globular shape, both at the level of the cranial bones and the underlying brain itself (Benitez-Burraco & Boeckx, 2015; Heide et al., 2020). Brains enlarge due to an increase in number of cells, includ-ing neurons. To get to sapiens' 86 billion neurons, an average of 12,400 neurons form per second during development (Ackerman, 1992; Cowan, 1979; Stiles & Jernigan, 2010). That's a tremendous rate! If this doesn't happen at the same rate everywhere, you get areas that get relatively larger, as does the prefrontal cortex, at the front of the brain. You get a more vertical forehead by the enlargement of that frontal brain area. This cortex is the conductor of the brain, directing how we plan and prepare actions (even simply bringing a cup of coffee to your mouth requires prior planning, if not the coffee risks ending up somewhere it doesn't belong). A change in the *TKLT1* gene already seems to have played an important role in bringing about this sphericity (Pinson et al., 2022). Indeed, sapiens has a variant of this gene that is different from that in all other primates, including Neanderthals. And the difference is very small, because the protein resulting from both gene variants differs only in one amino acid.

Experiments showed that the sapiens variant causes precursors of neurons to divide more frequently in certain parts of the brain, forming more neurons per precursor. This occurred mainly in the frontal lobe, where the prefrontal cortex is located, thus giving rise to a more bulbous front of the brain. In addition, brain growth, synapse formation and growth of neurons is controlled by variants of multiple genes that only got established in the sapiens evolutionary lineage (Neubauer et al., 2018).

**Healthy brains**

A healthy mind in a healthy body. A saying that effectively has a biological basis! In rats, voluntary physical exercise has been found to have a positive effect on the brain. Running in a wheel resulted in twice as many neurons in some parts of the brain (van Praag et al., 1999). Both short and vigorous efforts, and prolonged small efforts have a positive effect on the brain and even increase cognitive ability (Raichlen & Polk, 2013). When we do increased aerobic activity, our body produces more neurotrophins and growth factors, proteins that help maintain neurons in their construction and their activity. For example, the brain-derived neurotrophic factor (BDNF) is produced in the nervous system itself, while the liver produces Insulin-like Growth Factor 1 (IGF-1). This IGF also there increases the production of glucose, which is useful for the brain. But IGF also gets to the brain where it helps create new neurons. One might wonder what the strongest driving force was toward such a big brain? Was it the brain itself and what we could do with it that was under selective pressure? Or was it our walking ability that evolved through natural selection into something like prolonged running, and did the brain then become larger rather as a side effect of that? Ultimately, the most notable enlargement occurred in erectus, the first adapted to run for long periods of time.

## Technology ... let it be a bit more

### Acheulean technology on its last streaks

Teardrop-shaped hand axes, typical of erectus' Acheulean technology, have long been everyday tools. About 800,000 years ago, the early ancestors of European human species brought them into Europe, and 200,000 years later they had already reached as far as the United Kingdom (Moncel & Ashton, 2018). In the Arabian Peninsula, humans were a bit lazy and dull, though, as very few tools were made which were also very similar (Shipton et al., 2018). About 400,000 years ago, however, Neanderthals took a completely different approach in Europe: instead of ending up with one large stone with two cutting edges, it was the remaining core that was now further worked into finer tools. Thanks to the well-known Levallois technique, they thus obtained long and thin stone blades that were razor-sharp. This provided the entry of a new tool culture, the pre-Mousterian (Doronichev, 2016; Moncel et al., 2020; Picin et al., 2013). Tools were now no longer limited to fist axes, people now obtained sharp knives that were used to work other materials such as wood and bone (wooden spears over 400,000 years old are known from Clacton-on-Sea in the United Kingdom and Schöningen in Germany) (Conard et al., 2020; Wragg Sykes, 2020). Sapiens discovered already 300,000 years ago that this technique produced more and sharper stones, as the site at Jebel Irhoud in Morocco demonstrates (Hublin et al., 2017). Neanderthals refined their motor skills and discovered other materials, switching to Mousterian technology about 160,000 years ago, which they continued to use until they became extinct (Doronichev, 2016). Shortly before that, however, about 45,000 years ago, Neanderthals appeared to have rediscovered their creativity. New materials were combined in new ways to perform new functions, known as Chatelperronian technology. This new boost, which they did not get to enjoy for very long, may have been inspired after the contact with sapiens who introduced their own new technology to Europe 43,000 years ago: the Aurignacian technology (followed up 28,000 years ago by a lot of other technologies) (Higham et al., 2014). The Chatelperronian and Aurignacian formed the basis of a veritable cultural revolution among Neanderthals and sapiens, from user objects for hunting, working, eating, to objects for decorating, sleeping, heating, burying, protecting, making a roof over one's head, breeding, trading, making symbolic objects, etc. (Demay et al., 2012; Wadley et al., 2020; Wilkins et al., 2021; Willoughby, 2021). Neanderthals were already processing birds of prey's claws to make necklaces 160,000 years ago (Radovcic et al., 2015). More than 50,000 years ago, they even made whistles out of bones (Turk et al., 2018; Turk et al., 2020). A true cultural evolution only briefly successful in Neanderthals but one that would take over sapiens' biological evolution about 50,000 years ago (Lieberman, 2013).

## Handy Harry's

Some of us are more dexterous than others, but as a human species we stand out in our dexterity ... literally then. No other animal can use the hands in such a refined and powerful way as we do. We can grasp stones so firmly that we can knock off controlled pieces with another stone. But we are also very good at throwing objects both far and with great precision (Lieberman, 2013). A great ape throws with a straight arm. We fold our arm as we start throwing from behind our head, making a whip-like motion which makes us throw harder. Some even manage to throw a javelin 104.8m or a baseball at a speed of 170 km/h. At the same time, we also manage to produce very refined throws, as in darts (Wolfe et al., 2006).

These highly complex movements require an alignment between perceiving, thinking, planning and motor control. During our development, we go through six stages to finally reach the point where we can think and make symbolic movements with the hands, such as to depict objects (Parker, 2015). Surprisingly, great apes and humans are the only ones to get to that point (we reach that stage as early as two to five years of age). Prosimians are already stuck in the second stage, which is little more than repeating general movements and playing with the fingers. Other primates get as far as stage 4, as they can manipulate an object to get to another object. Capuchin monkeys get one step further and use sticks to fish in holes for food. But they can't put anything back together that was taken apart or portray objects with their hands. We, along with great apes, can do so, something very useful for creatively transforming objects into utensils.

Are our hands unique then? Sapiens hands can already perform three unique manipulations that even great apes fail to do. We need only one hand to reposition an object we grasp, while great apes need both hands to do so. We can grasp an object very forcefully and precisely with the inside of the thumb and the other fingers, whereas great apes use their fingertips for that and cannot grasp objects forcefully enough to, for example, hit them hard against another object. And we can hold an elongated object very forcefully while the fingers lie diagonally along the palm and so the object lies in line with the forearm, like the handle of a hammer. Great apes can grasp like this but not forcefully (Kivell et al., 2022). Even australopithecines did not yet have hands for a powerful precision grip, even though the thumb was easily opposable. So only *Homo* could do this (Karakostis et al., 2021; Lieberman, 2013; Tocheri et al., 2008). The hand in sapiens and Neanderthals (possibly also Denisova humans and naledi humans) is characterized by short fingers and a long, stout and strongly opposable thumb (the thumb is completely facing the palm with its underside). We have remarkably large muscles that forcefully close and open the thumb, as well as a large thumb muscle that flexes and holds the tip of the thumb steady (Bucchi et al., 2020; Rolian et al., 2011). Even our little finger, although our smallest finger, is firmly supported by a robust metacarpal bone (more substantial than the metacarpal bones of all other fingers). We may not realize it, but the little finger is very important when we hold an object to strike it with something. Great apes, on the other hand, have long and slimmer fingers, useful for climbing. Hominins have short fingers and long thumbs, more convenient for grasping and manipulating objects more forcefully (Williams-Hatala et al., 2020). So one can argue that our hand evolutionarily changed from a powerful hand for climbing vertically to a hand whose thumb remained long but became heftier and the fingers became short, enabling a powerful precision grip (Kivell et al., 2022; Tocheri et al., 2008). A thumb that remained long may even be associated with a big toe that remained larger than the other toes, as there appears to be similar development and evolution between anterior and posterior limbs (Rolian, 2014; Rolian, Liberman, et al., 2010). A unique mutation in the *HACNS1* gene may also help explain the origin of the human thumb (Prabhakar et al., 2008).

## Being creative with materials

Creative tool making and use manifested itself in the way different materials were used for different purposes, and how materials were even deliberately selected for their useful properties. The 400,000-year-old wooden spears were made of yew wood, known for its hardness (Wragg Sykes, 2020). As early as 500,000 years ago, people used fire in Europe and the Middle East, in part to make the tip of such spears harder and sharper, even though this came at the expense of the strength of the spear itself (Condemi & Savatier, 2019; Ennos & Chan, 2016; Goren-Inbar et al., 2004; Roebroeks & Villa, 2011). Like wood, bone and ivory have the property of being more finely workable, without breaking as quickly as stones do. Thanks to the sharp stone blades obtained through the Levallois technique, people were able to reduce bone and ivory to a variety of shapes. People were already manufacturing needles with an eye more than 50,000 years ago (Shunkov et al., 2020). This suggests that even then people were already working and binding hides together to protect themselves against the cold. But other small bone fragments or teeth were also perforated to make bracelets and necklaces, such as the recently found deer tooth with a hole in it (about 25,000 years old) (Essel et al., 2023). But bone was also worked more refined as early as 400,000 years ago, as evidenced by remains of an elephant bone from Casto di Guido in Italy (Villa et al., 2021). Neanderthals and sapiens also knew how to process raw materials to produce a dye, such as ochre (a red substance derived from manganese dioxide, also used to make fire) (Heyes et al., 2016). They used this 250,000 years ago to decorate their bodies, caves, shells or other objects (Condemi & Savatier, 2019; d'Errico et al., 2016; Mayer et al., 2009; Roebroeks et al., 2012; Suddendorf et al., 2020). Only sapiens succeeded, about 7000 years ago, in producing and crafting metal objects, ushering in the Copper Age (Radivojević et al., 2010; Wragg Sykes, 2020).

The dexterity and material to make small objects opened the way to now combining even small pieces into a complex tool, one that worked better than the individual pieces. As early as 500,000 years ago, pieces of stone or bone were combined with wooden sticks, using bitumen (a natural pitch that surfaces in some places) or plant resins to glue everything together (also used as a fuel for fire to heat plant material) (Hardy & Buckley, 2017; Hardy et al., 2012; Ochando et al., 2024; Picin et al., 2013; Schmidt et al., 2019; Schmidt et al., 2024; Shipton, 2022; Wilkins et al., 2012). Fine points were also attached to a wooden shaft via sinews of animals and plant fibers, so that 100,000 years and earlier arrows and pointed spears were made (Lieberman, 2013; Wragg Sykes, 2020). This solved the problem of the less sturdy tips of wooden spears, allowed the tips to be quickly replaced if damaged (without having to make a completely new spear), and provided weapons that could be used to hunt even more dangerous and larger mammals (Shea, 2010; Wilkins et al., 2012). Neanderthals and sapiens further refined their techniques thanks to an ability to plan, know the right materials and where to find them, provide spare parts, and much more (Langley & Suddendorf, 2022; Wragg Sykes, 2020; Wragg Sykes, 2017).

## Talking with hands and feet

### From tools to language

Having the dexterity to make tools is one thing. Being able to explain something figuratively with hands and feet to someone else is something else. It may come as a surprise but there is an evolutionary link between language and the dexterity for making tools, especially in the way both are controlled by our brain. We already talked about language centers, such as Broca's center and Wernicke's center, being larger on the left side of the brain. We also mentioned the fact that most people are right-handed, with the right hand being controlled from ... yep, the left side of the brain. And we also talked about the involvement of these language centers in sign language, a form of communication that combines handedness and language. We all use our hands now and then to explain something (Corballis, 2003).

An asymmetry in the brain indicates lateralization, where certain actions are controlled more from one side of the brain. For example, we have lateralization for controlling our hands and for producing language, and both develop in a coupled manner (people with pronounced lateralization for language are more pronounced right-handed) (Pritchard et al., 2019). Based on brain shape and thus skull shape, one can even infer whether a person is right-handed. In sapiens, as well as Neanderthals, this asymmetry is more pronounced than in other hominins, possibly due to more extensive use of the hands for sign language (Corballis, 2012; Holloway, 2015; Luef, 2018). Broca's center is located in a brain fold (the lower frontal gyrus) that is responsible for understanding and producing language, as well as manipulating objects. In fact, this language center is also activated when we manipulate tools, as in other primates (Higuchi et al., 2009). Even the temporal lobe, commonly thought of as the language lobe, contains a fold (the upper temporal gyrus) that responds to movements of hands, body and face, as well as to speech (Stout & Chaminade, 2012).

Communication through sign language is older than humans, as apes also make extensive use of it (Hobaiter et al., 2022). The area that forms our center of Broca is not responsible for speech production in monkeys but plays a role in sign language. This area contains so-called mirror neurons: neurons that are responsible both for performing hand movements and for recognizing hand movements in others (Acharya & Shukla, 2012; Corballis, 2003). In addition, the posterior part of Broca's center is involved in movements of the mouth, such as chewing. According to some, the source of speech movements must be found here, which from this brain center was linked to movements for communication through sign language. Whereas the predominance for communication in monkeys and apes lies primarily in gestures, in human species this has slowly shifted to a predominance in vocal communication (Corballis, 2003). One could argue that we slowly evolved from pantomime or body expressions to verbal communication with articulate speech, a form of communication that was much more diverse, efficient and brought about other advantages (Zywiczynski et al., 2021). For example, it freed up hands, useful for making better tools while communicating and allowed non-visual communication useful during group-hunting or power scavenging (Bickerton & Szathmary, 2011; Szamado & Szathmary, 2006). Yet, sign language remains important, just think about how you try to explain something to someone who speaks another language. This verbal ability only really became fully developed from Neanderthals and sapiens, about 200,000 years ago (Ferretti & Adornetti, 2021).

There is another aspect that tools and language share: the increase in complexity. Monkeys communicate verbally through simple shouts and grunts, mainly to elicit a response from a conspecific (Torti, 2021). But this is already functional, as this is how alarm cries can protect conspecifics. For example, green vervets use different grunts to alert groupmates when a predator is nearby, grunts that differ depending on the type of predator (Cheney & Seyfarth, 1982). If one plays the sound they make when they see a bird of prey in the sky, the monkeys rush into the bushes. If one plays the sound for a predator on the ground, the monkeys rush into the trees (Torti, 2021). Knowing where best to hide depending on the species of predator makes these very simple signals extremely functional. These simple grunts could somehow be compared to words: individual sounds that stand alone but are not combined. What Neanderthals and sapiens already did, however, was combine words into sentences, i.e. syntax. This gives a specific combination of words much more meaning than the individual words. The words 'lion', 'behind', 'me' and 'walk' take on slightly different meanings in 'I walk behind the lion' or 'the lion walks behind me'. Syntax thus opens doors for an almost infinite number of combinations and unlimited transmission of functional information (Griesser et al., 2018). The evolution of tools also experienced this: the composite tools of Neanderthals and sapiens are also a creative composition of parts (such as words) into a better working whole (such as a sentence). Both conversions required a mental change that allowed one to think in terms of composing rather than chopping into pieces, something that came about 200,000 years ago and was fully developed in sapiens when they made the move from Africa 70,000 years ago (Rosenberg & Trevathan, 2002; Rutherford, 2019). Brain scans show that the same brain areas are active when we use tools and syntax. Even by training language, tool use improves and vice versa (Thibault et al., 2021).

## The tools to talk

Producing articulated language requires a lot of functional units. Call it anatomical tools, like the language centers in the brain. Broca mainly provides the motor control to speak, and Wernicke provides the understanding of language. Also involved are the temporal cortex where sound information comes in, the visual cortex to see objects we can name (such as letters or symbols), the parietal cortex that converts visual information to motor information, and the frontal cortex to plan how we will process language information to respond to it (Ardesch et al., 2019). Compared to other primates, our language centers are more highly developed, with neurons making many more connections with each other (Palomero-Gallagher & Zilles, 2019).

Producing language obviously only works if all these centers exchange information with each other in a meaningful way. This is done through nerve pathways called 'streams'. Tool use is controlled via two such streams (Stout & Chaminade, 2012). An upper stream sends visual information from the occipital lobe to the parietal lobe, where it is translated into motor information for doing something with an object.

A lower stream exchanges that visual information with the temporal lobe, where it is converted to information about the object that is stored in memory. This information is also transmitted to the frontal lobe, where a plan is made of what to do with the object. By having people make different tools during experiments, brain scans showed that fewer areas are involved in making Oldowan than Acheulean tools (Stout & Chaminade, 2012). Surprising or not, but two streams also exist for language, albeit between other parts of the brain. There are more than two, but the main ones are an upper stream that maps sound and translates it into speech (runs from the center of Wernicke to the center of Broca) and a lower stream that translates sound into something meaningful (runs from the temporal lobe to the frontal lobe) (Alagoz et al., 2022; Catani, 2009; Kolb & Whishaw, 2014; Luef, 2018; Michon et al., 2019).

With all the information available today, it is safe to say that Neanderthals possessed a very similar ability as sapiens to use language (Dediu & Levinson, 2013). Genetically, they share a variant of the *FOXP-2* gene, previously thought to be unique to sapiens (Denisova humans also have that variant) (Spiteri et al., 2007; Vernes et al., 2007). This gene is partly responsible for the development of certain parts of the brain involved in language, including the control of facial muscles and speech (a negative mutation in this gene leads to language impairment in humans) (Atkinson et al., 2018; van 't Hoog, 2005). The gene is older than that, as also mice have it, where it also plays a role in communication (mice use ultrasonic communication) (Szamado & Szathmary, 2006). The anatomy of Neanderthals' brains, their respiratory system (such as the size of the rib cage) and their speaking apparatus also shows a system that allowed articulate speech (Condemi & Savatier, 2019; Jordan, 1999). In 1983, a 60,000-year-old hyoid (tongue bone) of a Neanderthal was found in a cave in Kebara, Israel (Arensburg et al., 1989). Our tongue is attached to the hyoid and its movement is essential for forming vowels (Lieberman, 1993). Both the general shape, muscle attachment sites and internal structure of this bone suggest that Neanderthals used it in a very similar way to sapiens (D'Anastasio et al., 2013). But this is not the only requirement, as the position of the hyoid bone within the pharynx is also crucial for producing different vowels. In infants, the hyoid bone, together with the larynx and the epiglottic valve that closes the trachea during swallowing, abut against the palate (Lieberman, 2013; Pagano et al., 2022). This allows them to continue breathing through the nose while drinking milk without choking. As we get older, the larynx descends. The advantage is that there is now an extended cavity in which sound vibrations are amplified and influenced in such a way that vowels can be formed. The disadvantage is that we choke more easily. It was long thought that the larynx was not yet so deeply descended in Neanderthals (Lieberman, 1993). However, this does not seem to be correct (Arensburg & Tillier, 1991; Dediu & Levinson, 2013). The auditory system also shows that Neanderthals had a similar ability to hear sounds and tones as is the case in sapiens (Conde-Valverde et al., 2021; Wragg Sykes, 2020).

Speaking of vowels, there is also something to be said about consonants. For example, our evolution to a softer farmer's diet had an impact on our jaw apparatus, with all sorts of negative effects (such as bad teeth, wisdom teeth that don't have enough space to break through and even difficult passage of air causing sleep problems) (Kahn et al., 2020). At the same time, it would also have had a positive effect on how we came to produce consonants. Looking at older sapiens, the teeth in the upper and lower jaws are much more aligned with the edges facing each other. Once agriculture and softer foods appeared, we see phenomena such as overbite (upper teeth overlapping frontally and vertically with the lower teeth) and overjet (upper teeth lying horizontally more forward than the lower teeth) appearing more frequently (Blasi et al., 2019). Having to bite less hard puts less pressure on to keep the teeth nicely opposed to each other. But this would have had an advantage for producing a new set of consonants, which only come about when the lower lip can be pushed against the upper teeth: the 'f' and the 'v'. By looking at the evolution of languages, one could even find out that these consonants are limited in so-called proto-languages that were used for 6000–8000 years (in only 3% of cases), compared to 76% in current languages.

## So ...

Inhabitants of the northern hemisphere, Neanderthals, Denisova humans and sapiens, managed to survive in cold regions, thanks to anatomical, physiological and technological novelties. Neanderthals were not inferior to sapiens in many areas, but only sapiens survived. Our whole evolution accelerated culturally when we decided to switch from hunting and gathering to agriculture and animal husbandry. However, this cultural shift also caused anatomical and genetic changes through natural selection, which are still manifest today. The variation in skin color, the refinement of articulate language, the ability to digest milk, these are just a few examples. The road was opened for an evolutionary highway to a new man, a modern man. A man who, by changing his environment himself, confronted his own evolutionary past. A sapiens adapted to make great physical efforts, to fight for food and endure periods of hunger, eventually bearing the consequences of this in an environment where sitting in a chair for hours and unbridled access to energy-rich food becomes the rule.

# 8

# What now?

Do we then know enough about our past so that we can now predict what the future sapiens will bring? We know a lot, but far from everything. We are even still discovering body parts whose existence or function were little or unknown, such as some muscles in the hips or ligaments in the knee (Benias et al., 2018; Breckling et al., 2022; Claes et al., 2013). Every day, studies appear that further expose our genetic evolution, including the genes we inherited from our close relatives. It is becoming increasingly clear that we should no longer speak of how natural selection and a changing environment directed 'our' genes and 'our' evolution as sapiens. Rather, it is a matter of how the exchange of our genes with those of other species has had an impact on how sapiens became sufficiently adapted to the environment. Also the 'environment' can no longer be limited to vegetation, climate or interaction with other organisms. We ourselves have primarily started to define our 'environment': sapiens who have come to co-determine its own ecology. Is this then what has made us so unique as a species?

## Are we truly unique then?

Sapiens has had an enormous impact on its environment through cultural and technological innovations, ignited with the rise of agriculture and specialized tool technology. Even behaviorally we have redirected our own evolution. For example, we domesticated ourselves, which facilitated living and working together in large communities. It's not that we made pets out of ourselves, but rather that we favored similar traits that were selected for in pets: less aggression. By experimentally selecting for more social and less aggressive silver foxes and continuing to breed with them, Dmitry Belyaev was able to show in 1959 that selection for these behaviors also brought about anatomical changes after only ten generations (males began to look more like females, ears were more drooping, they acquired more of a puppy look due to a shorter snout, ...) (Dugatkin, 2020; Perry, 2018). Extending this to sapiens, an increasing preference for less aggressive males in a group could explain why, for example, the snout shortened and the jaw apparatus became less robust (this, of course, in conjunction with a change in diet) or eyebrow arches virtually disappeared (Balcarcel et al., 2021; Gibbons, 2014). Self-domestication might even be partly behind how language came about or how our childhood lengthened (Benitez-Burraco & Progovac, 2021; Gibbons, 2014). Does this make us unique? Surely not, it may even be more common than initially thought (as in bonobos and recently observed in elephants) (Raviv et al., 2023).

DOI: 10.1201/9781003588795-9

If we look at individual traits and compare them to other animals, we are not unique in using and making tools, being able to solve problems, being able to follow the direction of someone's gaze (via the white in our eyes), being able to recognize ourselves in a mirror, being able to mimic others, being able to mentally project to how others think, having self-awareness, exhibiting social behavior or our language (Richerson et al., 2021; Roth & Dicke, 2019). Even getting drunk is not something only we do, something Charles Darwin already knew (Darwin, 1871). That our brains are big but not special for a primate with such a body size, we also know that by now. However, spherical brains are unique, but only to the later sapiens (Kuhlwilm & Boeckx, 2019). Of a top ten traits considered unique to sapiens, there is actually none that cannot be found in some form in other animals as well (Geggel, 2022).

Diving deeper into our biology, one would expect that we would have lots of unique genes or gene variants. But even there the question is mostly how many of these are unique to sapiens because until recently we only had the chimpanzee genome to compare with (McLean et al., 2011). This does tell us what happened before or after the split between the two, but not yet whether this was unique for sapiens (Almécija et al., 2021). With the Neanderthal genome included, this picture does become more and more clear. For example, it was possible to reduce the number of genetic differences between sapiens and chimpanzee from about 35 million to about 30,000 differences in 571 genes between sapiens and Neanderthals (Kuhlwilm & Boeckx, 2019). Of many of these, however, it is not clear what specific traits they brought about that would make sapiens unique. Then again, not everything is applicable to the entire species, as evidenced by a lot of regional differences resulting from crosses with other human species (Condemi & Savatier, 2019; Dannemann & Racimo, 2018; Peyregne et al., 2017). We can suspect some changes to be unique to sapiens, such as having many copies of amylase genes (three adjacent genes lie on the genome: *AMY1*, *AMY2A* and *AMY2B*) that express the enzymes in the salivary glands that break down starch (Bolognini et al., 2024; Inchley et al., 2016). The more variants one has, the more efficient one can digest starch. This genetic adaptation was useful after switching to a diet rich in grains, once we became farmers, which makes it no coincidence that the occurrence of these variants drastically increased since about 12,000 years ago. But analogous to the mutation for lactose tolerance, this is a sapiens trait that does not apply similarly to all sapiens populations. Extensive fading of the skin due to multiple mutations in the *SLC24A5* gene combined with a starchy diet is unique to European sapiens (Sabeti et al., 2007). Sleek and jet-black hair is unique to East Asian sapiens due to a mutation in the *EDAR* gene that appeared about 30,000 years ago (Hawks, 2016; Hlusko et al., 2018). Are you of the sweaty type and have sticky earwax? Then you probably have an original form of the *ABCC11* gene, which was altered about 30,000–20,000 years ago in many East Asians, causing them to produce less smelly sweat and rather dry earwax (Hawks, 2016). Just over 3,000 years ago, a mutation appeared among African and Indian sapiens that produced a variant of blood hemoglobin, making them more resistant to malaria. Another mutation did something analogous among East Asians.

Big difference: Africans and Indians are stuck with sickle cell anemia because of their mutation, East Asians are not.

Summarized, it is not our individual traits that make us unique as a species but rather the degree and combination in which they occur. No other species combines an ability to run marathons with very large brains, bare skin full of sweat glands, a social behavior and technical skill that led to industrialization and megacities, or a language so versatile that it can even be used to write songs and books. Mozarts and Hararis are found only among sapiens.

## Are we still evolving?

Where we are also unique is that we have managed to build a barrier to shield ourselves from natural selection. One might rightly ask whether natural selection still has any hold on us and thus whether we are still evolving. Nico Van Straalen wrote an entire book on that question (van Straalen & Roelofs, 2017). David Attenborough, the famous BBC documentary filmmaker, once stated that our evolution has passed (Hawks, 2016). Is this even true? Better is as Jonathan Pritchard, a population geneticist at Stanford University, formulated an answer to the question of whether we are still evolving: *"We certainly are. But the answer to the question of how we are changing is far more complicated"* (Pritchard, 2013).

The genetic and anatomical resources we are equipped with today were available when our sapiens ancestors had left Africa. The initial genetic divide between the sapiens who migrated to East Asia and the Anatolians who provided agriculture in Europe has since greatly blurred into a world population that is the result of subsequent migrations and remigrations, expansions and relapses of populations, and numerous admixtures between peoples (Rotival et al., 2021; Vallini et al., 2022). Even the skin color of people today is not only explained by the gene variants present that we talked about but also access to vitamin-rich diet and means to protect themselves from the sun play a role (Kim et al., 2021; Parra et al., 2003).

Yet the genome of the sapiens who left Africa provided us with adaptations that we still bear the fruits of today. At the same time, there are genetic changes that were advantageous to those early sapiens but still bring little benefit or are even rather disadvantages in our current environment. We even have adaptations that are the result of how we changed our environment. Just think about lactose tolerance, our wisdom teeth issues, or the major changes in the composition and diversity of our gut flora and the bacteria found in the oral cavity (Curry, 2021b; Fellows Yates et al., 2021; Kahn et al., 2020; Wibowo et al., 2021). During the last 40,000 years there has been an acceleration in the number of sweeps of positive selection on about 7% of our genes, due in part to the ever-increasing human population (Hawks et al., 2007). With the onset of the Neolithic, and thus agriculture, this accelerated even further and sapiens evolved 100 times faster than right before it (Ward, 2013). The peak in the number of variants selected was reached slightly earlier in Africa (about 8,000 years ago) than in Europe (about 5,000 years ago) (Hawks et al., 2007). Even since the start of the 20th century, changes in the skull and mandible have occurred, due to industrialization (Kilroy et al., 2020). In current sapiens, one can even see a link between diet, lifestyle, medical practices and the shape of the pelvis, all factors that affect a smooth birth (Mitteroecker et al., 2021). But not all adjustments are beneficial now. For example, agriculture greatly narrowed the diversity of our diet because it only paid off by investing time in those varieties of plants and animals that gave high yields. Agriculture did produce lots of cultivated varieties, but they are genetically much poorer than the original ones (such as European tomatoes, for example) that we had 1000–2000 years ago (Blanca et al., 2022; Stock et al., 2023). The modern human diet consists on average of 55–80% carbohydrates (Larbey et al., 2019).

All evidence directs to our evolution not having stopped at all, but that it no longer applies homogenously to the species as a whole. For that, we have become too large and too diverse a population that has assimilated into many different environments and that impacted its own natural selection. Recent and ongoing sapiens evolution and adaptations is thus mostly local. For example, Tibetans exhibit unique mutations in two genes (*EGLN1* and *EPAS1*) that make them better adapted to survive in low-oxygen altitudes, mutations that spread very rapidly about 3,000 years ago (Lorenzo et al., 2014; Pritchard, 2013; Xu et al., 2011). A recent study showed that the adaptations that come with these mutations also result in higher offspring and thus have a direct effect on their evolutionary fitness (Ye et al., 2024). The sea nomads from

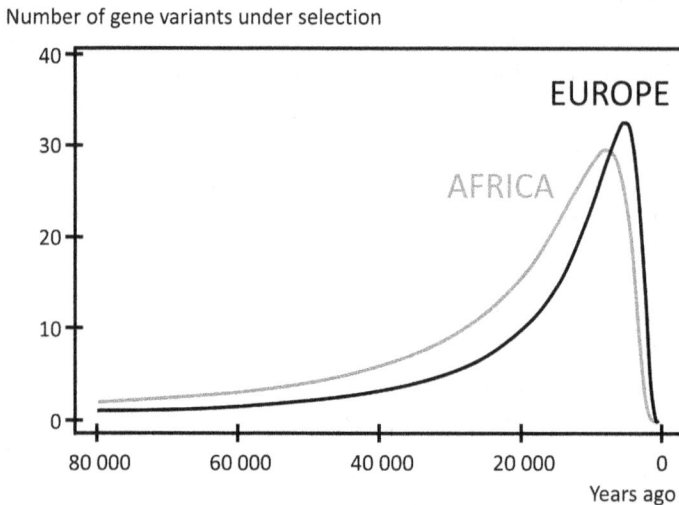

Figure 8.1    Humans have experienced an accelerated evolution over the last 20,000 years, as evidenced by the number of gene variants that have been the subject of positive selection. In Africa, the peak was reached 8,000 years ago, while it was 5250 years ago in Europe. (Modified from Hawks et al. (2007).)

Bajau in Southeast Asia experienced an evolution about a thousand years ago that has better adapted them to do what their workday consists of for more than 60% of the time: diving up to 70 m deep by holding their breath. Mutations in some genes have allowed their spleens to get bigger and to produce more red blood cells (the spleen is an organ that stores a lot of blood) (Ilardo et al., 2018). When they dive, the spleen contracts, allowing more blood to enter the organs and thus also 3–10% more oxygen. Very useful for collecting more food in a single dive. Even more recently, natural selection has spread a mutation that better protects people from the West African Lassa virus, a deadly virus to which more than 30 million people have been exposed. In fact, the Yoruba population in Nigeria has a mutation in two genes (*LARGE* and *IL2*) that better shields them from the virus (Andersen et al., 2012; Sabeti et al., 2007). When the mutation originated is not known, but the Lassa virus might have emerged only 153 years ago in New Guinea in West Africa (Fichet-Calvet et al., 2019).

# What will the future bring?

I often get asked about what we will look like in a hundred or a thousand years, and in what direction we will continue to evolve. Will we live even longer (life expectancy has increased from 70 years to 82 years since 1960 (Bank))? Will we become organisms with enormously large brains sitting on a body that is no longer adapted to great physical exertion? Will another new human species emerge? We walk, work, live in an industrialized society with a body that has largely been evolutionarily shaped when we were still walking around in grass savannahs, surviving ice ages and becoming farmers. An ever-increasing conflict between this ancestral body and the current environment manifests itself, among other things, in the numerous societal diseases (also called 'mismatch diseases') that are increasingly frequent (Leonard et al., 2010; Lieberman, 2013).

However, predicting our evolutionary future nevertheless remains speculation, and this is because we cannot predict the further course of two essential components of our evolution: future changes in the environment and how we will escape natural selection through new techniques we will develop ourselves. Will our brains permanently enlarge? Yes, if people with a smaller brain somehow have a lower chance of reproducing. Research already suggests that this is not the case, at least at the level of intelligence, quite the contrary. In fact, socioeconomic status and intelligence are inversely related to the number of children one has (Hopcroft, 2015). I also don't see a law being passed on anytime soon that prohibits anyone below a certain brain volume from procreating. Thankfully. And whether a larger head will be more attractive to partners also seems unlikely, when we know that babies with larger heads are more likely to have problems in childbirth. Our evolution has frequently shown that brain volume by itself is not an overly good indicator of how well or how poorly someone can function in an environment. Let alone that it is as reliable a signal for mate choice as a peacock's tail or an antelope's pronking to show off their vitality.

Can we not predict anything at all about our future? Assuming that the already prevailing trends will continue, we can make a few educated guesses about how things might progress. For example, our Y chromosome carries very few genes compared to other chromosomes (Hughes et al., 2005). Based on the rate at which this chromosome has lost genes in the past (45 genes remain of the original 1438 in mammals), it is predicted that the chromosome will no longer have a function within a decade or so (Aitken & Marshall Graves, 2002; Graves, 2004). Should we worry about that? Probably not (even though the chromosome contains the *SRY* gene that causes an embryo to develop a male phenotype). Most likely, by then we as a species will have faced bigger problems that will have caused a major shift in our genetic evolution: reduced fertility, global warming, overpopulation, biodiversity crisis, .... The irony is that where climate change was one of the main natural drivers that guided our evolution, we are now the first species to change this climate ourselves (Gavrilets, 2015; Muthukrishna et al., 2018). Is this what makes us unique as a species? Perhaps *Homo destructus* is a better name than *Homo sapiens* (Klein & Takahata, 2002). Let us hope that in the end, mental sapiens wisdom prevails, and we can continue our evolution on and with this planet and its inhabitants for millions more years to come. Or will we face an evolutionary race with artificial sapiens wisdom? It is up to sapiens man to keep common sense there.

# Glossary

**Acheulean**  A prehistoric stone tool industry characterized by large hand axes, typically processed on both sides (hence referred to as bifacies), manufactured by several species of early humans, starting from *Homo erectus*.

**Adipose tissue**  A type of body tissue that stores fat in specialized cells, providing energy reserves that can be used for metabolic activities in periods that there is a food shortage. It also functions as a thermoregulatory isolator.

**Amniota**  A clade of tetrapod vertebrates that includes reptiles, birds and mammals, characterized by having an amniotic egg.

**Anthropoids**  The group of primates that comprises monkeys, apes and humans.

**Apes**  The group of primates derived from Old World monkeys that are characterized by the absence of a tail. Recent taxa include gibbons, orangutans, gorillas, chimpanzees and bonobos but also humans that evolved from an ape ancestor.

**Arboreal**  Referring to animals that live and move in trees.

***Ardipithecus***  An early genus of hominins that lived in Africa between around 5.8 and 4.4 million years ago.

**Aurignacian**  An advanced stone tool and art culture from the Upper Paleolithic period (between about 43,000 and 26,000 years ago) associated with early European *Homo sapiens* when they dispersed out of Africa.

**Australopithecines**  Multiple species of early hominins within two main genera (*Australopithecus* and *Paranthropus*) that lived in East, Central and South Africa (between 3.8 and 0.3 million years ago). They are considered ancestors of the genus *Homo*.

***Australopithecus***  A genus with several species of early hominins, the oldest one being *A. anamensis* and the species *A. afarensis* to which the famous fossil Lucy has been assigned to. Fossil evidence is known from East and South Africa, from a period between 3.8 and 1.8 million years ago).

**Bent-hip-bent-knee**  A walking posture hypothesized to have been used by some early hominins, in which both the pelvis and the knees were not fully extended yet, resulting in a not as efficient way of walking that that of later *Homo* species.

**Bipedal**  Walking on two legs.

**Brachiator**  An animal that moves by swinging from branch to branch using its arms while hanging below branches. Typically, their arms and shoulders are more mobile and longer than their legs, such as in gibbons.

**Carbohydrates**  Organic compounds that are an important source for metabolic energy, found in foods like bread, rice and fruits.

**Catarrhini**   The group of primates that includes Old World monkeys, apes and humans. The name refers to their nostrils being direct downward. Their sister group are the Platyrrhini, the New World monkeys.

**Cellulose**   A complex, natural, carbohydrate polymer that forms the main component of plant cell walls but is hard to be digested by plant-eating mammals. Frequently symbiotic unicellular organisms (such as bacteria or ciliates) live in the intestinal canal of these mammals to digest the cellulose.

**Cerebellum**   The so-called small brain, which is the part of the brain that coordinates movement and balance (among other things). It lies behind the large brain or cerebrum.

**Cerebrum**   The largest part of the brain, responsible for higher cognitive functions like planning, problem-solving, designing, coordinating complex movement, generating language, memorizing etc.

**Chatelperronian**   A stone tool and art culture of the Upper Paleolithic period from Western Europe, associated with Neanderthals and early *Homo sapiens* (between 45,000 and 40,000 years ago). It was preceded by the Mousterian culture and followed in *Homo sapiens* by the Aurignacian.

**Clambering**   Way of climbing in trees used by some primates, including some apes and early hominins, where both hands (to grasp onto overhead branches) and feet (to grasp onto branches onto which they walk) are combined with an upright position of the body.

**Dendrites**   Branch-like extensions of neurons (nerve cells) that receive signals from other neurons. They are covered with synapses through which the connections with other neurons are established.

**Denisovans**   A late but extinct *Homo* species from Eastern Europe, known from DNA evidence and a few fossil remains. They formed the sister-species of Neanderthals.

**Dexterity**   Motor skills and ease in using the hands to manipulate objects and perform complex movements, such as needed to make stone tools.

**Diurnal**   Active during the day (opposite is nocturnal, active during the night).

**Enamel**   The hardest, outer layer in teeth that due to its very dense mineralization, makes teeth very common among fossil evidence.

**Encephalization**   The degree of brain size relative to body size, frequently quantified in so-called encephalization quotients, where numbers larger than one indicate that the brain is larger than what would be expected based on the body size.

**Endocasts**   Natural or artificial casts of the inside of a skull, showing imprints of the brain's surface folds or blood vessels.

**Endurance running**   The ability to sustain prolonged physical effort to keep on running for an extensive time and distance, a trait that evolved in hominins from *Homo erectus* on.

**Eumelanin**   A type of melanin pigment that gives hair, skin and eyes a brown to black color and is able to absorb UV-irradiation efficiently.

**Fallback food**   Alternative food sources used when preferred foods are scarce, generally implying food of a lower nutritional quality or harder to process and digest.

**Glacial and interglacial periods**   Glacial periods are ice-age periods, when the Earth experienced long periods of cold climate that resulted in the formation of large ice caps and glaciers (especially in Northern Europe), creating a harsh environment for Neanderthals and *Homo sapiens* to survive. Glacial periods were alternated by periods of better climatic conditions. These periods are referred to interglacial periods.

**Glial cells**   Supportive cells in the nervous system that protect, control and nourish neurons.

**Glucose**   A simple molecule of a carbohydrate, that is the most important energy source for the body that is taken up through the intestines. Cells use glucose to fuel their metabolic activities.

**Gyrus**   A ridge or fold in the cerebral cortex (large brain) and the cerebellum (small brain), that are especially well elaborately developed in human brains. Brains containing these folds are referred to as 'gyrencephalic'.

**Haplorrhini**   The group of primates that includes tarsiers, monkeys, apes and humans.

**Holocene**   A geological epoch that started about 11.7 thousand years ago, following after the Last Glacial Period.

**Hominidae**   The family of primates that comprises apes (both extinct and extant great apes) most closely related to and including the human lineage, with extant species being modern humans, chimpanzees and bonobos, gorillas and orangutans.

**Hominins**   Depending on the source, different interpretations can be given to this group of primates. When considering it as a synonym of the clade called 'Hominini', then it includes the genus *Pan* (chimpanzee and bonobo) and the complete human lineage. However, often and as done throughout this book, it refers only to the taxa that are part of the human lineage, thus after the split from the *Pan*-lineage.

**Hominoidea**   The group of primates that includes apes and humans.

*Homo*   The genus that includes multiple species of a lineage of hominins that originated from an australopithecine ancestor, starting with *Homo habilis* and leading to the origin of *Homo sapiens*.

**Hunter-gatherers**   People who obtain the majority of their food by hunting animals and gathering plants, rather than by cultivating plants or herding animals.

**Isotopes**   Variants of a chemical element with different numbers of neutrons. Some of these variants are stable ('stable isotopes') that can be present for a long period of time, allowing paleoanthropologists to use them to perform dating analyses of fossils or to make estimations of the diet of an extinct species.

**Joules**   A unit of energy, just like kilocalories (1 kilocalories equals 4184 Joules).

**Keratin**   A protein that is formed in the outer skin layer (epidermis) and that makes up hair and nails (but also feathers, scales or spines).

**Kilocalories**   A unit of energy commonly used to measure the energy content of food (also referred to as 'large Calories', written with a capital C, where 1 kilocalories equals 1 Calories and 1000 'small' calories).

**Lactase**   An enzyme that is needed to break down lactose, the sugar in milk. It is naturally produced in newborn mammals but its production stops after the lactation period. In early agricultural societies of *Homo sapiens*, a mutation enabled lactase production to continue, leading to lactose tolerance all the way into adulthood.

**Lactose**   A sugar found in milk, broken down in the intestine by the enzyme lactase.

**Larynx**   The voice box, located in the throat where vocal cords are attached to cartilaginous elements and controlled by muscles to produce sounds at different tones. A descended larynx originated late during human evolution and is linked to the origin of language using vowels.

**Levallois technique**   A specialized technique for stone tool production used by Neanderthals and *Homo sapiens*, where a large stone could be split into multiple blades with a good control over the size and shape of the blades.

**Lomekwi**   An archaeological site in Kenya known for some of the oldest stone tools of which australopithecines are considered to be the makers. This tool culture is then also referred to as Lomekwi culture.

**Lordosis**   A forward curve in the spine, where in a natural spine such a lordosis exists in the neck and the lower back. The latter is referred to as the lumbar lordosis, which is an adaptation to bipedal walking as it functions as a shock absorber.

**Melanin**   A pigment that gives color to skin, hair and eyes. It comes in two main forms: eumelanin and pheomelanin.

**Melanocytes**   Cells that produce melanin, which is stored in organelles called melanosomes.

**Miocene**   A geological epoch from about 23 to 5 million years ago, during which apes were very successful as they dispersed from Africa into Eurasia. During this epoch, the human lineage got split from that of the ape lineage leading to chimpanzee and bonobo.

**Mitochondria**   Organelles in cells that produce energy and that carry their own DNA (mtDNA). Both the organelles and their DNA are inherited only from the mother and thus allow to reconstruct so-called matrilineal evolutionary relationships.

**Mousterian**   A stone tool culture used by European Neanderthals and early African *Homo sapiens*, characterized by refined flakes production (from the period between 160,000 and 40,000 years ago).

**mtDNA**   Mitochondrial DNA, inherited from the mother.

**Mutation**   A change in one or more base pairs in the DNA string, that can be traced throughout evolution and can lead to adaptive evolution of novel traits.

**Nails (vs claws)**   Flat structures on the dorsal side of the tips of fingers and toes, made of dense keratin and where the underlying bone (distal phalanx) is cylindrical in shape. In contrast, claws are pointed and downward-directed keratinous sheets supported by a claw-shaped phalanx. All primates have nails, where prosimians and some New World monkeys still have some claws.

**nDNA**   Nuclear DNA, found in the cell nucleus. In contrast to mitochondrial DNA, the nDNA is inherited both from the father and the mother.

**Neanderthals**   An extinct species of hominins that lived in Europe and Asia, that split off from the Denisovans about 450,000 years ago and became extinct until about 30,000 years ago.

**Neocortex**   The outer layer of the mammalian large brain (cerebrum) that in many species, including primates, gets enlarged through the formation of folds and grooves.

**Neolithic**   The later phase of the Stone Age that started with the transition from hunter-gathering to farming by *Homo sapiens*, about 12,000 years ago.

**Neuron**   A nerve cell that transmits signals in the nervous system and stores memory by making synaptic connections with other neurons.

**Neurotransmitters**   Chemicals released by neurons to transmit signals between neurons.

**Nocturnal**   Active during the night.

**Oldowan**   Primitive stone tool technology produced by some australopithecines and the first *Homo* (such as *Homo habilis*), characterized by hand axes with flakes removed from a single side.

**Opposable digit**   A thumb or big toe that has an increased mobility in the joint at the base, allowing it to diverge and rotate so that it can touch the other fingers or toes, allowing for grasping and crucial for tool making.

*Orrorin*   An early hominin species that lived around 6 million years ago in Africa.

**Orthograde**   Upright posture.

**Palaeoanthropology**   The study of human evolution, especially focusing on human fossils and that of their relatives.

**Paleolithic**   The early phase of the Stone Age, characterized by the emergence of stone tool usage (about 3.3 million years ago during the Pliocene). It continues until the end of the Pleistocene, about 12,000 years ago (it is followed by the Neolithic).

*Paranthropus*   A genus of robust australopithecines, characterized by a very robust jaw apparatus and associated specializations in the skull, jaws and teeth.

**Pelvis**   The bony structure that attaches to the base of the spine through the sacroiliac joint and articulates with both legs at the level of the hip bones.

**Phalanges**   The bones of the fingers and toes.

**Pheomelanin**   A type of melanin pigment that gives hair and skin a reddish-yellow color and is less efficient in absorbing UV (compared to the darker eumelanin).

**Pleistocene**   A geological epoch, from about 2.6 million to 12,000 years ago, known for repeated glaciations.

**Plesiadapiformes**   An extinct group of primate-like mammals that lived during the Paleocene and Eocene epochs. They are considered to be the sister group to the primates.

**Pliocene**   A geological epoch from about 5.3 to 2.6 million years ago, preceding the Pleistocene.

**Prefrontal cortex**   The front part of the brain's frontal lobe, involved in complex behaviors like decision-making and social interactions.

**Prehensile**   Capable of grasping, using tails (like in some New World monkeys) or limbs (using specialized opposable thumb and large toe like in primates).

**Primates**   The order of mammals that is characterized by adaptations to climbing in trees, and that includes lemurs, monkeys, apes and humans.

**Pronograde**   Walking with the body parallel to the ground, typical of quadrupeds.

**Prosimians**   The group of primates that includes lemurs, lorises and tarsiers.

**Quadrupeds**   Animals that walk on four legs.

***Sahelanthropus***   An early hominin species that lived around 7 million years ago in Africa.

***Sapiens***   Species name of the modern human species, belonging to the genus *Homo*, hence the full species name is written as '*Homo sapiens*'.

**Starch**   Large polymers of carbohydrates found in plants, serving as a major energy source in the human diet.

**Stereoscopic vision**   The ability to perceive depth and three-dimensional structure, resulting from the combination of images from both eyes. It requires both eyes to have an overlapping field of view, as well as special neurological connections between the eye and the brain hemispheres.

**Strepsirrhini**   The group of primates that includes lemurs and lorises.

**Synapse**   The junction between two neurons where signals are transmitted, where the number of synapses depends on the amount of information being processed and memory being stored.

**Syntaxis**   The arrangement of words and phrases to create well-formed sentences in a language.

**Tetrapods**   Vertebrates with four limbs, including amphibians, reptiles, birds and mammals.

**UV-B**   Ultraviolet B radiation, a component of sunlight that can cause skin damage and break down vitamins, like folate, but also allows to produce other vitamins, like vitamin $D_3$. Skin pigments like eumelanin can absorb this radiation.

# Suggested reading per chapter

## Chapter 1 – Fragments of our existence

Berger, L. R., & Hilton-Barber, B. (2000). *In the footsteps of Eve. The mystery of human origins*, p. 323. National Geographic; Washington, DC.

Green, R. E., Krause, J., Briggs, A. W., Maricic, T., Stenzel, U., Kircher, M., Patterson, N., Li, H., Zhai, W. W., Fritz, M. H. Y., Hansen, N. F., Durand, E. Y., Malaspinas, A. S., Jensen, J. D., Marques-Bonet, T., Alkan, C., Prufer, K., Meyer, M., Burbano, H. A.,. . . & Paabo, S. (2010, May 7). A draft sequence of the neandertal genome. *Science, 328*(5979), 710–722. https://doi.org/10.1126/science.1188021

Johanson, D. C., & Wong, K. (2010). *Lucy's legacy. The quest for human origins*. Three Rivers Press.

Krause, J., & Trappe, T. (2019). *A short history of humanity: How migration made us who we are* (C. Waight, Trans.). Virgin Digital.

Pääbo, S. (2014). *Neanderthal man: In search of lost genomes*. Basic Books.

## Chapter 2 – What makes us unique?

Almécija, S., Hammond, A. S., Thompson, N. E., Pugh, K. D., Moyà-Solà, S., & Alba, D. M. (2021, May 7). Fossil apes and human evolution. *Science, 372*(6542), eabb4363. https://doi.org/10.1126/science.abb4363

Browne, J. (2021). Introduction. In J. M. DeSilva (Ed.), *A most interesting problem. What Darwin's descent of man got right and wrong about human evolution* (pp. 1–23). Princeton University Press.

Darwin, C. (1859). *The origin of species by means of natural selection, on the preservation of favoured races in the struggle for life* (1st ed.). John Murray. https://archive.org/stream/originofspecies00darwuoft?ref=ol

Darwin, C. (1871). *The descent of man, and selection in relation to sex* (1st ed.). John Murray. https://archive.org/stream/descentofmansele00darw?ref=ol#page/n6/mode/2up

Harari, Y. N. (2018). *Sapiens* (Nederlandse vertaling, 27e druk ed.). Uitgeverij Thomas Rap.

Herculano-Houzel, S. (2016). *The human advantage: How our brains became remarkable*. MIT Press.

Huxley, T. H. (1863). *Evidence as to man's place in nature*. D. Appleton and Company. https://archive.org/details/evidenceasto00huxl

Lieberman, D. (2013). *The story of the human body*. Penguin Books.

Shubin, N. H. (2008). *Your inner fish*. Pantheon Books.

Wragg Sykes, R. (2020). *Kindred Neanderthal life, love, death and art*. Bloomsbury Sigma.

Zimmer, C., & Emlen, D. J. (2013). *Evolution: Making sense of life*. Roberts and Company Publishers, Inc.

# Chapter 3 – The equatorial Eden

Aiello, L. C., & Dean, C. (1996). *An introduction to human evolutionary anatomy*. Academic Press.

Ankel-Simons, F. (2007). *Primate anatomy: An introduction* (3rd ed.). Academic Press. https://www.dropbox.com/s/w28col6sf8do23d/Primate%20Anatomy%20-%20An%20 Introduction.pdf?dl=0

Begun, D. R. (2003, 2003). Planet of the apes. *Scientific American, 289*(2), 64–73.

Bryson, B. (2020). *The body: A guide for occupants*. Transworld Publishers Ltd.

Chudler, E. H. (2020, 29/06/2020). *Brain facts and figures*. Retrieved 10/06/2022 from http://faculty.washington.edu/chudler/facts.html

Condemi, S., & Savatier, F. (2019). *Wie vermoordde de neanderthalers?*, p. 286. Paris: EPO vzw.

DeSilva, J. (2021). *First steps. How upright walking made us human*. Harper Collins Publishers.

Harcourt-Smith, W. E. H. (2013). The origins of bipedal locomotion. In *Handbook of paleoanthropology* (pp. 1–36). Springer-Verlag. https://doi.org/10.1007/978-3-642-27800-6_48-3

Herculano-Houzel, S. (2016). *The human advantage: How our brains became remarkable*. MIT Press.

Jablonski, N. G. (2012). *Living colors. The biological and social meaning of skin color*. University of California Press.

Kolb, B., & Whishaw, I. Q. (2014). *An introduction to brain and behavior* (4th ed.). Worth Publishers.

Lieberman, D. (2013). *The story of the human body*. Penguin Books.

Milton, K. (2012). Diet and primate evolution. *Scientific American, 12*, 22–29.

Rolian, C. (2014, May 6). Genes, development, and evolvability in primate evolution. *Evolutionary Anthropology, 23*(3), 93–104. https://doi.org/10.1002/evan.21409

Whiten, A., Goodall, J., McGrew, W. C., Nishida, T., Reynolds, V., Sugiyama, Y., Tutin, C. E., Wrangham, R. W., & Boesch, C. (1999, Jun 17). Cultures in chimpanzees. *Nature, 399* (6737), 682–685. https://doi.org/10.1038/21415

# Chapter 4 – The human cradle

Aiello, L. C., & Wheeler, P. (1995, Apr). The expensive-tissue hypothesis – The brain and the digestive system in human and primate evolution. *Current Anthropology, 36*(2), 199–221.

Almécija, S., Hammond, A. S., Thompson, N. E., Pugh, K. D., Moyà-Solà, S., & Alba, D. M. (2021, May 7). Fossil apes and human evolution. *Science, 372*(6542), eabb4363. https://doi.org/10.1126/science.abb4363

Andrews, P. (2020, Mar). Last common ancestor of apes and humans: Morphology and environment. *Folia Primatologica, 91*(2), 122–148. https://doi.org/10.1159/000501557

Coppens, Y. (1994, 1994). East side Story: The origin of humankind. *Scientific American, 270*(5), 62–69.

Domínguez-Rodrigo, M. (2014). Is the "Savanna hypothesis" a dead concept for explaining the emergence of the earliest hominins? *Current Anthropology, 55*(1), 59–81. https://doi.org/10.1086/674530

Fleagle, J. G., Stern, J. T., Jungers, W. L., Susman, R. L., Vangor, A. K., & Wells, J. P. (1981). Climbing: A biomechanical link with brachiation and with bipedalism. *Symposia of the Zoological Society of London, 48*, 359–375.

Lieberman, D. E. (2011). *The evolution of the human head*. The Belknap Press of Harvard University Press.

## Chapter 5 – The forest opens up

Berger, L. R., & Hilton-Barber, B. (2000). *In the footsteps of Eve. The mystery of human origins*, p. 323. National Geographic; Washington, DC.

Carey, T. S., & Crompton, R. H. (2005, Jan). The metabolic costs of 'bent-hip, bent-knee' walking in humans. *Journal of Human Evolution, 48*(1), 25–44. https://doi.org/10.1016/j.jhevol.2004.10.001

Collard, M. (2003). Grades and transitions in human evolution. *Speciation of Modern Homo Sapiens, 106*, 61–100.

Dawkins, R., & Wong, Y. (2004). *The ancestor's tale. A pilgrimage to the dawn of evolution* (1st ed.). Houghton Mifflin (US).

DeSilva, J. (2021). *First steps. How upright walking made us human*. Harper Collins Publishers.

DeSilva, J. (2022). Walks of life. *Scientific American, 327*(5), 72–81.

Galis, F. (1999). Why do almost all mammals have seven cervical vertebrae? Developmental constraints, Hox genes, and cancer. *Journal of Experimental Zoology, 285*, 19–26.

Jablonski, N. G., & Chaplin, G. (2000, 2000). The evolution of human skin coloration. *Journal of Human Evolution, 39*, 57–106.

Johanson, D. C., & Wong, K. (2010). *Lucy's legacy. The quest for human origins*. Three Rivers Press.

Laden, G., & Wrangham, R. (2005, Oct). The rise of the hominids as an adaptive shift in fallback foods: Plant underground storage organs (USOs) and australopith origins. *Journal of Human Evolution, 49*(4), 482–498. https://doi.org/10.1016/j.jhevol.2005.05.007

Levy, R. (2024, Mar 1). The prefrontal cortex: From monkey to man. *Brain, 147*(3), 794–815. https://doi.org/10.1093/brain/awad389

Wheeler, P. E. (1994, Apr). The thermoregulatory advantages of heat-storage and shade-seeking behavior to hominids foraging in equatorial Savanna environments. *Journal of Human Evolution, 26*(4), 339–350. https://doi.org/10.1006/jhev.1994.1021

Wrangham, R. W., Jones, J. H., Laden, G., Pilbeam, D., & Conklin-Brittain, N. (1999a, Dec). The raw and the stolen – Cooking and the ecology of human origins. *Current Anthropology, 40*(5), 567–594. https://doi.org/10.1086/300083

## Chapter 6 – Fully erect!

Allen, J. S. (2009). *The lives of the brain*. The Belknap Press of Harvard University Press.

Anton, S. C. (2003). Natural history of *Homo erectus*. *American Journal of Physical Anthropology, 122*(S37), 126–170. https://doi.org/10.1002/ajpa.10399

Ayala, F. J., & Cela-Conde, C. J. (2017). *Processes in human evolution*. Oxford University Press.

Bramble, D. M., & Lieberman, D. E. (2004). Endurance running and the evolution of *Homo*. *Nature, 432*(7015), 345–352. https://doi.org/10.1038/nature03052

Chaplin, G., Jablonski, N. G., & Cable, N. T. (1994, Dec). Physiology, thermoregulation and bipedalism. *Journal of Human Evolution, 27*(6), 497–510. https://doi.org/10.1006/jhev.1994.1066

Chazan, M. (2017, Aug). Toward a long prehistory of fire. *Current Anthropology, 58*, S351–S359. https://doi.org/10.1086/691988

Folk, G. E., & Semken, H. A. (1991). The evolution of sweat glands. *International Journal of Biometeorology, 35*(3), 180–186. https://doi.org/10.1007/Bf01049065

Fonseca-Azevedo, K., & Herculano-Houzel, S. (2012, Nov 6). Metabolic constraint imposes trade-off between body size and number of brain neurons in human evolution. *Proceedings of the National Academy of Sciences of the United States of America, 109*(45), 18571–18576. https://doi.org/10.1073/pnas.1206390109

Gibbons, A. (2011). Who was *Homo habilis*—And was it really *Homo? Science, 332*, 1370–1371.

Harcourt-Smith, W. E. H. (2013). The origins of bipedal locomotion. In *Handbook of paleoanthropology* (pp. 1–36). Springer-Verlag. https://doi.org/10.1007/978-3-642-27800-6_48-3

Herculano-Houzel, S. (2012). Neuronal scaling rules for primate brains: The primate advantage. In M. A. Hofman & D. Falk (Eds.), *Evolution of the primate brain: From neuron to behavior* (Vol. 195, pp. 325–340). Elsevier B.V. https://doi.org/10.1016/B978-0-444-53860-4.00015-5

Holloway, R. L. (2015). The evolution of the hominid brain. In W. Henke & I. Tattersall (Eds.), *Handbook of paleoanthropology* (Vol. 3, pp. 1961–1987). Springer-Verlag Berlin-Heidelberg. https://doi.org/10.1007/978-3-642-39979-4_81

Jablonski, N. G. (2017, 02/28/2017). *The real 'skin in the game': The history of naked, sweaty, and colorful skin in the human lineage.* UC Berkeley. https://www.uctv.tv/shows/The-Real-Skin-in-the-Game-The-History-of-Naked-Sweaty-and-Colorful-Skin-in-the-Human-Lineage-32129?s=09

Kimbel, W. H., & Villmoare, B. (2016, Jul 5). From *Australopithecus* to *Homo*: The transition that wasn't. *Philosophical Transactions of the Royal Society of London. Series B, Biological Sciences, 371*(1698). https://doi.org/10.1098/rstb.2015.0248

Smil, V. (2002). Eating meat: Evolution, patterns, and consequences. *Population and Development Review, 28*(4), 599–639.

Uchida, T. K., & Delp, S. L. (2020). *Biomechanics of movements. The science of sports, robotics, and rehabilitation.* The MIT Press.

# Chapter 7 – Conquer the world

Arensburg, B., & Tillier, A. M. (1991). Speech and the Neanderthals. *Endeavour, 15*(1), 26–28. https://doi.org/10.1016/0160-9327(91)90084-o

Bocquet-Appel, J. P., & Degioanni, A. (2013, Dec 1). Neanderthal demographic estimates. *Current Anthropology, 54*, S202–S213. https://doi.org/10.1086/673725

Chaplin, G. (2004, Nov). Geographic distribution of environmental factors influencing human skin coloration. *American Journal of Physical Anthropology, 125*(3), 292–302. https://doi.org/10.1002/ajpa.10263

Dannemann, M., & Kelso, J. (2017, Oct 5). The contribution of Neanderthals to phenotypic variation in modern humans. *American Journal of Human Genetics, 101*(4), 578–589. https://doi.org/10.1016/j.ajhg.2017.09.010

Gibbons, A. (2015, 06/02/2015). *How modern humans ate their way to world dominance.* Science Magazine. Retrieved 11/05/2021 from https://www.sciencemag.org/news/2015/02/how-modern-humans-ate-their-way-world-dominance

Gibbons, A. (2021b, Apr 9). When modern humans met Neanderthals. *Science, 372*(6538), 115–116. https://doi.org/10.1126/science.372.6538.115

Jordan, P. (1999). *Neanderthal.* Sutton Publishing Limited.

Leonard, W. R., Stock, J. T., & Valeggia, C. R. (2010, May-Jun). Evolutionary perspectives on human diet and nutrition. *Evolutionary Anthropology, 19*(3), 85–86. https://doi.org/10.1002/evan.20250

Mitteroecker, P., & Fischer, B. (2024, Mar). Evolution of the human birth canal. *American Journal of Obstetrics and Gynecology, 230*(3S), S841–S855. https://doi.org/10.1016/j.ajog.2022.09.010

Pääbo, S. (2014). *Neanderthal man: In search of lost genomes*. Basic Books.

Raff, J. (2021). Journey into the Americas. *Scientific American, 324*(5), 26–33.

Rolian, C. (2014, May 6). Genes, development, and evolvability in primate evolution. *Evolutionary Anthropology, 23*(3), 93–104. https://doi.org/10.1002/evan.21409

Rowley-Conwy, P. (2024, Mar 5). Hunter-gatherers and earliest farmers in western Europe. *Proceedings of the National Academy of Sciences of the United States of America, 121*(10), e2322683121. https://doi.org/10.1073/pnas.2322683121

Stout, D., & Chaminade, T. (2012, Jan 12). Stone tools, language and the brain in human evolution. *Philosophical Transactions of the Royal Society B-Biological Sciences, 367*(1585), 75–87. https://doi.org/10.1098/rstb.2011.0099

Wong, K. (2015). Neandertal minds. *Scientific American, 312*(2), 36–43.

Wragg Sykes, R. (2020). *Kindred Neanderthal life, love, death and art*. Bloomsbury Sigma.

# Chapter 8 – What now?

Dannemann, M., & Racimo, F. (2018, Dec). Something old, something borrowed: Admixture and adaptation in human evolution. *Current Opinion in Genetics & Development, 53*, 1–8. https://doi.org/10.1016/j.gde.2018.05.009

Geggel, L. (2022, 02/02/2022). *Top 10 things that make humans special* LiveScience. Retrieved 25/06/23 from https://www.livescience.com/15689-evolution-human-special-species.html

Gibbons, A. (2014, Oct 24). How we tamed ourselves–and became modern. *Science, 346*(6208), 405–406. https://doi.org/10.1126/science.346.6208.405

Hawks, J. (2016). Still evolving (after all these years). *Scientific American, 25*(4), 100–105.

Klein, J., & Takahata, N. (2002). *Where do we come from? The molecular evidence for human descent*. Springer-Verlag.

Leonard, W. R., Stock, J. T., & Valeggia, C. R. (2010, May-Jun). Evolutionary perspectives on human diet and nutrition. *Evolutionary Anthropology, 19*(3), 85–86. https://doi.org/10.1002/evan.20250

Lieberman, D. (2013). *The story of the human body*. Penguin Books.

Pritchard, J. K. (2013). How we are evolving? *Scientific American, 22*(1), 99–105.

Roth, G., & Dicke, U. (2019). Origin and evolution of human cognition. In M. A. Hofman (Ed.), *Progress in brain research* (2019/11/11 ed., Vol. 250, pp. 285–316). Elsevier B.V. https://doi.org/10.1016/bs.pbr.2019.02.004

Ward, P. (2013). What may become of *Homo sapiens*? *Scientific American, 22*(1), 107–111.

# References

Acharya, S., & Shukla, S. (2012). Mirror neurons: Enigma of the metaphysical modular brain. *Journal of Natural Science, Biology, and Medicine, 3*(2), 118–124. https://pmc.ncbi.nlm.nih.gov/articles/PMC3510904/

Ackerman, S. (1992). The development and shaping of the brain. In *Discovering the brain* (pp. 86–103). National Academies Press (US). https://doi.org/10.17226/1785

Adamantidis, A. (2022). How the gut talks to the brain. *Science, 376*(6590), 248–249.

Adegboyega, M. T., Stamos, P. A., Hublin, J. J., & Weaver, T. D. (2021). Virtual reconstruction of the Kebara 2 Neanderthal pelvis, *Journal of Human Evolution, 151*, 102922, https://doi.org/10.1016/j.jhevol.2020.102922

Adhikari, K., Mendoza-Revilla, J., Sohail, A., Fuentes-Guajardo, M., Lampert, J., Chacon-Duque, J. C., Hurtado, M., Villegas, V., Granja, V., Acuna-Alonzo, V., Jaramillo, C., Arias, W., Lozano, R. B., Everardo, P., Gomez-Valdes, J., Villamil-Ramirez, H., Silva de Cerqueira, C. C., Hunemeier, T., Ramallo, V., Schuler-Faccini, L., Salzano, F. M., Gonzalez-Jose, R., Bortolini, M. C., Canizales-Quinteros, S., Gallo, C., Poletti, G., Bedoya, G., Rothhammer, F., Tobin, D. J., Fumagalli, M., Balding, D., & Ruiz-Linares, A. (2019). A GWAS in Latin Americans highlights the convergent evolution of lighter skin pigmentation in Eurasia. *Nature Communications, 10*(1), 358. https://doi.org/10.1038/s41467-018-08147-0

Aerts, P., D'Aout, K., Thorpe, S., Berillon, G., & Vereecke, E. (2018). The gibbon's Achilles tendon revisited: consequences for the evolution of the great apes? *Proceedings. Biological sciences, 285*(1880). https://doi.org/10.1098/rspb.2018.0859

Ahern, J. C. (2005). Foramen magnum position variation in *Pan troglodytes*, Plio-Pleistocene hominids, and recent *Homo sapiens*: implications for recognizing the earliest hominids. *American Journal of Physical Anthropology, 127*(3), 267–276. https://doi.org/10.1002/ajpa.20082

Ahern, J. C. M. (2015). Archaic *Homo*. In *Basics in human evolution* (pp. 163–176). Elsevier Inc. https://doi.org/10.1016/b978-0-12-802652-6.00012-8

Aiello, L. C., & Dean, C. (1996). *An introduction to human evolutionary anatomy.* Academic Press.

Aiello, L. C., & Wheeler, P. (1995). The expensive-tissue hypothesis: The brain and the digestive-system in human and primate evolution. *Current Anthropology, 36*(2), 199–221. https://doi.org/10.1086/204350

Aitken, R. J., & Marshall Graves, J. A. (2002). The future of sex. *Nature, 415*(6875), 963. https://doi.org/10.1038/415963a

Akat, E., Yenmis, M., Pombal, M. A., Molist, P., Megias, M., Arman, S., Vesely, M., Anderson, R., & Ayaz, D. (2022). Comparison of vertebrate skin structure at class level: A review. *Anatomical Record (Hoboken, N.J.: 2007)*. https://doi.org/10.1002/ar.24908

Alagoz, G., Molz, B., Eising, E., Schijven, D., Francks, C., Stein, J. L., & Fisher, S. E. (2022). Using neuroimaging genomics to investigate the evolution of human brain structure. *Proceedings of the National Academy of Sciences of the United States of America, 119*(40), e2200638119. https://doi.org/10.1073/pnas.2200638119

Alba, D. M., Almecija, S., DeMiguel, D., Fortuny, J., Perez de los Rios, M., Pina, M., Robles, J. M., & Moya-Sola, S. (2015). Miocene small-bodied ape from Eurasia sheds light on hominoid evolution. *Science (New York, N.Y.), 350*(6260), aab2625. https://doi.org/10.1126/science.aab2625

Alemseged, Z., Spoor, F., Kimbel, W. H., Bobe, R., Geraads, D., Reed, D., & Wynn, J. G. (2006). A juvenile early hominin skeleton from Dikika, Ethiopia. *Nature, 443*(7109), 296–301. https://doi.org/10.1038/nature05047

Alexander, R. M. (1991). Energy-saving mechanisms in walking and running, *The Journal of Experimental Biology, 160*, 55–69. https://doi.org/10.1242/jeb.160.1.55

Alibardi, L. (2022). Vertebrate keratinization evolved into cornification mainly due to transglutaminase and sulfhydryl oxidase activities on epidermal proteins: An immunohisto-chemical survey. *Anatomical Record (Hoboken, N.J.: 2007), 305*(2), 333–358. https://doi.org/10.1002/ar.24705

Allen, J. S. (2009). *The lives of the brain*. The Belknap Press of Harvard University Press.

Allen, J. S., Bruss, J., & Damasio, H. (2006). Looking for the lunate sulcus: A magnetic resonance imaging study in modern humans. *Anatomical Record Part A-Discoveries in Molecular Cellular and Evolutionary Biology, 288a*(8), 867–876. https://doi.org/10.1002/ar.a.20362

Allen, K. L., & Kay, R. F. (2011). Dietary quality and encephalization in platyrrhine primates. *Proc Biol Sci*. https://doi.org/10.1098/rspb.2011.1311

Allentoft, M. E., Sikora, M., Fischer, A., Sjogren, K. G., Ingason, A., Macleod, R., Rosengren, A., Schulz Paulsson, B., Jorkov, M. L. S., Novosolov, M., Stenderup, J., Price, T. D., Fischer Mortensen, M., Nielsen, A. B., Ulfeldt Hede, M., Sorensen, L., Nielsen, P. O., Rasmussen, P., Jensen, T. Z. T., Refoyo-Martinez, A., Irving-Pease, E. K., Barrie, W., Pearson, A., Sousa da Mota, B., Demeter, F., Henriksen, R. A., Vimala, T., McColl, H., Vaughn, A., Vinner, L., Renaud, G., Stern, A., Johannsen, N. N., Ramsoe, A. D., Schork, A. J., Ruter, A., Gotfredsen, A. B., Henning Nielsen, B., Brinch Petersen, E., Kannegaard, E., Hansen, J., Buck Pedersen, K., Pedersen, L., Klassen, L., Meldgaard, M., Johansen, M., Uldum, O. C., Lotz, P., Lysdahl, P., Bangsgaard, P., Petersen, P. V., Maring, R., Iversen, R., Wahlin, S., Anker Sorensen, S., Andersen, S. H., Jorgensen, T., Lynnerup, N., Lawson, D. J., Rasmussen, S., Korneliussen, T. S., Kjaer, K. H., Durbin, R., Nielsen, R., Delaneau, O., Werge, T., Kristiansen, K., & Willerslev, E. (2024). 100 ancient genomes show repeated population turnovers in Neolithic Denmark. *Nature, 625*(7994), 329–337. https://doi.org/10.1038/s41586-023-06862-3

Almécija, S., Hammond, A. S., Thompson, N. E., Pugh, K. D., Moyà-Solà, S., & Alba, D. M. (2021). Fossil apes and human evolution. *Science, 372*(6542), eabb4363. https://doi.org/10.1126/science.abb4363

Almécija, S., Smaers, J. B., & Jungers, W. L. (2015). The evolution of human and ape hand proportions. *Nature Communications, 6*, 7717. https://doi.org/10.1038/ncomms8717

Altamura, F., Lehmann, J., Rodríguez-Álvarez, B., Urban, B., van Kolfschoten, T., Verheijen, I., Conard, N. J., & Serangeli, J. (2023). Fossil footprints at the late lower Paleolithic site of Schöningen (Germany): A new line of research to reconstruct animal and hominin paleo-ecology. *Quaternary Science Reviews*. https://doi.org/10.1016/j.quascirev.2023.108094

Amato, K. R., Chaves, O. M., Mallott, E. K., Eppley, T. M., Abreu, F., Baden, A. L., Barnett, A. A., Bicca-Marques, J. C., Boyle, S. A., Campbell, C. J., Chapman, C. A., De la Fuente, M. F., Fan, P., Fashing, P. J., Felton, A., Fruth, B., Fortes, V. B., Grueter, C. C., Hohmann, G., Irwin, M., Matthews, J. K., Mekonnen, A., Melin, A. D., Morgan, D. B., Ostner, J., Nguyen, N., Piel, A. K., Pinacho-Guendulain, B., Quintino-Aredes, E. P., Razanaparany, P. T., Schiel, N., Sanz, C. M., Schulke, O., Shanee, S., Souto, A., Souza-Alves, J. P., Stewart, F., Stewart,

K. M., Stone, A., Sun, B., Tecot, S., Valenta, K., Vogel, E. R., Wich, S., & Zeng, Y. (2021). Fermented food consumption in wild nonhuman primates and its ecological drivers. *American Journal of Physical Anthropology*, *175*(3), 513–530. https://doi.org/10.1002/ajpa.24257

Ami, O., Maran, J. C., Gabor, P., Whitacre, E. B., Musset, D., Dubray, C., Mage, G., & Boyer, L. (2019). Three-dimensional magnetic resonance imaging of fetal head molding and brain shape changes during the second stage of labor. *PloS One*, *14*(5), e0215721. https://doi.org/10.1371/journal.pone.0215721

An, N. A., Zhang, J., Mo, F., Luan, X., Tian, L., Shen, Q. S., Li, X., Li, C., Zhou, F., Zhang, B., Ji, M., Qi, J., Zhou, W. Z., Ding, W., Chen, J. Y., Yu, J., Zhang, L., Shu, S., Hu, B., & Li, C. Y. (2023). De novo genes with an lncRNA origin encode unique human brain developmental functionality. *Nature Ecology & Evolution*, *7*(2), 264–278. https://doi.org/10.1038/s41559-022-01925-6

Andersen, K. G., Shylakhter, I., Tabrizi, S., Grossman, S. R., Happi, C. T., & Sabeti, P. C. (2012). Genome-wide scans provide evidence for positive selection of genes implicated in Lassa fever. *Philosophical Transactions of the Royal Society B-Biological Sciences*, *367*(1590), 868–877. https://doi.org/10.1098/rstb.2011.0299

Andrews, P. (2020). Last common ancestor of apes and humans: Morphology and environment. *Folia Primatologica*, *91*(2), 122–148. https://doi.org/10.1159/000501557

Ankel-Simons, F. (2007). *Primate anatomy: An introduction* (3rd ed.). Academic Press.

Ankel-Simons, F., & Rasmussen, D. T. (2008). Diurnality, nocturnality, and the evolution of primate visual systems. *American Journal of Physical Anthropology*, *137*(S47), 100–117, https://doi.org/10.1002/ajpa.20957

Anton, S. C. (2003). Natural history of *Homo erectus*. *American Journal of Physical Anthropology*, *122*(S37), 126–170. https://doi.org/10.1002/ajpa.10399

Antón, S. C., Potts, R., & Aiello, L. C. (2014). Evolution of early *Homo*: An integrated biological perspective. *Science*, *345*(6192), 1236828. https://doi.org/10.1126/science.1236828

Aoki, K. (2002). Sexual selection as a cause of human skin colour variation: Darwin's hypothesis revisited. *Annals of Human Biology*, *29*(6), 589–608. https://doi.org/10.1080/03014460210000019144

Araiza, I., Meyer, M. R., & Williams, S. A. (2021). Is ulna curvature in the StW 573 ('Little Foot') *Australopithecus* natural or pathological? *Journal of Human Evolution*, *151*, 102927. https://doi.org/10.1016/j.jhevol.2020.102927

Arambourg, C., & Coppens, Y. (1968). Découverte d'un australopithecien nouveau dans les gisements de l'Omo (Ethiopie). *South African Journal of Science*, *64*(2), 58–59.

Ardesch, D. J., Scholtens, L. H., Li, L., Preuss, T. M., & Rilling, J. K. (2019). Evolutionary expansion of connectivity between multimodal association areas in the human brain compared with chimpanzees. *Proceedings of the National Academy of Sciences of the United States of America*, *116*(19), 7101–7106. https://doi.org/10.1073/pnas.1906107116

Arellano, C. J., & Kram, R. (2014). The metabolic cost of human running: Is swinging the arms worth it? *The Journal of Experimental Biology*, *217*(Pt 14), 2456–2461. https://doi.org/10.1242/jeb.100420

Arensburg, B., & Tillier, A. M. (1991). Speech and the Neanderthals. *Endeavour*, *15*(1), 26–28. https://doi.org/10.1016/0160-9327(91)90084-o

Arensburg, B., Tillier, A. M., Vandermeersch, B., Duday, H., Schepartz, L. A., & Rak, Y. (1989). A middle paleolithic human hyoid bone. *Nature*, *338*(6218), 758–760. https://doi.org/10.1038/338758a0

Arlegi, M., Pablos, A., & Lorenzo, C. (2023). Evolutionary selection and morphological integration in the foot of modern humans. *American Journal of Biological Anthropology.* https://
doi.org/10.1002/ajpa.24703

Armitage, S. J., Jasim, S. A., Marks, A. E., Parker, A. G., Usik, V. I., & Uerpmann, H. P. (2011). The
Southern route "Out of Africa": Evidence for an early expansion of modern humans into
Arabia. *Science, 331*(6016), 453–456. https://doi.org/10.1126/science.1199113

Arranz-Otaegui, A., Gonzalez Carretero, L., Ramsey, M. N., Fuller, D. Q., & Richter, T. (2018).
Archaeobotanical evidence reveals the origins of bread 14,400 years ago in northeastern
Jordan. *Proceedings of the National Academy of Sciences of the United States of America,
115*(31), 7925–7930. https://doi.org/10.1073/pnas.1801071115

Arsuaga, J. L., Martínez, I., Gracia, A., & Lorenzo, C. (1997). The Sima de los Huesos crania
(Sierra de Atapuerca, Spain). A comparative study. *Journal of Human Evolution, 33*(2-3),
219–281. https://doi.org/10.1006/jhev.1997.0133

Arzarello, M., De Weyer, L., & Peretto, C. (2016). The first European peopling and the Italian
case: Peculiarities and "opportunism". *Quaternary International, 393,* 41–50. https://doi.
org/10.1016/j.quaint.2015.11.005

Arzarello, M., & Peretto, C. (2010). Out of Africa: The first evidence of Italian peninsula occupation, *Quaternary International, 223,* 65–70, https://doi.org/10.1016/j.quaint.2010.01.006

Asfaw, B., White, T. D., Lovejoy, O., Latimer, B., Simpson, S., & Suwa, G. (1999, 1999).
*Australopithecus garhi:* a new species of early hominid from Ethiopia. *Science, 284,*
629–635.

Athreya, S., & Hopkins, A. (2021). Conceptual issues in hominin taxonomy: *Homo heidelbergensis* and an ethnobiological reframing of species. *American Journal of Physical
Anthropology, 175*(Suppl 72), 4–26. https://doi.org/10.1002/ajpa.24330

Atkinson, E. G., Audesse, A. J., Palacios, J. A., Bobo, D. M., Webb, A. E., Ramachandran, S., &
Henn, B. M. (2018). No evidence for recent selection at *FOXP2* among diverse human
populations. *Cell, 174*(6), 1424–1435.e15. https://doi.org/10.1016/j.cell.2018.06.048

Ayala, F. J., & Cela-Conde, C. J. (2017). *Processes in human evolution.* Oxford University Press.

Azevedo, F. A. C., Carvalho, L. R. B., Grinberg, L. T., Farfel, J. M., Ferretti, R. E. L., Leite, R. E. P.,
Jacob, W., Lent, R., & Herculano-Houzel, S. (2009). Equal numbers of neuronal and non-
neuronal cells make the human brain an isometrically scaled-up primate brain. *Journal of
Comparative Neurology, 513*(5), 532–541. https://doi.org/10.1002/cne.21974

Baab, K. L. (2008a). A re-evaluation of the taxonomic affinities of the early *Homo* cranium
KNM-ER 42700. *Journal of Human Evolution, 55*(4), 741–746. https://doi.org/10.1016/j.
jhevol.2008.02.013

Baab, K. L. (2008b). The taxonomic implications of cranial shape variation in *Homo erectus.*
*Journal of Human Evolution, 54*(6), 827–847. https://doi.org/10.1016/j.jhevol.2007.11.003

Baab, K. L. (2015). Defining *Homo erectus.* In W. Henke & I. Tattersall (Eds.), *Handbook of paleoanthropology* (pp. 2189–2219). Springer. https://doi.org/10.1007/978-3-642-39979-4_73

Bae, C. J., & Wu, X. (2024). Making sense of eastern Asian Late Quaternary hominin variability.
*Nature Communications, 15*(1), 9479. https://doi.org/10.1038/s41467-024-53918-7

Baker, L. B. (2019). Physiology of sweat gland function: The roles of sweating and sweat composition in human health. *Temperature (Austin, Tex.), 6*(3), 211–259. https://doi.org/10.
1080/23328940.2019.1632145

Balcarcel, A. M., Veitschegger, K., Clauss, M., & Sanchez-Villagra, M. R. (2021). Intensive
human contact correlates with smaller brains: Differential brain size reduction in cattle

types. *Proceedings. Biological Sciences, 288*(1952), 20210813. https://doi.org/10.1098/rspb.2021.0813

Balolia, K. L., Soligo, C., & Wood, B. (2017). Sagittal crest formation in great apes and gibbons. *Journal of Anatomy, 230*(6), 820–832. https://doi.org/10.1111/joa.12609

Balter, M. (2010). Archaeology. The tangled roots of agriculture. *Science (New York, N.Y.), 327*(5964), 404–406. https://doi.org/10.1126/science.327.5964.404

Balzeau, A., Buck, L. T., Albessard, L., Becam, G., Grimaud-Herve, D., Rae, T. C., & Stringer, C. (2017). The Internal Cranial Anatomy of the Middle Pleistocene Broken Hill 1 Cranium. *PaleoAnthropology*, 107–138. https://paleoanthro.org/media/journal/content/PA20170107.pdf

Bank, T. W. *World Development Indicators.* Retrieved 26/12/2021 from https://datatopics.world-bank.org/world-development-indicators/

Bard, E., Heaton, T. J., Talamo, S., Kromer, B., Reimer, R. W., & Reimer, P. J. (2020). Extended dilation of the radiocarbon time scale between 40,000 and 48,000 y BP and the overlap between Neanderthals and *Homo sapiens. Proceedings of the National Academy of Sciences of the United States of America,* https://doi.org/10.1073/pnas.2012307117

Bardo, A., Borel, A., Meunier, H., Guery, J. P., & Pouydebat, E. (2016). Behavioral and functional strategies during tool use tasks in bonobos. *American Journal of Physical Anthropology, 161*(1), 125–140. https://doi.org/10.1002/ajpa.23015

Bardo, A., Pouydebat, E., & Meunier, H. (2015). Do bimanual coordination, tool use, and body posture contribute equally to hand preferences in bonobos? *Journal of Human Evolution, 82*, 159–169. https://doi.org/10.1016/j.jhevol.2015.02.015

Bardo, A., Vigouroux, L., Kivell, T. L., & Pouydebat, E. (2018). The impact of hand proportions on tool grip abilities in humans, great apes and fossil hominins: A biomechanical analysis using musculoskeletal simulation. *Journal of Human Evolution, 125*, 106–121. https://doi.org/10.1016/j.jhevol.2018.10.001

Barnes, K. R., & Kilding, A. E. (2015). Running economy: Measurement, norms, and determining factors. *Sports Medicine - Open, 1*(1), 8. https://doi.org/10.1186/s40798-015-0007-y

Barr, W. A., & Wood, B. (2024). Spatial sampling bias influences our understanding of early hominin evolution in eastern Africa. *Nat Ecol Evol, 8*(11), 2113–2120. https://doi.org/10.1038/s41559-024-02522-5

Barrett, G., Revell, D., Harding, L., Mills, I., Jorcin, A., & Stiefel, K. (2019). Tool use by four species of Indo-Pacific Sea Urchins. *Journal of Marine Science and Engineering, 7*(3), 69. https://doi.org/10.3390/jmse7030069

Barrett, P. (2003). Palaeoclimatology: Cooling a continent. *Nature, 421*(6920), 221–223. https://doi.org/10.1038/421221a

Barrickman, N. L., Bastian, M. L., Isler, K., & van Schaik, C. P. (2008). Life history costs and benefits of encephalization: A comparative test using data from long-term studies of primates in the wild. *Journal of Human Evolution, 54*(5), 568–590. https://doi.org/10.1016/j.jhevol.2007.08.012

Bartlett, J. L., Sumner, B., Ellis, R. G., & Kram, R. (2014). Activity and functions of the human gluteal muscles in walking, running, sprinting, and climbing. *American Journal of Physical Anthropology, 153*(1), 124–131. https://doi.org/10.1002/ajpa.22419

Barton, R. A., & Harvey, P. H. (2000). Mosaic evolution of brain structure in mammals. *Nature, 405*(6790), 1055–1058. https://doi.org/10.1038/35016580

Bartsiokas, A., & Arsuaga, J.-L. (2020). Hibernation in hominins from Atapuerca, Spain half a million years ago. *L'Anthropologie, 124*(5), 102797. https://doi.org/10.1016/j.anthro.2020.102797

Bastir, M., Garcia-Martinez, D., Torres-Tamayo, N., Palancar, C. A., Beyer, B., Barash, A., Villa, C., Sanchis-Gimeno, J. A., Riesco-Lopez, A., Nalla, S., Torres-Sanchez, I., Garcia-Rio, F., Been, E., Gomez-Olivencia, A., Haeusler, M., Williams, S. A., & Spoor, F. (2020). Rib cage anatomy in *Homo erectus* suggests a recent evolutionary origin of modern human body shape. *Nat Ecol Evol*. https://doi.org/10.1038/s41559-020-1240-4

Bastir, M., Sanz-Prieto, D., & Burgos, M. (2021). Three-dimensional form and function of the nasal cavity and nasopharynx in humans and chimpanzees. *Anatomical Record (Hoboken, N.J.: 2007)*. https://doi.org/10.1002/ar.24790

Basu, S., Ramegowda, V., Kumar, A., & Pereira, A. (2016). Plant adaptation to drought stress. *F1000Research, 5*, 1554. https://doi.org/10.12688/f1000research.7678.1

Batchelor, C. L., Margold, M., Krapp, M., Murton, D. K., Dalton, A. S., Gibbard, P. L., Stokes, C. R., Murton, J. B., & Manica, A. (2019). The configuration of Northern Hemisphere ice sheets through the Quaternary. *Nature Communications, 10*(1), 3713. https://doi.org/10.1038/s41467-019-11601-2

Bates, K. T., McCormack, S., Donald, E., Coatham, S., Brassey, C. A., Charles, J., O'Mahoney, T., van Bijlert, P. A., & Sellers, W. I. (2025). Running performance in *Australopithecus afarensis*. *Curr Biol, 35*(1), 224–230.e224. https://doi.org/10.1016/j.cub.2024.11.025

Beaudet, A., Clarke, R. J., Bruxelles, L., Carlson, K. J., Crompton, R., de Beer, F., Dhaene, J., Heaton, J. L., Jakata, K., Jashashvili, T., Kuman, K., McClymont, J., Pickering, T. R., & Stratford, D. (2019). The bony labyrinth of StW 573 ("Little Foot"): Implications for early hominin evolution and paleobiology. *Journal of Human Evolution, 127*, 67–80. https://doi.org/10.1016/j.jhevol.2018.12.002

Beaudet, A., Clarke, R. J., de Jager, E. J., Bruxelles, L., Carlson, K. J., Crompton, R., de Beer, F., Dhaene, J., Heaton, J. L., Jakata, K., Jashashvili, T., Kuman, K., McClymont, J., Pickering, T. R., & Stratford, D. (2019). The endocast of StW 573 ("Little Foot") and hominin brain evolution. *Journal of Human Evolution, 126*, 112–123. https://doi.org/10.1016/j.jhevol.2018.11.009

Beaudet, A., Clarke, R. J., Heaton, J. L., Pickering, T. R., Carlson, K. J., Crompton, R. H., Jashashvili, T., Bruxelles, L., Jakata, K., Bam, L., Van Hoorebeke, L., Kuman, K., & Stratford, D. (2020). The atlas of StW 573 and the late emergence of human-like head mobility and brain metabolism. *Scientific Reports, 10*(1), 4285. https://doi.org/10.1038/s41598-020-60837-2

Beaudet, A., Du, A., & Wood, B. (2019). Evolution of the modern human brain. In M. A. Hofman (Ed.), *Progress in brain research* (2019/11/11 ed., Vol. 250, pp. 219–250). Elsevier B.V. https://doi.org/10.1016/bs.pbr.2019.01.004

Been, E., Gomez-Olivencia, A., & Kramer, P. A. (2012). Lumbar lordosis of extinct hominins. *American Journal of Physical Anthropology, 147*(1), 64–77. https://doi.org/10.1002/ajpa.21633

Begun, D. R. (2003). Planet of the apes. *Scientific American, 289*(2), 64–73.

Begun, D. R., Nargolwalla, M. C., & Kordos, L. (2012). European Miocene Hominids and the origin of the African ape and human clade. *Evolutionary Anthropology, 21*(1), 10–23. https://doi.org/10.1002/evan.20329

Bell, A., & Jedrzejczak, W. (2021). Why three ossicles are better than one. *Acoustics Australia, 49*(2), 1–26. https://doi.org/10.1007/s40857-021-00246-1

Ben-Dor, M., Gopher, A., & Barkai, R. (2016). Neandertals' large lower thorax may represent adaptation to high protein diet. *American Journal of Physical Anthropology, 160*(3), 367–378. https://doi.org/10.1002/ajpa.22981

Ben-Dor, M., Sirtoli, R., & Barkai, R. (2021). The evolution of the human trophic level during the Pleistocene. *American Journal of Physical Anthropology*, *175*(Suppl 72), 27–56. https://doi.org/10.1002/ajpa.24247

Benias, P. C., Wells, R. G., Sackey-Aboagye, B., Klavan, H., Reidy, J., Buonocore, D., Miranda, M., Kornacki, S., Wayne, M., Carr-Locke, D. L., & Theise, N. D. (2018). Structure and distribution of an unrecognized interstitium in human tissues. *Scientific Reports*, *8*(1). https://doi.org/10.1038/s41598-018-23062-6

Benitez-Burraco, A., & Boeckx, C. (2015). Possible functional links among brain- and skull-related genes selected in modern humans. *Frontiers in Psychology*, *6*, 794. https://doi.org/10.3389/fpsyg.2015.00794

Benitez-Burraco, A., & Progovac, L. (2021). Language evolution: Examining the link between cross-modality and aggression through the lens of disorders. *Philosophical Transactions of the Royal Society of London. Series B, Biological Sciences*, *376*(1824), 20200188. https://doi.org/10.1098/rstb.2020.0188

Bennett, M. R., Bustos, D., Pigati, J. S., Springer, K. B., Urban, T. M., Holliday, V. T., Reynolds, S. C., Budka, M., Honke, J. S., Hudson, A. M., Fenerty, B., Connelly, C., Martinez, P. J., Santucci, V. L., & Odess, D. (2021). Evidence of humans in North America during the Last Glacial Maximum. *Science (New York, N.Y.)*, *373*(6562), 1528–1531. https://doi.org/10.1126/science.abg7586

Bennett, M. R., Harris, J. W. K., Richmond, B. G., Braun, D. R., Mbua, E., Kiura, P., Olago, D., Kibunjia, M., Omuombo, C., Behrensmeyer, A. K., Huddart, D., & Gonzalez, S. (2009). Early hominin foot morphology based on 1.5-million-year-old footprints from Ileret, Kenya. *Science*, *323*(5918), 1197–1201. https://doi.org/10.1126/science.1168132

Bennett, M. R., Reynolds, S. C., Morse, S. A., & Budka, M. (2016). Footprints and human evolution: Homeostasis in foot function? *Palaeogeography Palaeoclimatology Palaeoecology*, *461*, 214–223. https://doi.org/10.1016/j.palaeo.2016.08.026

Berge, C., Penin, X., & Pelle, E. (2006). New interpretation of Laetoli footprints using an experimental approach and Procrustes analysis: Preliminary results. *Comptes Rendus Palevol*, *5*(3-4), 561–569. https://doi.org/10.1016/j.crpv.2005.09.001

Berger, L. R., de Ruiter, D. J., Churchill, S. E., Schmid, P., Carlson, K. J., Dirks, P. H., & Kibii, J. M. (2010). *Australopithecus sediba*: A new species of *Homo*-like australopith from South Africa. *Science*, *328*(5975), 195–204. https://doi.org/10.1126/science.1184944

Berger, L. R., Hawks, J., de Ruiter, D. J., Churchill, S. E., Schmid, P., Delezene, L. K., Kivell, T. L., Garvin, H. M., Williams, S. A., DeSilva, J. M., Skinner, M. M., Musiba, C. M., Cameron, N., Holliday, T. W., Harcourt-Smith, W., Ackermann, R. R., Bastir, M., Bogin, B., Bolter, D., Brophy, J., Cofran, Z. D., Congdon, K. A., Deane, A. S., Dembo, M., Drapeau, M., Elliott, M. C., Feuerriegel, E. M., Garcia-Martinez, D., Green, D. J., Gurtov, A., Irish, J. D., Kruger, A., Laird, M. F., Marchi, D., Meyer, M. R., Nalla, S., Negash, E. W., Orr, C. M., Radovcic, D., Schroeder, L., Scott, J. E., Throckmorton, Z., Tocheri, M. W., VanSickle, C., Walker, C. S., Wei, P. P., & Zipfel, B. (2015). *Homo naledi*, a new species of the genus *Homo* from the Dinaledi Chamber, South Africa. *eLife*, *4*, 1–35. https://doi.org/10.7554/eLife.09560

Berger, L. R., & Hilton-Barber, B. (2000). *In the footsteps of Eve. The mystery of human origins*. National Geographic Washington DC, ISBN: 0-7922-7682-5.

Bergström, A., Stringer, C., Hajdinjak, M., Scerri, E. M. L., & Skoglund, P. (2021). Origins of modern human ancestry. *Nature*, *590*(7845), 229–237. https://doi.org/10.1038/s41586-021-03244-5

Bermúdez de Castro, J. M., & Martinon-Torres, M. (2019). What does *Homo antecessor* tell us about the origin of the "emergent humanity" that gave rise to *Homo sapiens*? *Journal of Anthropological Sciences, 97*, 209–213. https://doi.org/10.4436/jass.97012

Bermúdez de Castro, J. M., & Martinon-Torres, M. (2020). Another interpretation of *Homo antecessor* - Reply. *Journal of Anthropological Sciences, 98*, 171–179. https://doi.org/10.4436/jass.97015

Berna, F., Goldberg, P., Horwitz, L. K., Brink, J., Holt, S., Bamford, M., & Chazan, M. (2012). Microstratigraphic evidence of in situ fire in the Acheulean strata of Wonderwerk Cave, Northern Cape province, South Africa. *Proceedings of the National Academy of Sciences of the United States of America, 109*(20), E1215–E1220. https://doi.org/10.1073/pnas.1117620109

Bertram, J. E., & Biewener, A. A. (1988). Bone curvature: Sacrificing strength for load predictability? *Journal of Theoretical Biology, 131*(1), 75–92. https://doi.org/10.1016/s0022-5193(88)80122-x

Bertrand, O. C., Shelley, S. L., Williamson, T. E., Wible, J. R., Chester, S. G. B., Flynn, J. L., Holbrook, L. T., Lyson, T. R., Meng, J., Miller, I. M., Püschel, H. P., Smith, T., Spaulding, M., Tseng, Z. J., & Brusatte, S. L. (2022). Brawn before brains in placental mammals after the end-Cretaceous extinction. *Science, 376*(6588), 80–85.

Best, A., & Kamilar, J. M. (2018). The evolution of eccrine sweat glands in human and non-human primates. *Journal of Human Evolution, 117*, 33–43. https://doi.org/10.1016/j.jhevol.2017.12.003

Best, A. W., Lieberman, D. E., Gerson, A. R., Holt, B. M., & Kamilar, J. M. (2023). Variation in human functional eccrine gland density and its implications for the evolution of human sweating. *American Journal of Biological Anthropology*. https://doi.org/10.1002/ajpa.24723

Bethlehem, R. A. I., Seidlitz, J., White, S. R., Vogel, J. W., Anderson, K. M., Adamson, C., Adler, S., Alexopoulos, G. S., Anagnostou, E., Areces-Gonzalez, A., Astle, D. E., Auyeung, B., Ayub, M., Bae, J., Ball, G., Baron-Cohen, S., Beare, R., Bedford, S. A., Benegal, V.,. . . Alexander-Bloch, A. F (2022). Brain charts for the human lifespan. *Nature*. https://doi.org/10.1038/s41586-022-04554-y

Bezanson, M. (2012). The ontogeny of prehensile-tail use in *Cebus capucinus* and *Alouatta palliata*. *American Journal of Primatology, 74*(8), 770–782. https://doi.org/10.1002/Ajp.22028

Bianconi, E., Piovesan, A., Facchin, F., Beraudi, A., Casadei, R., Frabetti, F., Vitale, L., Pelleri, M. C., Tassani, S., Piva, F., Perez-Amodio, S., Strippoli, P., & Canaider, S. (2013). An estimation of the number of cells in the human body. *Annals of Human Biology, 40*(6), 463–471. https://doi.org/10.3109/03014460.2013.807878

Bickerton, D., & Szathmary, E. (2011). Confrontational scavenging as a possible source for language and cooperation. *BMC Evolutionary Biology, 11*, 261. https://doi.org/10.1186/1471-2148-11-261

Black, D. (1929). *Sinanthropus Pekinensis*: The recovery of further fossil remains of this early Hominid from the Chou Kou Tien deposit. *Science, 69*(1800), 674–676.

Blanca, J., Pons, C., Montero-Pau, J., Sanchez-Matarredona, D., Ziarsolo, P., Fontanet, L., Fisher, J., Plazas, M., Casals, J., Rambla, J. L., Riccini, A., Pombarella, S., Ruggiero, A., Sulli, M., Grillo, S., Kanellis, A., Giuliano, G., Finkers, R., Cammareri, M., Grandillo, S., Mazzucato, A., Causse, M., Diez, M. J., Prohens, J., Zamir, D., Canizares, J., Monforte, A.

J., & Granell, A. (2022). European traditional tomatoes galore: A result of farmers' selection of a few diversity-rich loci. *J Exp Bot, 73*(11), 3431–3445. https://doi.org/10.1093/jxb/erac072

Blasi, D. E., Moran, S., Moisik, S. R., Widmer, P., Dediu, D., & Bickel, B. (2019). Human sound systems are shaped by post-Neolithic changes in bite configuration. *Science, 363*(eaav3218), 1–10. https://doi.org/10.1126/science.aav3218

Blaszczyk, M. B., & Vaughan, C. L. (2007). Re-interpreting the evidence for bipedality in *Homo floresiensis*. *South African Journal of Science, 103*(9-10), 409–414.

Bleasdale, M., Richter, K. K., Janzen, A., Brown, S., Scott, A., Zech, J., Wilkin, S., Wang, K., Schiffels, S., Desideri, J., Besse, M., Reinold, J., Saad, M., Babiker, H., Power, R. C., Ndiema, E., Ogola, C., Manthi, F. K., Zahir, M., Petraglia, M., Trachsel, C., Nanni, P., Grossmann, J., Hendy, J., Crowther, A., Roberts, P., Goldstein, S. T., & Boivin, N. (2021). Ancient proteins provide evidence of dairy consumption in eastern Africa. *Nature Communications, 12*(1), 632. https://doi.org/10.1038/s41467-020-20682-3

Boaretto, E., Wu, X. H., Yuan, J. R., Bar-Yosef, O., Chu, V., Pan, Y., Liu, K. X., Cohen, D., Jiao, T. L., Li, S. C., Gu, H. B., Goldberg, P., & Weiner, S. (2009). Radiocarbon dating of charcoal and bone collagen associated with early pottery at Yuchanyan Cave, Hunan Province, China. *Proceedings of the National Academy of Sciences of the United States of America, 106*(24), 9595–9600. https://doi.org/10.1073/pnas.0900539106

Boaz, N. T., Ciochon, R. L., Xu, Q., & Liu, J. (2004). Mapping and taphonomic analysis of the *Homo erectus* loci at locality 1 Zhoukoudian, China. *Journal of Human Evolution, 46*(5), 519–549. https://doi.org/10.1016/j.jhevol.2004.01.007

Bobe, R., Manthi, F. K., Ward, C. V., Plavcan, J. M., & Carvalho, S. (2020). The ecology of *Australopithecus anamensis* in the early Pliocene of Kanapoi, Kenya. *Journal of Human Evolution, 140*, 102717. https://doi.org/10.1016/j.jhevol.2019.102717

Bocherens, H. (2009). Neanderthal Dietary Habits: Review of the Isotopic Evidence. In J. J. Hublin & M. P. Richards (Eds.), *Evolution of hominin diets: integrating approaches to the study of palaeolithic subsistence* (pp. 241–250). Springer Science & Business Media.

Bocherens, H., Billiou, D., Mariotti, A., Patou-Mathis, M., Otte, M., Bonjean, D., & Toussaint, M. (1999). Palaeoenvironmental and palaeodietary implications of isotopic biogeochemistry of last interglacial Neanderthal and mammal bones in Scladina Cave (Belgium). *Journal of Archaeological Science, 26*(6), 599–607. https://doi.org/10.1006/jasc.1998.0377

Bocherens, H., Billiou, D., Mariotti, A., Toussaint, M., Patou-Mathis, M., Bonjean, D., & Otte, M. (2001). New isotopic evidence for dietary habits of Neandertals from Belgium. *Journal of Human Evolution, 40*(6), 497–505. https://doi.org/10.1006/jhev.2000.0452

Bocherens, H., Drucker, D. G., Billiou, D., Patou-Mathis, M., & Vandermeersch, B. (2005). Isotopic evidence for diet and subsistence pattern of the Saint-Cesaire I Neanderthal: Review and use of a multi-source mixing model. *Journal of Human Evolution, 49*(1), 71–87. https://doi.org/10.1016/j.jhevol.2005.03.003

Bocquet-Appel, J. P., & Degioanni, A. (2013). Neanderthal demographic estimates. *Current Anthropology, 54*, S202–S213. https://doi.org/10.1086/673725

Bohme, M., Spassov, N., Fuss, J., Troscher, A., Deane, A. S., Prieto, J., Kirscher, U., Lechner, T., & Begun, D. R. (2019). A new Miocene ape and locomotion in the ancestor of great apes and humans. *Nature*. https://doi.org/10.1038/s41586-019-1731-0

Bohmer, C., & Werneburg, I. (2017). Deep time perspective on turtle neck evolution: Chasing the Hox code by vertebral morphology. *Sci Rep, 7*(1), 8939. https://doi.org/10.1038/s41598-017-09133-0

Bolanowski, S. J., & Pawson, L. (2003). Organization of Meissner corpuscles in the glabrous skin of monkey and cat. *Somatosensory and Motor Research, 20*(3-4), 223–231.

Bolognini, D., Halgren, A., Lou, R. N., Raveane, A., Rocha, J. L., Guarracino, A., Soranzo, N., Chin, C. S., Garrison, E., & Sudmant, P. H. (2024). Recurrent evolution and selection shape structural diversity at the amylase locus. *Nature, 634*(8034), 617–625. https://doi.org/10.1038/s41586-024-07911-1

Bond, M., Tejedor, M. F., Campbell, K. E., Jr., Chornogubsky, L., Novo, N., & Goin, F. (2015). Eocene primates of South America and the African origins of New World monkeys. *Nature, 520*(7548), 538–541. https://doi.org/10.1038/nature14120

Brace, C. L. (1999). Comment to 'The raw and the stolen - Cooking and the ecology of human origins'. *Current Anthropology, 40*(5), 577–579. https://doi.org/10.1086/300083

Bradbury, J. (2005). Molecular insights into human brain evolution. *Plos Biology, 3*(3), 367–370. https://doi.org/10.1371/journal.pbio.0030050

Brain, C., & Mitchell, D. (1999). Body temperature changes in free-ranging baboons (*Papio hamadryas ursinus*) in the Namib Desert, Namibia. *International Journal of Primatology, 20*(4), 585–598. https://doi.org/10.1023/A:1020394824547

Bramble, D. M., & Lieberman, D. E. (2004). Endurance running and the evolution of *Homo*. *Nature, 432*(7015), 345–352. https://doi.org/10.1038/nature03052

Brant, R. (2019). *Aap op schoenen met een buikje. Een evolutionaire kijk op gezondheid*. Scriptum.

Brassey, C. A., O'Mahoney, T. G., Chamberlain, A. T., & Sellers, W. I. (2018). A volumetric technique for fossil body mass estimation applied to *Australopithecus afarensis*. *Journal of Human Evolution, 115*, 47–64. https://doi.org/10.1016/j.jhevol.2017.07.014

Braun, D. R., Aldeias, V., Archer, W., Arrowsmith, J. R., Baraki, N., Campisano, C. J., Deino, A. L., DiMaggio, E. N., Dupont-Nivet, G., Engda, B., Feary, D. A., Garello, D. I., Kerfelew, Z., McPherron, S. P., Patterson, D. B., Reeves, J. S., Thompson, J. C., & Reed, K. E. (2019). Earliest known Oldowan artifacts at >2.58 Ma from Ledi-Geraru, Ethiopia, highlight early technological diversity. *Proceedings of the National Academy of Sciences of the United States of America, 116*(24) 11712–11717. https://doi.org/10.1073/pnas.1820177116

Breckling, A. C. M., Katrikh, A. Z., Jones, M. W., & Ferrigno, C. (2022). Iliocapsularis: An exploration of the muscle and its omission in education. *Journal of Morphology, 283*(7), 899–907. https://doi.org/10.1002/jmor.21468

Briggs, A. W., Good, J. M., Green, R. E., Krause, J., Maricic, T., Stenzel, U., Lalueza-Fox, C., Rudan, P., Brajkovic, D., Kucan, Z., Gusic, I., Schmitz, R., Doronichev, V. B., Golovanova, L. V., de la Rasilla, M., Fortea, J., Rosas, A., & Paabo, S. (2009). Targeted retrieval and analysis of five Neandertal mtDNA genomes. *Science, 325*(5938), 318–321. https://doi.org/10.1126/science.1174462

Brinkerhoff, S. A., Sanchez, N., Culver, M. N., Murrah, W. M., Robinson, A. T., McCullough, J. D., Miller, M. W., & Roper, J. A. (2024). The dual timescales of gait adaptation: Initial stability adjustments followed by subsequent energetic cost adjustments. *Journal of Experimental Biology, 227*(23). https://doi.org/10.1242/jeb.249217

Briscoe, S. D., & Ragsdale, C. W. (2019). Evolution of the chordate telencephalon. *Current Biology, 29*(13), R647–R662. https://doi.org/10.1016/j.cub.2019.05.026

Brodmann, K. (1909). *Vergleichende Localisationslehre der Grosshirnrinde*. Verlag von Johan Ambrosius Barth.

Broom, R. (1925). Some notes on the taungs skull. *Nature, 115*(2894), 569–571.

Broom, R. (1938). The pleistocene anthropod apes of South Africa. *Nature, 163*(3591), 377–379.

Brown, C. (2012). Tool use in fishes. *Fish and Fisheries*, *13*(1), 105–115. https://doi.org/10.1111/j.1467-2979.2011.00451.x

Brown, F., Harris, J., Leakey, R., & Walker, A. (1985). Early *Homo erectus* skeleton from west Lake Turkana, Kenya. *Nature*, *316*(6031), 788–792. https://www.ncbi.nlm.nih.gov/pubmed/3929141

Brown, P., Sutikna, T., Morwood, M. J., Soejono, R. P., Jatmiko, Saptomo, E. W., & Due, R. A. (2004). A new small-bodied hominin from the Late Pleistocene of Flores, Indonesia. *Nature*, 431(7012), 1055–1061. https://doi.org/10.1038/nature02999

Brown, S., Massilani, D., Kozlikin, M. B., Shunkov, M. V., Derevianko, A. P., Stoessel, A., Jope-Street, B., Meyer, M., Kelso, J., Paabo, S., Higham, T., & Douka, K. (2021). The earliest Denisovans and their cultural adaptation. *Nature Ecology & Evolution*, 6, 28–35. https://doi.org/10.1038/s41559-021-01581-2

Browne, J. (2021). Introduction. In J. M. DeSilva (Ed.), *A most interesting problem. What Darwin's descent of man got right and wrong about human evolution* (pp. 1–23). Princeton University Press.

Bruner, E., & Beaudet, A. (2023). The brain of *Homo habilis*: Three decades of paleoneurology. *Journal of Human Evolution*, *174*, 103281. https://doi.org/10.1016/j.jhevol.2022.103281

Bruner, E., Mantini, S., Musso, F., De La Cuetara, J. M., Ripani, M., & Sherkat, S. (2011). The evolution of the meningeal vascular system in the human genus: From brain shape to thermoregulation. *American Journal of Human Biology: the Official Journal of the Human Biology Council*, *23*(1), 35–43. https://doi.org/10.1002/ajhb.21123

Brunet, M. (2002). *Sahelanthropus* or 'Sahelpithecus'? Reply. *Nature*, *419*(6907), 582–582. https://doi.org/10.1038/419582a

Brunet, M., Beauvilain, A., Coppens, Y., Heintz, E., Moutaye, A. H. E., & Pilbeam, D. (1996). *Australopithecus bahrelghazali*, une nouvelle espèce d'hominidé ancien de la région de Koro Toro (Tchad). *Comptes Rendus de l'Académie des Sciences*, *322*(IIa), 907–913.

Brunet, M., Guy, F., Pilbeam, D., Mackay, H. T., Likius, A., Ahounta, D., Beauvilain, A., Blondel, C., Bocherens, H., Boisserie, J.-R., De Bonis, L., Coppens, Y., Dejax, J., Denys, C., Duringer, P., Eisenmann, V., Fanone, G., Fronty, P., Geraads, D., Lehmann, T., Lihoreau, F., Louchart, A., Mahamat, A., Merceron, G., Mouchelin, G., Otero, O., Campomanes, P. P., De Leon, M. P., Rage, J.-C., Sapanet, M., Schuster, M., Sudre, J., Tassy, P., Valentin, X., Vignaud, P., Viriot, L., Zazzo, A., & Zollikofer, C. (2002). A new hominid from the Upper Miocene of Chad, Central Africa. *Nature*, *418*(6894), 145–151. https://doi.org/10.1038/nature00879

Brunet, M., Guy, F., Pilbeam, D., Lieberman, D. E., Likius, A., Mackaye, H. T., Ponce de Leon, M. S., Zollikofer, C. P., & Vignaud, P. (2005). New material of the earliest hominid from the Upper Miocene of Chad. *Nature*, *434*(7034), 752–755. https://doi.org/10.1038/nature03392

Bryson, B. (2019). *Het lichaam: een reisgids* (J. W. Reitsma & A. Witteveen, Trans.). Atlas Contact.

Bucchi, A., Luengo, J., Del Bove, A., & Lorenzo, C. (2020). Insertion sites in manual proximal phalanges of African apes and modern humans. *American Journal of Physical Anthropology*, *173*(3), 1–12. https://doi.org/10.1002/ajpa.24127

Bunn, H. T. (1999). Comment to 'The raw and the stolen - Cooking and the ecology of human origins'. *Current Anthropology*, *40*(5), 579–580. https://doi.org/10.1086/300083

Burger, J., Kirchner, M., Bramanti, B., Haak, W., & Thomas, M. G. (2007). Absence of the lactase-persistence-associated allele in early Neolithic Europeans. *Proceedings of the National*

*Academy of Sciences of the United States of America, 104*(10), 3736–3741. https://doi.org/10.1073/pnas.0607187104

Burger, J., Link, V., Blocher, J., Schulz, A., Sell, C., Pochon, Z., Diekmann, Y., Zegarac, A., Hofmanova, Z., Winkelbach, L., Reyna-Blanco, C. S., Bieker, V., Orschiedt, J., Brinker, U., Scheu, A., Leuenberger, C., Bertino, T. S., Bollongino, R., Lidke, G., Stefanovic, S., Jantzen, D., Kaiser, E., Terberger, T., Thomas, M. G., Veeramah, K. R., & Wegmann, D. (2020). Low prevalence of lactase persistence in Bronze Age Europe indicates ongoing strong selection over the last 3,000 years. *Current Biology, 30*(21), 4307–4315.e13. https://doi.org/10.1016/j.cub.2020.08.033

Burgin, C. J., Wilson, D. E., Mittermeier, R. A., Rylands, A. B., Lacher, T. E., & Sechrest, W. (2020). *Illustrated checklist of the mammals of the world* (A. Martinez-Vilalta, M. Olivé, & D. M. Leslie, Jr., Eds.). Lynx Edicions.

Butler, D. (2001). The battle of Tugen Hills. *Nature, 410*(6828), 508–509. https://doi.org/10.1038/35069232

Byrne, R. W. (2001). Social and technical forms of primate intelligence. In F. B. M. De Waal (Ed.), *What primate behavior can tell us about human social evolution* (pp. 147–172). President and Fellows of Harvard College.

Callaway, E. (2015). Oldest stone tools raise questions about their creators. *Nature, 520*(7548), 421–421. https://doi.org/10.1038/520421a

Callaway, E. (2021, 7/4/21). Oldest DNA from a Homo sapiens reveals surprisingly recent Neanderthal ancestry. *Nature Magazine.* Retrieved 8/4/2021 from https://www.nature.com/articles/d41586-021-00916-0

Cameron, D. W., & Groves, P. C. (2004). *Bones, stones and molecules. "Out of Africa" and human origins.* Elsevier Academic Press.

Campbell, M. C., & Ranciaro, A. (2021). Human adaptation, demography and cattle domestication: an overview of the complexity of lactase persistence in Africa. *Human Molecular Genetics, 30*(R1), R98–R109. https://doi.org/10.1093/hmg/ddab027

Cann, R. L., Stoneking, M., & Wilson, A. C. (1987). Mitochondrial DNA and human evolution. *Nature, 325*(6099), 31–36. https://doi.org/10.1038/325031a0

Cantalupo, C., & Hopkins, W. D. (2001). Asymmetric Broca's area in great apes. *Nature, 414*(6863), 505. https://doi.org/10.1038/35107134

Carey, T. S., & Crompton, R. H. (2005). The metabolic costs of 'bent-hip, bent-knee' walking in humans. *Journal of human evolution, 48*(1), 25–44. https://doi.org/10.1016/j.jhevol.2004.10.001

Carlson, K. J., Green, D. J., Jashashvili, T., Pickering, T. R., Heaton, J. L., Beaudet, A., Stratford, D., Crompton, R., Kuman, K., Mcclymont, J., Bruxelles, L., & Clarke, R. J. (2021). The pectoral girdle of StW 573 ('Little Foot') and its implications for hominin evolution. *American Journal of Physical Anthropology, 174,* 15–16.

Carlson, K. J., Stout, D., Jashashvili, T., de Ruiter, D. J., Tafforeau, P., Carlson, K., & Berger, L. R. (2011). The endocast of MH1, *Australopithecus sediba. Science, 333*(6048), 1402–1407. https://doi.org/10.1126/science.1203922

Carmody, R. N., Dannemann, M., Briggs, A. W., Nickel, B., Groopman, E. E., Wrangham, R. W., & Kelso, J. (2016). Genetic evidence of human adaptation to a cooked diet. *Genome Biology and Evolution, 8*(4), 1091–1103. https://doi.org/10.1093/gbe/evw059

Carrier, D. R. (1984). The energetic paradox of human running and hominid evolution. *Current Anthropology, 25*(4), 483–495. https://doi.org/10.1086/203165

Carrier, D. R. (2007). The short legs of great apes: Evidence for aggressive behavior in australopiths. *Evolution; International Journal of Organic Evolution, 61*(3), 596–605. https://doi.org/10.1111/j.1558-5646.2007.00061.x

Carrier, D. R. (2011). The advantage of standing up to fight and the evolution of habitual bipedalism in hominins. *PloS One, 6*(5), e19630. https://doi.org/10.1371/journal.pone.0019630

Carrión, J. S., Rose, J., & Stringer, C. (2011). Early human evolution in the Western palaearctic: Ecological scenarios. *Quaternary Science Reviews, 30*(11-12), 1281–1295. https://doi.org/10.1016/j.quascirev.2011.04.003

Carroll, S. B. (2003). Genetics and the making of *Homo sapiens. Nature, 422*(6934), 849–857.

Cartmill, M. (1974). Rethinking primate origins. *Science, 184*(4135), 436–443. https://doi.org/10.1126/science.184.4135.436

Cartmill, M. (1983). 4 legs good, 2 legs bad - mans place (if any) in nature. *Natural History, 92*(11), 64.

Carvalho, S., Biro, D., Cunha, E., Hockingss, K., McGrew, W. C., Richmond, B. G., & Matsuzawa, T. (2012). Chimpanzee carrying behaviour and the origins of human bipedality. *Current Biology, 22*(6), R180–R181. https://doi.org/10.1016/j.cub.2012.01.052

Catani, M. (2009). The connectional anatomy of language: Recent contributions from diffusion tensor tractography. In H. Johansen-Berg, & T. E. J. Behrens (Eds.), *Diffusion MRI* (pp. 403–413). Elsevier Inc. https://doi.org/10.1016/B978-0-12-374709-9.00018-3

Cavener, D. R., Bond, M. L., Wu-Cavener, L., Lohay, G. G., Cavener, M. W., Hou, X., Pearce, D. L., & Lee, D. E. (2024). Sexual dimorphisms in body proportions of Masai giraffes and the evolution of the giraffe's neck. *Mammalian Biology.* https://doi.org/10.1007/s42991-024-00424-4

Cazenave, M., Oettle, A., Pickering, T. R., Heaton, J. L., Nakatsukasa, M., Francis Thackeray, J., Hoffman, J., & Macchiarelli, R. (2021). Trabecular organization of the proximal femur in *Paranthropus robustus*: Implications for the assessment of its hip joint loading conditions. *Journal of Human Evolution, 153*, 102964. https://doi.org/10.1016/j.jhevol.2021.102964

Cazenave, M., Pina, M., Hammond, A. S., Bohme, M., Begun, D. R., Spassov, N., Gazabon, A. V., Zanolli, C., Bergeret-Medina, A., Marchi, D., Macchiarelli, R., & Wood, B. (2025). Postcranial evidence does not support habitual bipedalism in *Sahelanthropus tchadensis*: A reply to Daver et al. (2022). *Journal of Human Evolution, 198*, 103557. https://doi.org/10.1016/j.jhevol.2024.103557

Cerling, T. E., Manthi, F. K., Mbua, E. N., Leakey, L. N., Leakey, M. G., Leakey, R. E., Brown, F. H., Grine, F. E., Hart, J. A., Kaleme, P., Roche, H., Uno, K. T., & Wood, B. A. (2013). Stable isotope-based diet reconstructions of Turkana Basin hominins. *Proceedings of the National Academy of Sciences of the United States of America, 110*(26), 10501–10506. https://doi.org/10.1073/pnas.1222568110

Cerling, T. E., Mbua, E., Kirera, F. M., Manthi, F. K., Grine, F. E., Leakey, M. G., Sponheimer, M., & Uno, K. T. (2011). Diet of *Paranthropus boisei* in the early Pleistocene of East Africa. *Proceedings of the National Academy of Sciences of the United States of America, 108*(23), 9337–9341. https://doi.org/10.1073/pnas.1104627108

Cerling, T. E., Wynn, J. G., Andanje, S. A., Bird, M. I., Korir, D. K., Levin, N. E., Mace, W., Macharia, A. N., Quade, J., & Remien, C. H. (2011). Woody cover and hominin environments in the past 6 million years. *Nature, 476*(7358), 51–56. https://doi.org/10.1038/nature10306

Chaney, M. E., Ruiz, C. A., Meindl, R. S., & Lovejoy, C. O. (2021). The foot of the human-chimpanzee last common ancestor was not African ape-like: A response to Prang (2019). *Journal of Human Evolution, 102940*. https://doi.org/10.1016/j.jhevol.2020.102940

Changizi, M., Weber, R., Kotecha, R., & Palazzo, J. (2011). Are wet-induced wrinkled fingers primate rain treads? *Brain, Behavior and Evolution, 77*(4), 286–290. https://doi.org/10.1159/000328223

Channon, A. J., Usherwood, J. R., Crompton, R. H., Gunther, M. M., & Vereecke, E. E. (2011). The extraordinary athletic performance of leaping gibbons. *Biology Letters.* https://doi.org/10.1098/rsbl.2011.0574

Chaplin, G. (2004). Geographic distribution of environmental factors influencing human skin coloration. *American Journal of Physical Anthropology, 125*(3), 292–302. https://doi.org/10.1002/ajpa.10263

Chaplin, G., Jablonski, N. G., & Cable, N. T. (1994). Physiology, thermoregulation and bipedalism. *Journal of Human Evolution, 27*(6), 497–510. https://doi.org/10.1006/jhev.1994.1066

Chappell, J., & Kacelnik, A. (2002). Tool selectivity in a non-primate, the New Caledonian crow (*Corvus moneduloides*). *Animal Cognition, 5*(2), 71–78. https://doi.org/10.1007/s10071-002-0130-2

Charles, J. P., Grant, B., D'Aout, K., & Bates, K. T. (2021). Foot anatomy, walking energetics, and the evolution of human bipedalism. *Journal of Human Evolution, 156*, 103014. https://doi.org/10.1016/j.jhevol.2021.103014

Chatterjee, H. J., Ho, S. Y. W., Barnes, I., & Groves, C. (2009). Estimating the phylogeny and divergence times of primates using a supermatrix approach *BMC Evolutionary Biology, 9*, 259. https://doi.org/10.1186/1471-2148-9-259

Chazan, M. (2017). Toward a long prehistory of fire. *Current Anthropology, 58*, S351–S359. https://doi.org/10.1086/691988

Chen, F., Welker, F., Shen, C.-C., Bailey, S. E., Bergmann, I., Davis, S., Xia, H., Wang, H., Fischer, R., Freidline, S. E., Yu, T.-L., Skinner, M. M., Stelzer, S., Dong, G., Fu, Q., Dong, G., Wang, J., Zhang, D., & Hublin, J.-J. (2019). A late middle Pleistocene Denisovan mandible from the Tibetan Plateau. *Nature, 364*(6439), 409–412. https://doi.org/10.1038/s41586-019-1139-x

Chen, J. Y., Huang, D. Y., & Li, C. W. (1999). An early Cambrian craniate-like chordate. *Nature, 402*(6761), 518–522. https://doi.org/10.1038/990080

Chen, L., Wolf, A. B., Fu, W. Q., Li, L. M., & Akey, J. M. (2020). Identifying and interpreting apparent Neanderthal ancestry in African individuals. *Cell, 180*(4), 677. https://doi.org/10.1016/j.cell.2020.01.012

Cheney, D. L., & Seyfarth, R. M. (1982). How vervet monkeys perceive their grunts - field playback experiments. *Animal Behaviour, 30*(Aug), 739–751. https://doi.org/10.1016/S0003-3472(82)80146-2

Chivers, D. J. (1998). Measuring food intake in wild animals: Primates. *The Proceedings of the Nutrition Society, 57*(2), 321–332. https://doi.org/10.1079/pns19980047

Christmas, M. J., Kaplow, I. M., Genereux, D. P., Dong, M. X., Hughes, G. M., Li, X., Sullivan, P. F., Hindle, A. G., Andrews, G., Armstrong, J. C., Bianchi, M., Breit, A. M., Diekhans, M., Fanter, C., Foley, N. M., Goodman, D. B., Goodman, L., Keough, K. C., Kirilenko, B.,... & Zhang, X. (2023). Evolutionary constraint and innovation across hundreds of placental mammals. *Science (New York, N.Y.), 380*(6643), eabn3943. https://doi.org/10.1126/science.abn3943

Chudler, E. H. (2020, 29/06/2020). *Brain facts and figures.* Retrieved 10/06/2022 from http://faculty.washington.edu/chudler/facts.html

Ciurana, N., Artells, R., Munoz, C., Arias-Martorell, J., Bello-Hellegouarch, G., Casado, A., Cuesta, E., Perez-Perez, A., Pastor, J. F., & Potau, J. M. (2017). Expression of myosin heavy chain isoforms mRNA transcripts in the temporalis muscle of common chimpanzees (*Pan troglodytes*). *Annals of Anatomy, 214*, 80–85. https://doi.org/10.1016/j.aanat.2017.08.001

Ciurana, N., Artells, R., Munoz, C., Arias-Martorell, J., Bello-Hellegouarch, G., Perez-Perez, A., Pastor, J. F., & Potau, J. M. (2017). Expression of MyHC isoforms mRNA transcripts in different regions of the masseter and medial pterygoid muscles in chimpanzees. *Archives of Oral Biology, 83*, 63–67. https://doi.org/10.1016/j.archoralbio.2017.07.003

Claes, S., Vereecke, E., Maes, M., Victor, J., Verdonk, P., & Bellemans, J. (2013). Anatomy of the anterolateral ligament of the knee. *Journal of Anatomy, 223*(4), 321–328. https://doi.org/10.1111/joa.12087

Clarke, R. J. (1998). First ever discovery of a well-preserved skull and associated skeleton of *Australopithecus*. *South African Journal of Science, 94*(10), 460–463.

Clarke, R. J. (2002). Newly revealed information on the Sterkfontein member 2 *Australopithecus* skeleton. *South African Journal of Science, 98*(11-12), 523–526.

Clarke, R. J., & Tobias, P. V. (1995). Sterkfontein-member-2 foot bones of the oldest South-African hominid. *Science, 269*(5223), 521–524. https://doi.org/10.1126/science.7624772

Code, C. (2021). The prehistory of speech and language is revealed in brain damage. *Philosophical Transactions of the Royal Society of London. Series B, Biological Sciences, 376*(1824), 20200191. https://doi.org/10.1098/rstb.2020.0191

Collard, M. (2003). Grades and transitions in human evolution. *Speciation of Modern Homo Sapiens, 106*, 61–100.

Conard, N. J., Serangeli, J., Bigga, G., & Rots, V. (2020). A 300,000-year-old throwing stick from Schoningen, northern Germany, documents the evolution of human hunting. *Nature Ecology & Evolution, 4*(5), 690–693. https://doi.org/10.1038/s41559-020-1139-0

Condemi, S., & Savatier, F. (2019). *Wie vermoordde de neanderthalers?* EPO vzw.

Conde-Valverde, M., Martínez, I., Quam, R. M., Rosa, M., Velez, A. D., Lorenzo, C., Jarabo, P., Bermúdez de Castro, J. M., Carbonell, E., & Arsuaga, J. L. (2021). Neanderthals and *Homo sapiens* had similar auditory and speech capacities. *Nature Ecology & Evolution, 5*(5), 609. https://doi.org/10.1038/s41559-021-01391-6

Constantino, P., & Wood, B. (2007). The evolution of *Zinjanthropus boisei*. *Evolutionary Anthropology: Issues, News, and Reviews, 16*(2), 49–62. https://doi.org/10.1002/evan.20130

Coppens, Y. (1994). East side Story: The origin of humankind. *Scientific American, 270*(5), 62–69.

Corballis, M. C. (2003). From mouth to hand: Gesture, speech, and the evolution of right-handedness. *Behavioral and Brain Sciences, 26*(2), 199. https://doi.org/10.1017/S0140525x03000062

Corballis, M. C. (2012). Lateralization of the human brain. In M. A. Hofman & D. Falk (Eds.), *Evolution of the primate brain* (Vol. 195, pp. 103–121). Elsevier B.V. https://doi.org/10.1016/B978-0-444-53860-4.00006-4

Covarrubias, B. V., Kamminga, J. M., Muchlinski, M. N., Munds, R. A., Villero Nunez, V., Bauman Surratt, S., Cayo Biobank Research, U., Martinez, M. I., Montague, M. J., Higham, J. P., Melin, A. D., & Veilleux, C. C. (2024). Investigating mechanoreceptor variability and morphometric proxies in Rhesus Macaques: Implications for primate precision touch studies. *Anatomical Record (Hoboken, N.J.: 2007)*. https://doi.org/10.1002/ar.25587

Cowan, C. S. M., Dinan, T. G., & Cryan, J. F. (2020). Annual research review: Critical windows - the microbiota-gut-brain axis in neurocognitive development. *Journal of Child Psychology and Psychiatry, 61*(3), 353–371. https://doi.org/10.1111/jcpp.13156

Cowan, W. M. (1979). The development of the brain. *Scientific American, 241*(3), 112–133. https://www.ncbi.nlm.nih.gov/pubmed/493917

Coyne, J. (2011, 28/05/2011). *The longest cell in the history of life*. Retrieved 10/06/2022 from https://whyevolutionistrue.com/2011/05/28/the-longest-cell-in-the-history-of-life/

Crittenden, A. N., & Schnorr, S. L. (2017). Current views on hunter-gatherer nutrition and the evolution of the human diet. *American Journal of Physical Anthropology, 162*(Suppl 63), 84–109. https://doi.org/10.1002/ajpa.23148

Crompton, R. H., Li, Y., Wang, W. J., Gunther, M., & Savage, R. (1998). The mechanical effectiveness of erect and "bent-hip, bent-knee" bipedal walking in *Australopithecus afarensis*. *Journal of Human Evolution, 35*(1), 55–74. https://doi.org/10.1006/jhev.1998.0222

Crompton, R. H., Pataky, T. C., Savage, R., D'Aout, K., Bennett, M. R., Day, M. H., Bates, K., Morse, S., & Sellers, W. I. (2012). Human-like external function of the foot, and fully upright gait, confirmed in the 3.66 million year old laetoli hominin footprints by topographic statistics, experimental footprint-formation and computer simulation. *Journal of the Royal Society Interface, 9*(69), 707–719. https://doi.org/10.1098/rsif.2011.0258

Crompton, R. H., Sellers, W. I., & Thorpe, S. K. S. (2010). Arboreality, terrestriality and bipedalism. *Philosophical Transactions of the Royal Society B-Biological Sciences, 365*(1556), 3301–3314. https://doi.org/10.1098/rstb.2010.0035

Cueva-Temprana, A., Lombao, D., Soto, M., Itambu, M., Bushozi, P., Boivin, N., Petraglia, M., & Mercader, J. (2022). Oldowan technology amid shifting environments ~2.03–1.83 million years ago. *Frontiers in Ecology and Evolution, 10*. https://doi.org/10.3389/fevo.2022.788101

Cummings, J. H. (1984). Cellulose and the human gut. *Gut, 25*(8), 805–810. https://doi.org/10.1136/gut.25.8.805

Cunningham, D. L., Graves, R. R., Wescott, D. J., & McCarthy, R. C. (2018). The effect of ontogeny on estimates of KNM-WT 15000's adult body size. *Journal of Human Evolution, 121*, 119–127. https://doi.org/10.1016/j.jhevol.2018.04.002

Curnoe, D. (2010). A review of early *Homo* in southern Africa focusing on cranial, mandibular and dental remains, with the description of a new species (*Homo gautengensis* sp nov.). *Homo-Journal of Comparative Human Biology, 61*(3), 151–177. https://doi.org/10.1016/j.jchb.2010.04.002

Currat, M., & Excoffier, L. (2004). Modern humans did not admix with Neanderthalsneanderthals during their range expansion into. *Europe. PLoS, 2*(12), 1–11.

Currie, P. (2004). Muscling in on hominid evolution. *Nature, 428*(6981), 373–374.

Curry, A. (2021a). The ancient carb revolution. *Nature, 594*(7864), 488–491.

Curry, A. (2021b). Ancient poop reveals extinction in gut bacteria. *Science, 372*(6543), 671. https://doi.org/10.1126/science.372.6543.671

Cutts, R. B., Hlubik, S., Campbell, R., Muschinski, J., Akuku, P., Braun, D. R., Patterson, D. B., O'Brien, J. J., Garrison, E., & Harris, J. W. K. (2019). Thermal curved-fragments: A method for identifying anthropogenic fire in the archaeological record. *Journal of Archaeological Science, 106*, 10–22. https://doi.org/10.1016/j.jas.2019.03.006

D'Anastasio, R., Wroe, S., Tuniz, C., Mancini, L., Cesana, D. T., Dreossi, D., Ravichandiran, M., Attard, M., Parr, W. C., Agur, A., & Capasso, L. (2013). Micro-biomechanics of the Kebara 2 hyoid and its implications for speech in Neanderthals. *PloS One, 8*(12), e82261. https://doi.org/10.1371/journal.pone.0082261

D'Antona, G., & Burtscher, M. (2022). Muscle endurance: Is bipedalism the cause? *Frontiers in Ecology and Evolution, 10*. https://doi.org/10.3389/fevo.2022.1067396

d'Errico, F., Dayet Bouillot, L., García-Diez, M., Pitarch Martí, A., Garrido Pimentel, D., & Zilhão, J. (2016). The technology of the earliest European Cave paintings: El Castillo Cave, Spain. *Journal of Archaeological Science, 70*, 48–65. https://doi.org/10.1016/j.jas.2016.03.007

Daeschler, E. B., Shubin, N. H., & Jenkins, F. A. (2006). A Devonian tetrapod-like fish and the evolution of the tetrapod body plan. *Nature, 440*(7085), 757–763.

Dainton, M. (2001). Did our ancestors knuckle-walk? *Nature, 410*(6826), 324–325.

Damjanovic, L., Roberts, S. G. B., & Roberts, A. I. (2022). Language as a tool for social bonding: Evidence from wild chimpanzee gestural, vocal and bimodal signals. *Philosophical Transactions of the Royal Society of London. Series B, Biological Sciences, 377*(1860), 20210311. https://doi.org/10.1098/rstb.2021.0311

Dannemann, M., & Kelso, J. (2017). The contribution of Neanderthals to phenotypic variation in modern humans. *American Journal of Human Genetics, 101*(4), 578–589. https://doi.org/10.1016/j.ajhg.2017.09.010

Dannemann, M., & Racimo, F. (2018). Something old, something borrowed: Admixture and adaptation in human evolution. *Current Opinion in Genetics & Development, 53*, 1–8. https://doi.org/10.1016/j.gde.2018.05.009

Dart, R. A. (1925). *Australopithecus africanus*: The Man-Ape of South Africa. *Nature, 115*(2884), 195–199.

Darwin, C. (1859). *The origin of species by means of natural selection, on the preservation of favoured races in the struggle for life* (1st ed.). John Murray. https://archive.org/stream/originofspecies00darwuoft?ref=ol

Darwin, C. (1871). *The descent of man, and selection in relation to sex* (1st ed.). John Murray. https://archive.org/stream/descentofmansele00darw?ref=ol#page/n6/mode/2up

Daver, G., Guy, F., Mackaye, H. T., Likius, A., Boisserie, J. R., Moussa, A., Pallas, L., Vignaud, P., & Clarisse, N. D. (2022). Postcranial evidence of late Miocene hominin bipedalism in Chad. *Nature*. https://doi.org/10.1038/s41586-022-04901-z

David, L. A., Maurice, C. F., Carmody, R. N., Gootenberg, D. B., Button, J. E., Wolfe, B. E., Ling, A. V., Devlin, A. S., Varma, Y., Fischbach, M. A., Biddinger, S. B., Dutton, R. J., & Turnbaugh, P. J. (2014). Diet rapidly and reproducibly alters the human gut microbiome. *Nature, 505*(7484), 559–563. https://doi.org/10.1038/nature12820

Davis, B. M. (2011). Evolution of the tribosphenic molar pattern in early mammals, with comments on the "dual-origin" hypothesis. *Journal of Mammalian Evolution, 18*(4), 227–244. https://doi.org/10.1007/s10914-011-9168-8

Dawkins, R., & Wong, Y. (2004). *The ancestor's tale. A pilgrimage to the dawn of evolution* (1st ed.). Houghton Mifflin.

Day, M. H., & Wickens, E. H. (1980). Laetoli Pliocene hominid footprints and bipedalism. *Nature, 286*(5771), 385–387. https://doi.org/10.1038/286385a0

Dean, M. C. (2006). Tooth microstructure tracks the pace of human life-history evolution. *Proceedings of the Royal Society B-Biological Sciences, 273*(1603), 2799–2808. https://doi.org/10.1098/rspb.2006.3583

Deaner, R. O., Isler, K., Burkart, J., & van Schaik, C. (2007). Overall brain size, and not encephalization quotient, best predicts cognitive ability across non-human primates. *Brain Behavior and Evolution, 70*(2), 115–124. https://doi.org/10.1159/000102973

de Azevedo, S., Gonzalez, M. F., Cintas, C., Ramallo, V., Quinto-Sanchez, M., Marquez, F., Hunemeier, T., Paschetta, C., Ruderman, A., Navarro, P., Pazos, B. A., Silva de Cerqueira, C. C., Velan, O., Ramirez-Rozzi, F., Calvo, N., Castro, H. G., Paz, R. R., & Gonzalez-Jose,

R. (2017). Nasal airflow simulations suggest convergent adaptation in Neanderthals and modern humans. *Proceedings of the National Academy of Sciences of the United States of America, 114*(47), 12442–12447. https://doi.org/10.1073/pnas.1703790114

DeCasien, A. R., Williams, S. A., & Higham, J. P. (2017). Primate brain size is predicted by diet but not sociality. *Nature Ecology & Evolution, 1*(5), 0112. https://doi.org/10.1038/s41559-017-0112

Dediu, D., & Levinson, S. C. (2013). On the antiquity of language: The reinterpretation of Neandertal linguistic capacities and its consequences. *Frontiers in Psychology, 4,* 397. https://doi.org/10.3389/fpsyg.2013.00397

De Groote, I. (2008). *A comprehensive analysis of long bone curvature in Neanderthals and modern humans using 3D morphometrics.* University College of London.

De Groote, I. (2011). Femoral curvature in Neanderthals and modern humans: A 3D geometric morphometric analysis. *Journal of Human Evolution, 60*(5), 540–548. https://doi.org/10.1016/j.jhevol.2010.09.009

De Groote, I., Flink, L. G., Abbas, R., Bello, S. M., Burgia, L., Buck, L. T., Dean, C., Freyne, A., Higham, T., Jones, C. G., Kruszynski, R., Lister, A., Parfitt, S. A., Skinner, M. M., Shindler, K., & Stringer, C. B. (2016). New genetic and morphological evidence suggests a single hoaxer created 'Piltdown man. *Royal Society Open Science, 3*(8), 160328. https://doi.org/10.1098/rsos.160328

Dehay, C., & Kennedy, H. (2020). Evolution of the human brain. *Science, 369*(6503), 506–507. https://doi.org/10.1126/science.abd1840

de Heinzelin, J., Clark, J. D., White, T., Hart, W., Renne, P., WoldeGabriel, G., Beyene, Y., & Vrba, E. (1999). Environment and behavior of 2.5-million-year-old Bouri hominids. *Science, 284*(5414), 625–629.

de la Torre, I. (2011). The origins of stone tool technology in Africa: A historical perspective. *Philosophical Transactions of the Royal Society of London. Series B, Biological Sciences, 366*(1567), 1028–1037. https://doi.org/10.1098/rstb.2010.0350

de la Torre, I. (2016). The origins of the Acheulean: Past and present perspectives on a major transition in human evolution. *Philosophical Transactions of the Royal Society of London. Series B, Biological Sciences, 371*(1698). https://doi.org/10.1098/rstb.2015.0245

Del Bino, S., Ito, S., Sok, J., Nakanishi, Y., Bastien, P., Wakamatsu, K., & Bernerd, F. (2015). Chemical analysis of constitutive pigmentation of human epidermis reveals constant eumelanin to pheomelanin ratio. *Pigment Cell Melanoma Res, 28*(6), 707–717. https://doi.org/10.1111/pcmr.12410

Delezene, L. K. (2015). Modularity of the anthropoid dentition: Implications for the evolution of the hominin canine honing complex. *Journal of Human Evolution, 86,* 1–12. https://doi.org/10.1016/j.jhevol.2015.07.001

de Lussanet, M. H. E., & Osse, J. W. M. (2012). An ancestral axial twist explains the contralateral forebrain and the optic chiasm in vertebrates. *Animal Biology, 62*(2), 193–216. https://doi.org/10.1163/157075611x617102

Demay, L., Péan, S., & Patou-Mathis, M. (2012). Mammoths used as food and building resources by Neanderthals: Zooarchaeological study applied to layer 4, Molodova I (Ukraine). *Quaternary International, 276-277,* 212–226. https://doi.org/10.1016/j.quaint.2011.11.019

DeMenocal, P. B. (2004). African climate change and faunal evolution during the Pliocene-Pleistocene. *Earth and Planetary Science Letters, 220*(1-2), 3–24. https://doi.org/10.1016/S0012-821x(04)00003-2

Demeter, F., Zanolli, C., Westaway, K. E., Joannes-Boyau, R., Duringer, P., Morley, M. W., Welker, F., Rüther, P. L., Skinner, M. M., McColl, H., Gaunitz, C., Vinner, L., Dunn, T. E., Olsen, J. V., Sikora, M., Ponche, J.-L., Suzzoni, E., Frangeul, S., Boesch, Q., . . . & Shackelford, L. (2022). A middle pleistocene denisovan molar from the annamite chain of northern Laos. *Nature Communications, 13*(1), 2557. https://doi.org/10.1038/s41467-022-29923-z

Deng, Y., Misselwitz, B., Dai, N., & Fox, M. (2015). Lactose intolerance in adults: Biological mechanism and dietary management. *Nutrients, 7*(9), 8020–8035. https://doi.org/10.3390/nu7095380

Dennell, R. (2019). Dating of hominin discoveries at Denisova. *Nature, 565*(7741), 571–572. https://doi.org/10.1038/d41586-019-00264-0

Deol, M. S. (1975). Racial differences in pigmentation and natural-selection. *Annals of Human Genetics, 38*(May), 501–503. https://doi.org/10.1111/j.1469-1809.1975.tb00640.x

Desfarges, S., & Ciuffi, A. (2012). Viral integration and consequences on host Gene expression. In G. Witzany (Ed.), *Viruses: Essential agents of life* (pp. 147–175). https://doi.org/10.1007/978-94-007-4899-6_7

Desilva, J. M., Laudicina, N. M., Rosenberg, K. R., & Trevathan, W. R. (2017). Neonatal shoulder width suggests a semirotational, oblique birth mechanism in *Australopithecus afarensis*. *Anatomical Record-Advances in Integrative Anatomy and Evolutionary Biology, 300*(5), 890–899. https://doi.org/10.1002/ar.23573

DeSilva, J. (2021). *First steps. How upright walking made us human*. Harper Collins Publishers.

DeSilva, J. (2022). Walks of life. *Scientific American, 327*(5), 72–81.

DeSilva, J., & Lesnik, J. (2006). Chimpanzee neonatal brain size: Implications for brain growth in *Homo erectus*. *Journal of Human Evolution, 51*(2), 207–212. https://doi.org/10.1016/j.jhevol.2006.05.006

DeSilva, J. M. (2009). Functional morphology of the ankle and the likelihood of climbing in early hominins. *Proceedings of the National Academy of Sciences of the United States of America, 106*(16), 6567–6572. https://doi.org/10.1073/pnas.0900270106

DeSilva, J. M. (2011). A shift toward birthing relatively large infants early in human evolution. *Proceedings of the National Academy of Sciences of the United States of America, 108*(3), 1022–1027. https://doi.org/10.1073/pnas.1003865108

DeSilva, J. M., Gill, C. M., Prang, T. C., Bredella, M. A., & Alemseged, Z. (2018). A nearly complete foot from Dikika, Ethiopia and its implications for the ontogeny and function of *Australopithecus afarensis*. *Science Advances, 4*(7). https://doi.org/10.1126/sciadv.aar7723

DeSilva, J. M., & Lesnik, J. J. (2008). Brain size at birth throughout human evolution: A new method for estimating neonatal brain size in hominins. *Journal of Human Evolution, 55*(6), 1064–1074. https://doi.org/10.1016/j.jhevol.2008.07.008

DeSilva, J. M., Traniello, J. F. A., Claxton, A. G., & Fannin, L. D. (2021). When and why did human brains decrease in size? A new change-point analysis and insights from brain evolution in ants. *Frontiers in Ecology and Evolution, 9*. https://doi.org/10.3389/fevo.2021.742639

Détroit, F., Mijares, A. S., Corny, J., Daver, G., Zanolli, C., Dizon, E., Robles, E., Grün, R., & Piper, P. J. (2019). A new species of *Homo* from the late pleistocene of the Philippines. *Nature, 568*(7751), 181–186. https://doi.org/10.1038/s41586-019-1067-9

Dietrich, L., & Haibt, M. (2020). Bread and porridge at early Neolithic Gobekli Tepe: A new method to recognize products of cereal processing using quantitative functional analyses on grinding stones. *Journal of Archaeological Science-Reports, 33*. https://doi.org/10.1016/j.jasrep.2020.102525

Dietrich, L., Meister, J., Dietrich, O., Notroff, J., Kiep, J., Heeb, J., Beuger, A., & Schutt, B. (2019). Cereal processing at early Neolithic Gobekli Tepe, southeastern Turkey. *PloS One*, *14*(5), e0215214. https://doi.org/10.1371/journal.pone.0215214

Dingwall, H. L., Hatala, K. G., Wunderlich, R. E., & Richmond, B. G. (2013). Hominin stature, body mass, and walking speed estimates based on 1.5 million-year-old fossil footprints at Ileret, Kenya. *Journal of Human Evolution*, *64*(6), 556–568. https://doi.org/10.1016/j.jhevol.2013.02.004

Dobbing, J., & Sands, J. (1973). Quantitative growth and development of human brain. *Archives of Disease in Childhood*, *48*, 757–767.

Dobzhansky, T. (1967). Changing man: Modern evolutionary biology justifies an optimistic view of man's biological future. *Science*, *155*(3761), 409–415.

Dolbunova, E., Lucquin, A., McLaughlin, T. R., Bondetti, M., Courel, B., Oras, E., Piezonka, H., Robson, H. K., Talbot, H., Adamczak, K., Andreev, K., Asheichyk, V., Charniauski, M., Czekaj-Zastawny, A., Ezepenko, I., Grechkina, T., Gunnarssone, A., Gusentsova, T. M., Haskevych, D.,. . . & Craig, O. E. (2022). The transmission of pottery technology among prehistoric European hunter-gatherers. *Nature Human Behaviour*. https://doi.org/10.1038/s41562-022-01491-8

Dominguez-Rodrigo, M., & Pickering, T. R. (2017). The meat of the matter: An evolutionary perspective on human carnivory. *Azania-Archaeological Research in Africa*, *52*(1), 4–32. https://doi.org/10.1080/0067270x.2016.1252066

Dominguez-Rodrigo, M., Pickering, T. R., Semaw, S., & Rogers, M. J. (2005). Cutmarked bones from Pliocene archaeological sites at Gona, Afar, Ethiopia: Implications for the function of the world's oldest stone tools. *Journal of Human Evolution*, *48*(2), 109–121. https://doi.org/10.1016/j.jhevol.2004.09.004

Domínguez-Rodrigo, M. (2014). Is the "Savanna hypothesis" a dead concept for explaining the emergence of the earliest hominins? *Current Anthropology*, *55*(1), 59–81. https://doi.org/10.1086/674530

Dominy, N. J. (2004). Color as an indicator of food quality to anthropoid primates: Ecological evidence and an evolutionary scenario. In C. F. Ross, & R. F. Kay (Eds.), *Anthropoid origins: New visions* (pp. 615–644). Kluwer Academic/Plenum Publishers.

Donnelly, M. P., Paschou, P., Grigorenko, E., Gurwitz, D., Barta, C., Lu, R. B., Zhukova, O. V., Kim, J. J., Siniscalco, M., New, M., Li, H., Kajuna, S. L., Manolopoulos, V. G., Speed, W. C., Pakstis, A. J., Kidd, J. R., & Kidd, K. K. (2012). A global view of the *OCA2-HERC2* region and pigmentation. *Human Genetics*, *131*(5), 683–696. https://doi.org/10.1007/s00439-011-1110-x

Doronichev, V. (2016). The Pre-Mousterian industrial complex in Europe between 400 and 300 ka: Interpreting its origin and spatiotemporal variability. *Quaternary International*, *409*, 222–240. https://doi.org/10.1016/j.quaint.2015.05.063

Douka, K., Slon, V., Jacobs, Z., Ramsey, C. B., Shunkov, M. V., Derevianko, A. P., Mafessoni, F., Kozlikin, M. B., Li, B., Grun, R., Comeskey, D., Deviese, T., Brown, S., Viola, B., Kinsley, L., Buckley, M., Meyer, M., Roberts, R. G., Paabo, S., Kelso, J., & Higham, T. (2019). Age estimates for hominin fossils and the onset of the Upper Palaeolithic at Denisova Cave. *Nature*, *565*(7741), 640–644. https://doi.org/10.1038/s41586-018-0870-z

Drummond-Clarke, R. C., Kivell, T. L., Sarringhaus, L., Stewart, F. B., Humle, T., & Piel, A. K. (2022). Wild chimpanzee behavior suggests that a savanna-mosaic habitat did not support the emergence of hominin terrestrial bipedalism. *Science Advances*, *8*(50), eadd9752.

Dubois, E. (1896). On *Pithecanthropus eshenrectus*: A transitional form between man and the apes. *The Journal of the Anthropological Institute of Great Britain and Ireland, 25*, 240–255.

Dugatkin, L. A. (2020). The silver fox domestication experiment how to tame a fox and build a dog. *Resonance-Journal of Science Education, 25*(7), 987–1000. https://doi.org/10.1007/s12045-020-1014-y

Dunmore, C. J., Skinner, M. M., Bardo, A., Berger, L. R., Hublin, J. J., Pahr, D. H., Rosas, A., Stephens, N. B., & Kivell, T. L. (2020). The position of *Australopithecus sediba* within fossil hominin hand use diversity. *Nature Ecology & Evolution, 4*, 911–918. https://doi.org/10.1038/s41559-020-1207-5

Dunsworth, H. (2021). This view of wife. In J. M. DeSilva (Ed.), *A most interesting problem. What Darwin's descent of man got right and wrong about human evolution* (pp. 183–203). Princeton University Press.

Dunsworth, H. M., Warrener, A. G., Deacon, T., Ellison, P. T., & Pontzer, H. (2012). Metabolic hypothesis for human altriciality. *Proceedings of the National Academy of Sciences of the United States of America, 109*(38), 15212–15216. https://doi.org/10.1073/pnas.1205282109

Dupressoir, A., Lavialle, C., & Heidmann, T. (2012). From ancestral infectious retroviruses to bona fide cellular genes: Role of the captured syncytins in placentation. *Placenta, 33*(9), 663–671. https://doi.org/10.1016/j.placenta.2012.05.005

Edwards, M., Bigham, A., Tan, J., Li, S., Gozdzik, A., Ross, K., Jin, L., & Parra, E. J. (2010). Association of the *OCA2* polymorphism His615Arg with melanin content in East Asian populations: Further evidence of convergent evolution of skin pigmentation. *PLoS Genetics, 6*(3), e1000867. https://doi.org/10.1371/journal.pgen.1000867

Elad, D., Wolf, M., & Keck, T. (2008). Air-conditioning in the human nasal cavity. *Respiratory Physiology & Neurobiology, 163*(1-3), 121–127. https://doi.org/10.1016/j.resp.2008.05.002

Endicott, P., Ho, S. Y. W., & Stringer, C. (2010). Using genetic evidence to evaluate four palaeoanthropological hypotheses for the timing of Neanderthal and modern human origins. *Journal of Human Evolution, 59*(1), 87–95. https://doi.org/10.1016/j.jhevol.2010.04.005

Eng, C. M., Lieberman, D. E., Zink, K. D., & Peters, M. A. (2013). Bite force and occlusal stress production in hominin evolution. *American Journal of Physical Anthropology, 151*(4), 544–557. https://doi.org/10.1002/ajpa.22296

Ennos, A. R., & Chan, T. L. (2016). 'Fire hardening' spear wood does slightly harden it, but makes it much weaker and more brittle. *Biology Letters, 12*(5), 20160174. https://doi.org/10.1098/rsbl.2016.0174

Ernest, S. K. M. (2003). Life history characteristics of placental nonvolant mammals. *Ecology, 84*(12), 3402–3402. https://doi.org/10.1890/02-9002

Essel, E., Zavala, E. I., Schulz-Kornas, E., Kozlikin, M. B., Fewlass, H., Vernot, B., Shunkov, M. V., Derevianko, A. P., Douka, K., Barnes, I., Soulier, M. C., Schmidt, A., Szymanski, M., Tsanova, T., Sirakov, N., Endarova, E., McPherron, S. P., Hublin, J. J., Kelso, J., Paabo, S., Hajdinjak, M., Soressi, M., & Meyer, M. (2023). Ancient human DNA recovered from a Palaeolithic pendant. *Nature, 618*(7964), 328–332. https://doi.org/10.1038/s41586-023-06035-2

Evrard, H. C. (2018). Von Economo and fork neurons in the monkey insula, implications for evolution of cognition. *Current Opinion in Behavioral Sciences, 21*, 182–190. https://doi.org/10.1016/j.cobeha.2018.05.006

Faith, J. T., Du, A., Behrensmeyer, A. K., Davies, B., Patterson, D. B., Rowan, J., & Wood, B. (2021). Rethinking the ecological drivers of hominin evolution. *Trends in Ecology & Evolution, 36*(9), 797–807. https://doi.org/10.1016/j.tree.2021.04.011

Falk, D. (1980). A reanalysis of the South-African Australopithecine Natural Endocasts. *American Journal of Physical Anthropology, 53*(4), 525–539. https://doi.org/10.1002/ajpa.1330530409

Falk, D. (1985). Hadar AI 162-28 endocast as evidence that brain enlargement preceded cortical reorganization in hominid evolution. *Nature, 313*(5997), 45–47. https://doi.org/10.1038/313045a0

Falk, D. (1990). Brain evolution in *Homo* – The radiator theory. *Behavioral and Brain Sciences, 13*(2), 333–343. https://doi.org/10.1017/S0140525x00078973

Falk, D. (2009). The natural endocast of Taung (*Australopithecus africanus*): Insights from the unpublished papers of Raymond Arthur Dart. *American Journal of Physical Anthropology, 140*(Suppl 49), 49–65. https://doi.org/10.1002/ajpa.21184

Falk, D. (2014). Interpreting sulci on hominin endocasts: Old hypotheses and new findings. *Frontiers in Human Neuroscience, 8*, 134. https://doi.org/10.3389/fnhum.2014.00134

Falk, D., & Clarke, R. (2007). Brief communication: New reconstruction of the Taung endocast. *American Journal of Physical Anthropology, 134*(4), 529–534. https://doi.org/10.1002/ajpa.20697

Falk, D., Zollikofer, C. P., Morimoto, N., & Ponce de Leon, M. S. (2012). Metopic suture of Taung (*Australopithecus africanus*) and its implications for hominin brain evolution. *Proceedings of the National Academy of Sciences of the United States of America, 109*(22), 8467–8470. https://doi.org/10.1073/pnas.1119752109

Falk, D., Zollikofer, C. P. E., Ponce de Leon, M., Semendeferi, K., Alatorre Warren, J. L., & Hopkins, W. D. (2018). Identification of in vivo Sulci on the external surface of eight adult chimpanzee brains: Implications for interpreting early hominin endocasts. *Brain, Behavior and Evolution, 91*(1), 45–58. https://doi.org/10.1159/000487248

Falotico, T., & Ottoni, E. B. (2023). Greater tool use diversity is associated with increased terrestriality in wild capuchin monkeys. *American Journal of Biological Anthropology.* https://doi.org/10.1002/ajpa.24740

Fang, R., Xia, C., Close, J. L., Zhang, M., He, J., Huang, Z., Halpern, A. R., Long, B., Miller, J. A., Lein, E. S., & Zhuang, X. (2022). Conservation and divergence of cortical cell organization in human and mouse revealed by MERFISH. *Science, 377*(6601), 56–62. https://doi.org/10.1126/science.abm1741

Fellows Yates, J. A., Velsko, I. M., Aron, F., Posth, C., Hofman, C. A., Austin, R. M., Parker, C. E., Mann, A. E., Nägele, K., Arthur, K. W., Arthur, J. W., Bauer, C. C., Crevecoeur, I., Cupillard, C., Curtis, M. C., Dalén, L., Díaz-Zorita Bonilla, M., Díez Fernández-Lomana, J. C., Drucker, D. G.,... & Warinner, C. (2021). The evolution and changing ecology of the African hominid oral microbiome. *Proceedings of the National Academy of Sciences of the United States of America, 118*(20), e2021655118. https://doi.org/10.1073/pnas.2021655118

Feng, Y. Y., & Walsh, C. A. (2004). Mitotic spindle regulation by *Nde1* controls cerebral cortical size. *Neuron, 44*(2), 279–293. https://doi.org/10.1016/j.neuron.2004.09.023

Ferraro, J. V., Plummer, T. W., Pobiner, B. L., Oliver, J. S., Bishop, L. C., Braun, D. R., Ditchfield, P. W., Seaman, J. W., III, Binetti, K. M., Seaman, J. W., Jr., Hertel, F. K., & Potts, R. (2013). Earliest archaeological evidence of persistent hominin carnivory. *PloS One, 8*(4), e62174. https://doi.org/10.1371/journal.pone.0062174.g001

Ferretti, F., & Adornetti, I. (2021). Persuasive conversation as a new form of communication in *Homo sapiens*. *Philosophical Transactions of the Royal Society of London. Series B, Biological Sciences, 376*(1824), 20200196. https://doi.org/10.1098/rstb.2020.0196

Fichet-Calvet, E., Magassouba, N., & Günther, S. (2019). Estimation of Lassa virus emergence in Upper Guinea through a time-calibrated phylogeny. *Virus Evolution, 5*, 12–13.

Fiers, P., De Clercq, D., Segers, V., & Aerts, P. (2013). Biomechanics of human bipedal gallop: Asymmetry dictates leg function. *The Journal of Experimental Biology, 216*(Pt 7), 1338–1349. https://doi.org/10.1242/jeb.074690

Finlayson, C., Carrion, J., Brown, K., Finlayson, G., Sanchez-Marco, A., Fa, D., Rodriguez-Vidal, J., Fernandez, S., Fierro, E., Bernal-Gomez, M., & Giles-Pacheco, F. (2011). The *Homo* habitat niche: Using the avian fossil record to depict ecological characteristics of Palaeolithic Eurasian hominins. *Quaternary Science Reviews, 30*(11-12), 1525–1532. https://doi.org/10.1016/j.quascirev.2011.01.010

Finlayson, C., Pacheco, F. G., Rodriguez-Vidal, J., Fa, D. A., Lopez, J. M. G., Perez, A. S., Finlayson, G., Allue, E., Preysler, J. B., Caceres, I., Carrion, J. S., Fernandez-Jalvo, Y., Gleed-Owen, C. P., Espejo, F. J. J., Lopez, P., Saez, J. A. L., Cantal, J. A. R., Marco, A. S., Guzman, F. G., Brown, K., Fuentes, N., Valarino, C. A., Villalpando, A., Stringer, C. B., Ruiz, F. M., & Sakamoto, T. (2006). Late survival of Neanderthals at the southernmost extreme of Europe. *Nature, 443*(7113), 850–853. https://doi.org/10.1038/nature05195

Finstermeier, K., Zinner, D., Brameier, M., Meyer, M., Kreuz, E., Hofreiter, M., & Roos, C. (2013). A mitogenomic phylogeny of living primates. *PloS One, 8*(7), e69504. https://doi.org/10.1371/journal.pone.0069504

Fiorenza, L., Benazzi, S., Tausch, J., Kullmer, O., Bromage, T. G., & Schrenk, F. (2011). Molar macrowear reveals Neanderthal eco-geographic dietary variation. *PloS One, 6*(3), e14769. https://doi.org/10.1371/journal.pone.0014769

Fischer, B., Grunstra, N. D. S., Zaffarini, E., & Mitteroecker, P. (2021). Sex differences in the pelvis did not evolve de novo in modern humans. *Nature Ecology & Evolution, 5*(5), 625–630. https://doi.org/10.1038/s41559-021-01425-z

Fitts, R. H., & Widrick, J. J. (1996). Muscle mechanics: Adaptations with exercise-training. *Exercise and Sport Sciences Reviews, 24*(1), 427–474.

Fleagle, J. G., Stern, J. T., Jungers, W. L., Susman, R. L., Vangor, A. K., & Wells, J. P. (1981). Climbing: A biomechanical link with brachiation and with bipedalism. *Symposia of the Zoological Society of London, 48*, 359–375.

Fleming, T. H., & Kress, W. J. (2011). A brief history of fruits and frugivores. *Acta Oecologica-International Journal of Ecology, 37*(6), 521–530. https://doi.org/10.1016/j.actao.2011.01.016

Folk, G. E., & Semken, H. A. (1991). The evolution of sweat glands. *International Journal of Biometeorology, 35*(3), 180–186. https://doi.org/10.1007/Bf01049065

Fonseca-Azevedo, K., & Herculano-Houzel, S. (2012). Metabolic constraint imposes tradeoff between body size and number of brain neurons in human evolution. *Proceedings of the National Academy of Sciences of the United States of America, 109*(45), 18571–18576. https://doi.org/10.1073/pnas.1206390109

Font, E., Garcia-Roa, R., Pincheira-Donoso, D., & Carazo, P. (2019). Rethinking the effects of body size on the study of brain size evolution. *Brain, Behavior and Evolution, 93*(4), 182–195. https://doi.org/10.1159/000501161

Franciscus, R. G., & Trinkaus, E. (1988). Nasal morphology and the emergence of *Homo erectus*. *American Journal of Physical Anthropology, 75*(4), 517–527. https://doi.org/10.1002/ajpa.1330750409

Frank, H. E. R., Amato, K., Trautwein, M., Maia, P., Liman, E. R., Nichols, L. M., Schwenk, K., Breslin, P. A. S., & Dunn, R. R. (2022). The evolution of sour taste. *Proceedings. Biological Sciences, 289*(1968), 20211918. https://doi.org/10.1098/rspb.2021.1918

Fraser, C. M. (2021). A genome to celebrate. *Science, 371*(6529), 545. https://doi.org/10.1126/science.abg8615

Frayer, D. W., Radovcic, J., & Radovcic, D. (2020). Krapina and the case for Neandertal symbolic behavior. *Current Anthropology, 61*(6), 713–731. https://doi.org/10.1086/712088

Freidline, S. E., Westaway, K. E., Joannes-Boyau, R., Duringer, P., Ponche, J. L., Morley, M. W., Hernandez, V. C., McAllister-Hayward, M. S., McColl, H., Zanolli, C., Gunz, P., Bergmann, I., Sichanthongtip, P., Sihanam, D., Boualaphane, S., Luangkhoth, T., Souksavatdy, V., Dosseto, A., Boesch, Q., . . . & Demeter, F. (2023). Early presence of *Homo sapiens* in Southeast Asia by 86-68 kyr at Tam Pa Ling, Northern Laos. *Nature Communications, 14*(1), 3193. https://doi.org/10.1038/s41467-023-38715-y

Frelat, M. A., Shaw, C. N., Sukhdeo, S., Hublin, J. J., Benazzi, S., & Ryan, T. M. (2017). Evolution of the hominin knee and ankle. *Journal of Human Evolution, 108*, 147–160. https://doi.org/10.1016/j.jhevol.2017.03.006

Friis, E. M., Pedersen, K. R., & Crane, P. R. (2010). Diversity in obscurity: Fossil flowers and the early history of angiosperms. *Philosophical Transactions of the Royal Society of London. Series B, Biological Sciences, 365*(1539), 369–382. https://doi.org/10.1098/rstb.2009.0227

Frumin, I., Perl, O., Endevelt-Shapira, Y., Eisen, A., Eshel, N., Heller, I., Shemesh, M., Ravia, A., Sela, L., Arzi, A., & Sobel, N. (2015). A social chemosignaling function for human handshaking. *eLife, 4*, e05154. https://doi.org/10.7554/eLife.05154

Fu, Q., Hajdinjak, M., Moldovan, O. T., Constantin, S., Mallick, S., Skoglund, P., Patterson, N., Rohland, N., Lazaridis, I., Nickel, B., Viola, B., Prufer, K., Meyer, M., Kelso, J., Reich, D., & Paabo, S. (2015). An early modern human from Romania with a recent Neanderthal ancestor. *Nature, 524*, 216–219. https://doi.org/10.1038/nature14558

Fu, Q., Li, H., Moorjani, P., Jay, F., Slepchenko, S. M., Bondarev, A. A., Johnson, P. L. F., Aximu-Petri, A., Prüfer, K., de Filippo, C., Meyer, M., Zwyns, N., Salazar-García, D. C., Kuzmin, Y. V., Keates, S. G., Kosintsev, P. A., Razhev, D. I., Richards, M. P., Peristov, N. V., . . . & Pääbo, S. (2014). Genome sequence of a 45,000-year-old modern human from western Siberia. *Nature, 514*(7523), 445–449. https://doi.org/10.1038/nature13810

Fuchs, J., Garcia-Tabernero, A., Rosas, A., Camus, H., Metz, L., Slimak, L., & Zanolli, C. (2024). The dentition of a new adult Neanderthal individual from Grotte Mandrin, France. *Journal of Human Evolution, 196*, 103599. https://doi.org/10.1016/j.jhevol.2024.103599

Fuentes, A. (2021). "On the races of man": Race, racism, science and hope. In J. M. DeSilva (Ed.), *A most interesting problem. What Darwin's descent of man got right and wrong about human evolution* (pp. 144–161). Princeton University Press.

Gabi, M., Neves, K., Masseron, C., Ribeiro, P. F. M., Ventura-Antunes, L., Torres, L., Mota, B., Kaas, J. H., & Herculano-Houzel, S. (2016). No relative expansion of the number of prefrontal neurons in primate and human evolution. *Proceedings of the National Academy of Sciences of the United States of America, 113*(34), 9617–9622. https://doi.org/10.1073/pnas.1610178113

Gabounia, L., de Lumley, M. A., Vekua, A., Lordkipanidze, D., & de Lumley, H. (2002). Discovery of a new hominid at Dmanisi (Transcaucasia, Georgia). *Comptes Rendus Palevol, 1*(4), 243–253. https://doi.org/10.1016/S1631-0683(02)00032-5

Gabunia, L., & Vekua, A. (1995). A Plio-Pleistocene hominid from Dmanisi, East Georgia, Caucasus. *Nature, 373*(6514), 509–512. https://doi.org/10.1038/373509a0

Galis, F. (1999). Why do almost all mammals have seven cervical vertebrae? Developmental constraints, Hox genes, and cancer. *Journal of Experimental Zoology, 285*, 19–26.

Galis, F., Schut, P. C., Cohen-Overbeek, T. E., & ten Broek, C. M. A. (2021). Evolutionary and developmental issues of cervical ribs/evolutionary issues of cervical ribs. In K. A. Illig, et al. (Eds.), *Thoracic outlet syndrome* (pp. 23–35). Springer. https://doi.org/10.1007/978-3-030-55073-8_4

Galis, F., Van Dooren, T. J., Feuth, J. D., Metz, J. A., Witkam, A., Ruinard, S., Steigenga, M. J., & Wijnaendts, L. C. (2006). Extreme selection in humans against homeotic transformations of cervical vertebrae. *Evolution; International Journal of Organic Evolution, 60*(12), 2643–2654. https://www.ncbi.nlm.nih.gov/pubmed/17263123

Galis, F., Van Dooren, T. J. M., & van der Geer, A. A. E. (2022). Breaking the constraint on the number of cervical vertebrae in mammals: On homeotic transformations in lorises and pottos. *Evolution & Development.* https://doi.org/10.1111/ede.12424

Galletta, L., Stephens, N. B., Bardo, A., Kivell, T. L., & Marchi, D. (2019). Three-dimensional geometric morphometric analysis of the first metacarpal distal articular surface in humans, great apes and fossil hominins. *Journal of Human Evolution, 132*, 119–136. https://doi.org/10.1016/j.jhevol.2019.04.008

Gallup, A. C. (2022). The causes and consequences of yawning in animal groups, *Animal Behaviour, 187*, 209–219. https://doi.org/10.1016/j.anbehav.2022.03.011

Gannon, P. J., Holloway, R. L., Broadfield, D. C., & Braun, A. R. (1998). Asymmetry of chimpanzee planum temporale: Humanlike pattern of Wernicke's brain language area homolog. *Science (New York, N.Y.), 279*(5348), 220–222. https://doi.org/10.1126/science.279.5348.220

Gannon, P. J., Kheck, N. M., Braun, A. R., & Holloway, R. L. (2005). Planum parietale of chimpanzees and orangutans: A comparative resonance of human-like planum temporale asymmetry. *Anatomical Record Part a-Discoveries in Molecular Cellular and Evolutionary Biology, 287a*(1), 1128–1141. https://doi.org/10.1002/ar.a.20256

Garba, R., Usyk, V., Yla-Mella, L., Kamenik, J., Stubner, K., Lachner, J., Rugel, G., Veselovsky, F., Gerasimenko, N., Herries, A. I. R., Kucera, J., Knudsen, M. F., & Jansen, J. D. (2024). East-to-west human dispersal into Europe 1.4 million years ago. *Nature, 627*(8005), 805–810. https://doi.org/10.1038/s41586-024-07151-3

Garber, P. A., & Rehg, J. A. (1999). The ecological role of the prehensile tail in white-faced capuchins (*Cebus capucinus*). *American Journal of Physical Anthropology, 110*(3), 325–339. https://doi.org/10.1002/(Sici)1096-8644(199911)110:3<325::Aid-Ajpa5>3.0.Co;2-D

Garcea, E. A. A. (2022). Critical overview of the Middle Stone Age in North Africa. *Anthropologie, 126*(2), 103022. https://doi.org/10.1016/j.anthro.2022.103022

Garcia-Martinez, D., Bastir, M., Huguet, R., Estalrrich, A., Garcia-Tabernero, A., Rios, L., Cunha, E., Rasilla, M., & Rosas, A. (2017). The costal remains of the El Sidron Neanderthal site (Asturias, northern Spain) and their importance for understanding Neanderthal thorax morphology. *Journal of Human Evolution, 111*, 85–101. https://doi.org/10.1016/j.jhevol.2017.06.003

Gardiner, B. G. (2003). The piltdown forgery: A re-statement of the case against Hinton. *Zoological Journal of the Linnean Society, 139*(3), 315–335. https://doi.org/10.1046/j.1096-3642.2003.00079.x

Gavrilets, S. (2015). Collective action and the collaborative brain. *Journal of the Royal Society, Interface, 12*(102), 20141067. https://doi.org/10.1098/rsif.2014.1067

Gebo, D. L. (1996). Climbing, brachiation, and terrestrial quadrupedalism: Historical precursors of hominid bipedalism. *American Journal of Physical Anthropology*, *101*(1), 55–92.

Gebo, D. L., Smith, T., & Dagosto, M. (2012). New postcranial elements for the earliest Eocene fossil primate Teilhardina belgica. *Journal of Human Evolution*, *63*(1), 205–218. https://doi.org/10.1016/j.jhevol.2012.03.010

Geggel, L. (2022, 02/02/2022). Top 10 things that make humans special. *LiveScience*. Retrieved 25/06/23 from https://www.livescience.com/15689-evolution-human-special-species.html

Geissmann, T., & Orgeldinger, M. (1995). Neonatal weight in Gibbons (*Hylobates* Spp). *American Journal of Primatology*, *37*(3), 179–189. https://doi.org/10.1002/ajp.1350370302

Germonpré, M., Udrescu, M., & Fiers, E. (2014). Possible evidence of mammoth hunting at the Neanderthal site of Spy (Belgium). *Quaternary International*, *337*, 28–42. https://doi.org/10.1016/j.quaint.2012.10.035

Giaobini, G. (2006). Not only burials. History of ideas on other mortuary practices attributed to Neandertals. *Comptes Rendus Palevol*, *5*(1-2), 177–182. https://doi.org/10.1016/j.crpv.2005.09.026

Gibbons, A. (2007). Food for thought. *Science*, *317*(5831), 1558–1560. https://doi.org/10.1126/science.316.5831.1558

Gibbons, A. (2011). Who was *Homo habilis*—And was it really *Homo*? *Science*, *332*, 1370–1371.

Gibbons, A. (2014). How we tamed ourselves - and became modern. *Science*, *346*(6208), 405–406. https://doi.org/10.1126/science.346.6208.405

Gibbons, A. (2015, 06/02/2015). How modern humans ate their way to world dominance. *Science Magazine*. Retrieved 11/05/2021 from https://www.sciencemag.org/news/2015/02/how-modern-humans-ate-their-way-world-dominance

Gibbons, A. (2021a). 'Dragon Man' may be an elusive Denisovan. *Science*, *373*(6550), 11–12.

Gibbons, A. (2021b, 25/06/21). Stunning 'Dragon Man' skull may be an elusive Denisovan—or a new species of human. *Science Magazine*. Retrieved 27/06/21 from https://www.sciencemag.org/news/2021/06/stunning-dragon-man-skull-may-be-elusive-denisovan-or-new-species-human

Gibbons, A. (2021c). When modern humans met Neanderthals. *Science*, *372*(6538), 115–116. https://doi.org/10.1126/science.372.6538.115

Gibbons, A. (2023). Should an also-ran in human evolution get more respect? *Science*, *379*(6632), 522–523.

Gilad, Y., Przeworski, M., & Lancet, D. (2004). Loss of olfactory receptor genes coincides with the acquisition of full trichromatic vision in primates. *Plos Biology*, *2*(1), E5. https://doi.org/10.1371/journal.pbio.0020005

Gilbert, C. C., Ortiz, A., Pugh, K. D., Campisano, C. J., Patel, B. A., Singh, N. P., Fleagle, J. G., & Patnaik, R. (2020). New middle miocene ape (Primates: Hylobatidae) from Ramnagar, India fills major gaps in the hominoid fossil record. *Proceedings. Biological Sciences*, *287*(1934), 20201655. https://doi.org/10.1098/rspb.2020.1655

Gilby, I. C., Machanda, Z. P., O'Malley, R. C., Murray, C. M., Lonsdorf, E. V., Walker, K., Mjungu, D. C., Otali, E., Muller, M. N., Thompson, M. E., Pusey, A. E., & Wrangham, R. W. (2017). Predation by female chimpanzees: Toward an understanding of sex differences in meat acquisition in the last common ancestor of *Pan* and *Homo*. *Journal of Human Evolution*, *110*, 82–94. https://doi.org/10.1016/j.jhevol.2017.06.015

Gittelman, R. M., Schraiber, J. G., Vernot, B., Mikacenic, C., Wurfel, M. M., & Akey, J. M. (2016). Archaic hominin admixture facilitated adaptation to out-of-Africa environments. *Current Biology, 26*(24), 3375–3382. https://doi.org/10.1016/j.cub.2016.10.041

Glazier, D. S. (2008). Effects of metabolic level on the body size scaling of metabolic rate in birds and mammals. *Proceedings. Biological Sciences, 275*(1641), 1405–1410. https://doi.org/10.1098/rspb.2008.0118

Goffette, Q., Rots, V., Abrams, G., Pirson, S., Di Modica, K., Bray, F., Cnuts, D., Bonjean, D., & Amos, L. (2024). Neanderthal exploitation of birds in north-western Europe: Avian remains from Scladina Cave (Belgium). *Frontiers in Environmental Archaeology, 3*. https://doi.org/10.3389/fearc.2024.1441926

Gokhman, D., Mishol, N., de Manuel, M., de Juan, D., Shuqrun, J., Meshorer, E., Marques-Bonet, T., Rak, Y., & Carmel, L. (2019). Reconstructing Denisovan anatomy using DNA methylation maps. *Cell, 179*(1), 180–192.e110. https://doi.org/10.1016/j.cell.2019.08.035

Gomez-Olivencia, A., Barash, A., Garcia-Martinez, D., Arlegi, M., Kramer, P., Bastir, M., & Been, E. (2018). 3D virtual reconstruction of the Kebara 2 Neandertal thorax. *Nature Communications, 9*(1), 4387. https://doi.org/10.1038/s41467-018-06803-z

Gomez-Olivencia, A., Quam, R., Sala, N., Bardey, M., Ohman, J. C., & Balzeau, A. (2018). La Ferrassie 1: New perspectives on a "classic" Neandertal. *Journal of Human Evolution, 117*, 13–32. https://doi.org/10.1016/j.jhevol.2017.12.004

Gonder, M. K., Locatelli, S., Ghobrial, L., Mitchell, M. W., Kujawski, J. T., Lankester, F. J., Stewart, C. B., & Tishkoff, S. A. (2011). Evidence from Cameroon reveals differences in the genetic structure and histories of chimpanzee populations. *Proceedings of the National Academy of Sciences of the United States of America, 108*(12), 4766–4771. https://doi.org/10.1073/pnas.1015422108

Gordon, K. E., Ferris, D. P., & Kuo, A. D. (2009). Metabolic and mechanical energy costs of reducing vertical center of mass movement during gait. *Archives of Physical Medicine and Rehabilitation, 90*(1), 136–144. https://doi.org/10.1016/j.apmr.2008.07.014

Goren-Inbar, N., Alperson, N., Kislev, M. E., Simchoni, O., Melamed, Y., Ben-Nun, A., & Werker, E. (2004). Evidence of hominin control of fire at Gesher Benot Ya'aqov, Israel. *Science, 304*, 725–727. https://doi.org/10.1126/science.1095443

Gosling, W. D., Scerri, E. M. L., & Kaboth-Bahr, S. (2022). The climate and vegetation backdrop to hominin evolution in Africa. *Philosophical Transactions of the Royal Society of London. Series B, Biological Sciences, 377*(1849), 20200483. https://doi.org/10.1098/rstb.2020.0483

Gottschall, J. (2012). *The storytelling animal*. Mariner Books.

Gowlett, J. A. J. (2016). The discovery of fire by humans: A long and convoluted process. *Philosophical Transactions: Biological Sciences, 317*(1696). https://doi.org/10.1098/rstb.2015.0164

Grabowski, M., Hatala, K. G., Jungers, W. L., & Richmond, B. G. (2015). Body mass estimates of hominin fossils and the evolution of human body size. *Journal of Human Evolution, 85*, 75–93. https://doi.org/10.1016/j.jhevol.2015.05.005

Grant, P. R., & Grant, B. R. (2008). *How and why species multiply. The radiation of Darwin's finches*. Princeton University Press.

Graves, J. A. M. (2004). The degenerate Y chromosome - can conversion save it? *Reproduction Fertility and Development, 16*(5), 527–534. https://doi.org/10.1071/Rd03096

Graves, R. R., Lupo, A. C., McCarthy, R. C., Wescott, D. J., & Cunningham, D. L. (2010). Just how strapping was KNM-WT 15000? *Journal of Human Evolution, 59*(5), 542–554. https://doi.org/10.1016/j.jhevol.2010.06.007

Green, R. E., Krause, J., Briggs, A. W., Maricic, T., Stenzel, U., Kircher, M., Patterson, N., Li, H., Zhai, W. W., Fritz, M. H. Y., Hansen, N. F., Durand, E. Y., Malaspinas, A. S., Jensen, J. D., Marques-Bonet, T., Alkan, C., Prufer, K., Meyer, M., Burbano, H. A., . . . & Paabo, S. (2010). A draft sequence of the Neandertal genome. *Science, 328*(5979), 710–722. https://doi.org/10.1126/science.1188021

Green, R. E., Malaspinas, A. S., Krause, J., Briggs, A. W., Johnson, P. L. F., Uhler, C., Meyer, M., Good, J. M., Maricic, T., Stenzel, U., Prufer, K., Siebauer, M., Burbano, H. A., Ronan, M., Rothberg, J. M., Egholm, M., Rudan, P., Brajkovic, D., Kucan, Z., . . . & Paabo, S. (2008). A complete neandertal mitochondrial genome sequence determined by high-throughput sequencing. *Cell, 134*(3), 416–426. https://doi.org/10.1016/j.cell.2008.06.021

Griesser, M., Wheatcroft, D., & Suzuki, T. N. (2018). From bird calls to human language: Exploring the evolutionary drivers of compositional syntax. *Current Opinion in Behavioral Sciences, 21*, 6–12. https://doi.org/10.1016/j.cobeha.2017.11.002

Grine, F. E., Jungers, W. L., & Schultz, J. (1996). Phenetic affinities among early *Homo* crania from East and South Africa. *Journal of Human Evolution, 30*(3), 189–225. https://doi.org/10.1006/jhev.1996.0019

Grine, F. E., Sponheimer, M., Ungar, P. S., Lee-Thorp, J., & Teaford, M. F. (2012). Dental microwear and stable isotopes inform the paleoecology of extinct hominins. *American Journal of Physical Anthropology, 148*(2), 285–317. https://doi.org/10.1002/ajpa.22086

Groscurth, P. (2002). Anatomy of sweat glands. *Current Problems in Dermatology, 30*, 1–9.

Groucutt, H. S., White, T. S., Scerri, E. M. L., Andrieux, E., Clark-Wilson, R., Breeze, P. S., Armitage, S. J., Stewart, M., Drake, N., Louys, J., Price, G. J., Duval, M., Parton, A., Candy, I., Carleton, W. C., Shipton, C., Jennings, R. P., Zahir, M., Blinkhorn, J., . . . & Petraglia, M. D. (2021). Multiple hominin dispersals into Southwest Asia over the past 400,000 years. *Nature*. https://doi.org/10.1038/s41586-021-03863-y

Grün, R., Brink, J. S., Spooner, N. A., Taylor, L., Stringer, C. B., Franciscus, R. G., & Murray, A. S. (1996). Direct dating of Florisbad hominid. *Nature, 382*(6591), 500–501. https://doi.org/10.1038/382500a0

Grün, R., Pike, A., McDermott, F., Eggins, S., Mortimer, G., Aubert, M., Kinsley, L., Joannes-Boyau, R., Rumsey, M., Denys, C., Brink, J., Clark, T., & Stringer, C. (2020). Dating the skull from Broken Hill, Zambia, and its position in human evolution. *Nature, 580*(7803), 372–375. https://doi.org/10.1038/s41586-020-2165-4

Gruss, L. T., Gruss, R., & Schmitt, D. (2017). Pelvic breadth and locomotor kinematics in human evolution. *Anatomical Record (Hoboken), 300*(4), 739–751. https://doi.org/10.1002/ar.23550

Gruss, L. T., & Schmitt, D. (2015). The evolution of the human pelvis: Changing adaptations to bipedalism, obstetrics and thermoregulation. *Philosophical Transactions of the Royal Society B-Biological Sciences, 370*(1663), 20140063. https://doi.org/10.1098/rstb.2014.0063

Guatelli-Steinberg, D. (2024). Growing up slowed down for an early *Homo* individual. *Nature, 635*(8040), 820–822. https://doi.org/10.1038/d41586-024-03547-3

Gunz, P. (2012). Evolutionary relationships among robust and gracile australopiths: An "Evo-devo" perspective. *Evolutionary Biology, 39*(4), 472–487. https://doi.org/10.1007/s11692-012-9185-4

Gunz, P., Neubauer, S., Falk, D., Tafforeau, P., Le Cabec, A., Smith, T. M., Kimbel, W. H., Spoor, F., & Alemseged, Z. (2020). *Australopithecus afarensis* endocasts suggest ape-like brain organization and prolonged brain growth. *Science Advances, 6*(14), eaaz4729. https://doi.org/10.1126/sciadv.aaz4729

Gunz, P., Neubauer, S., Golovanova, L., Doronichev, V., Maureille, B., & Hublin, J. J. (2012). A uniquely modern human pattern of endocranial development. Insights from a new cranial reconstruction of the Neandertal newborn from Mezmaiskaya. *Journal of Human Evolution, 62*(2), 300–313. https://doi.org/10.1016/j.jhevol.2011.11.013

Gunz, P., Neubauer, S., Maureille, B., & Hublin, J. J. (2010). Brain development after birth differs between Neanderthals and modern humans. *Current Biology, 20*(21), R921–R922. https://doi.org/10.1016/j.cub.2010.10.018

Gunz, P., Neubauer, S., Maureille, B., & Hublin, J.-J. (2011). Virtual reconstruction of the Le Moustier 2 newborn skull. *Paléo, 22*(22), 155–172. https://doi.org/10.4000/paleo.2107

Guran, S. H., Yousefi, M., Kafash, A., & Ghasidian, E. (2024). Reconstructing contact and a potential interbreeding geographical zone between Neanderthals and anatomically modern humans. *Scientific Reports, 14*(1), 20475. https://doi.org/10.1038/s41598-024-70206-y

Guy, F., Lieberman, D. E., Pilbeam, D., de Leon, M. P., Likius, A., Mackaye, H. T., Vignaud, P., Zollikofer, C., & Brunet, M. (2005). Morphological affinities of the *Sahelanthropus tchadensis* (Late Miocene hominid from Chad) cranium. *Proceedings of the National Academy of Sciences of the United States of America, 102*(52), 18836–18841. https://doi.org/10.1073/pnas.0509564102

Haeusler, M., Grunstra, N. D. S., Martin, R. D., Krenn, V. A., Fornai, C., & Webb, N. M. (2021). The obstetrical dilemma hypothesis: there's life in the old dog yet. *Biological Reviews of the Cambridge Philosophical Society, 96*(5), 2031–2057. https://doi.org/10.1111/brv.12744

Haeusler, M., Martelli, S. A., & Boeni, T. (2002). Vertebrae numbers of the early hominid lumbar spine. *Journal of Human Evolution, 43*(5), 621–643. https://doi.org/10.1006/jhev.2002.0595

Haeusler, M., Schiess, R., & Boeni, T. (2011). New vertebral and rib material point to modern bauplan of the Nariokotome *Homo erectus* skeleton. *Journal of Human Evolution, 61*(5), 575–582. https://doi.org/10.1016/j.jhevol.2011.07.004

Hagen, E. H., Blackwell, A. D., Lightner, A. D., & Sullivan, R. J. (2023). *Homo medicus*: The transition to meat eating increased pathogen pressure and the use of pharmacological plants in *Homo*. *American Journal of Biological Anthropology*. https://doi.org/10.1002/ajpa.24718

Hahn, C. A., & O'Toole, A. J. (2017). Recognizing approaching walkers: Neural decoding of person familiarity in cortical areas responsive to faces, bodies, and biological motion. *Neuroimage, 146*, 859–868. https://doi.org/10.1016/j.neuroimage.2016.10.042

Haile-Selassie, Y. (2001). Late Miocene hominids from the Middle Awash, Ethiopia. *Nature, 412*(6843), 178–181.

Haile-Selassie, Y. (2021). Charles Darwin and the fossil evidence for human evolution. In J. M. DeSilva (Ed.), *A most interesting problem. What Darwin's descent of man got right and wrong about human evolution* (pp. 82–102). Princeton University Press.

Haile-Selassie, Y., Gibert, L., Melillo, S. M., Ryan, T. M., Alene, M., Deino, A., Levin, N. E., Scott, G., & Saylor, B. Z. (2015). New species from Ethiopia further expands Middle Pliocene hominin diversity. *Nature, 521*(7553), 483–U500. https://doi.org/10.1038/nature14448

Haile-Selassie, Y., Melillo, S. M., Vazzana, A., Benazzi, S., & Ryan, T. M. (2019). A 3.8-million-year-old hominin cranium from Woranso-Mille, Ethiopia. *Nature*. https://doi.org/10.1038/s41586-019-1513-8

Haile-Selassie, Y., Suwa, G., & White, T. D. (2004). Late Miocene teeth from Middle Awash, Ethiopia, and early hominid dental evolution. *Science, 303*(5663), 1503–1505. https://doi.org/10.1126/science.1092978

Hain, D., Gallego-Flores, T., Klinkmann, M., Macias, A., Ciirdaeva, E., Arends, A., Thum, C., Tushev, G., Kretschmer, F., Tosches, M. A., & Laurent, G. (2022). Molecular diversity and evolution of neuron types in the amniote brain. *Science (New York, N.Y.), 377*(6610), eabp8202. https://doi.org/10.1126/science.abp8202

Hajdinjak, M., Mafessoni, F., Skov, L., Vernot, B., Hübner, A., Fu, Q., Essel, E., Nagel, S., Nickel, B., Richter, J., Moldovan, O. T., Constantin, S., Endarova, E., Zahariev, N., Spasov, R., Welker, F., Smith, G. M., Sinet-Mathiot, V., Paskulin, L., . . . & Pääbo, S. (2021). Initial upper palaeolithic humans in Europe had recent Neanderthal ancestry. *Nature, 592*(7853), 253–257. https://doi.org/10.1038/s41586-021-03335-3

Hammer, M. F., Karafet, T., Rasanayagam, A., Wood, E. T., Altheide, T. K., Jenkins, T., Griffiths, R. C., Templeton, A. R., & Zegura, S. L. (1998). Out of Africa and back again: Nested cladistic analysis of human Y chromosome variation. *Molecular Biology and Evolution, 15*(4), 427–441. https://doi.org/10.1093/oxfordjournals.molbev.a025939

Hammond, A. S., Mavuso, S. S., Biernat, M., Braun, D. R., Jinnah, Z., Kuo, S., Melaku, S., Wemanya, S. N., Ndiema, E. K., Patterson, D. B., Uno, K. T., & Palcu, D. V. (2021). New hominin remains and revised context from the earliest *Homo erectus* locality in East Turkana, Kenya. *Nature Communications, 12*(1), 1939. https://doi.org/10.1038/s41467-021-22208-x

Hammond, A. S., Rook, L., Anaya, A. D., Cioppi, E., Costeur, L., Moyà-Solà, S., & Almécija, S. (2020). Insights into the lower torso in late Miocene hominoid *Oreopithecus bambolii. Proceedings of the National Academy of Science, 117*(1), 278–284.

Harari, Y. N. (2018). *Sapiens* (Nederlandse vertaling, 27e druk ed.). Uitgeverij Thomas Rap.

Harcourt-Smith, W. (2016). Early hominin diversity and the emergence of the genus *Homo, J Anthropol Sci, 94*, 19–27, https://pubmed.ncbi.nlm.nih.gov/27124766/

Harcourt-Smith, W. E. (2005). Did *Australopithecus afarensis* make the Laetoli footprint trail? - New insights into an old problem. *American Journal of Physical Anthropology* (S40), 112–112.

Harcourt-Smith, W. E. H. (2013). The origins of bipedal locomotion. In W. Henke & I. Tattersall (Eds.), *Handbook of paleoanthropology* (pp. 1–36). Springer-Verlag. https://doi.org/10.1007/978-3-642-27800-6_48-3

Hardy, K., & Buckley, S. (2017). Earliest evidence of bitumen from *Homo* sp teeth is from El Sidron. *American Journal of Physical Anthropology, 164*(1), 212–213. https://doi.org/10.1002/ajpa.23255

Hardy, K., Buckley, S., Collins, M. J., Estalrrich, A., Brothwell, D., Copeland, L., Garcia-Tabernero, A., Garcia-Vargas, S., de la Rasilla, M., Lalueza-Fox, C., Huguet, R., Bastir, M., Santamaria, D., Madella, M., Wilson, J., Cortes, A. F., & Rosas, A. (2012). Neanderthal medics? Evidence for food, cooking, and medicinal plants entrapped in dental calculus. *Naturwissenschaften, 99*(8), 617–626. https://doi.org/10.1007/s00114-012-0942-0

Harmand, S., Lewis, J. E., Feibel, C. S., Lepre, C. J., Prat, S., Lenoble, A., Boes, X., Quinn, R. L., Brenet, M., Arroyo, A., Taylor, N., Clement, S., Daver, G., Brugal, J. P., Leakey, L., Mortlock, R. A., Wright, J. D., Lokorodi, S., Kirwa, C., Kent, D. V., & Roche, H. (2015). 3.3-million-year-old stone tools from Lomekwi 3, West Turkana, Kenya. *Nature, 521*(7552), 310–315. https://doi.org/10.1038/nature14464

Harper, C. M., Ruff, C. B., & Sylvester, A. D. (2021). Calcaneal shape variation in humans, non-human primates, and early hominins. *Journal of Human Evolution, 159,* 103050. https://doi.org/10.1016/j.jhevol.2021.103050

Harris, E. E. (2010). Nonadaptive processes in primate and human evolution. *American Journal of Physical Anthropology, 143*(Suppl 51), 13–45. https://doi.org/10.1002/ajpa.21439

Harris-Tryon, T. A., & Grice, E. A. (2022). Microbiota and maintenance of skin barrier function. *Science, 376,* 940–945.

Harvati, K., & Ackermann, R. R. (2022). Merging morphological and genetic evidence to assess hybridization in Western Eurasian late Pleistocene hominins. *Nature Ecology & Evolution, 6*(10), 1573–1585. https://doi.org/10.1038/s41559-022-01875-z

Harvati, K., & Reyes-Centeno, H. (2022). Evolution of *Homo* in the Middle and Late Pleistocene. *Journal of Human Evolution, 173,* 103279. https://doi.org/10.1016/j.jhevol.2022.103279

Harvati, K., Roding, C., Bosman, A. M., Karakostis, F. A., Grun, R., Stringer, C., Karkanas, P., Thompson, N. C., Koutoulidis, V., Moulopoulos, L. A., Gorgoulis, V. G., & Kouloukoussa, M. (2019). Apidima Cave fossils provide earliest evidence of *Homo sapiens* in Eurasia. *Nature, 571*(7766), 500–504. https://doi.org/10.1038/s41586-019-1376-z

Harvey, P. H., & Cluttonbrock, T. H. (1985). Life-history variation in primates. *Evolution, 39*(3), 559–581. https://doi.org/10.2307/2408653

Haslam, M. (2019). The other tool users. *Scientific American, 320*(3), 64–69.

Hatala, K. G., & Boyle, E. K. (2022). The feet of fossil *Homo*. In L. Barrett (Ed.), *The evolution of the primate foot. Anatomy, function and palaeontological evidence* (pp. 387–413). Springer. https://doi.org/10.1007/978-3-031-06436-4

Hatala, K. G., Demes, B., & Richmond, B. G. (2016). Laetoli footprints reveal bipedal gait biomechanics different from those of modern humans and chimpanzees. *Proceedings of the Royal Society B-Biological Sciences, 283*(1836). https://doi.org/10.1098/rspb.2016.0235

Hatala, K. G., Gatesy, S. M., & Falkingham, P. L. (2023). Arched footprints preserve the motions of fossil hominin feet. *Nature Ecology & Evolution, 7,* 32–41.

Hatala, K. G., Roach, N. T., Behrensmeyer, A. K., Falkingham, P. L., Gatesy, S. M., Williams-Hatala, E. M., Feibel, C. S., Dalacha, I., Kirinya, M., Linga, E., Loki, R., Longaye, A. A., Longaye, M., Lonyericho, E., Loyapan, I., Nakudo, N., Nyete, N., & Leakey, L. N. (2024). Footprint evidence for locomotor diversity andshared habitats among early Pleistocene hominins. *Science, 386*(6725), 1004–1010.

Hatten, M. E. (2020). Adding cognitive connections to the cerebellum. *Science (New York, N.Y.), 370*(6523), 1411–1412. https://doi.org/10.1126/science.abf4483

Hausler, M., & Schmid, P. (1995). Comparison of the pelves of Sts-14 and Al-288-1 - Implications for birth and sexual dimorphism in australopithecines. *Journal of Human Evolution, 29*(4), 363–383. https://doi.org/10.1006/jhev.1995.1063

Hawkes, K., O'Connell, J. F., & Blurton Jones, N. G. (1991). Hunting income patterns among the Hadza: Big game, common goods, foraging goals and the evolution of the human diet. *Philosophical Transactions of the Royal Society of London. Series B, Biological Sciences, 334*(1270), 243–250; discussion 250–241. https://doi.org/10.1098/rstb.1991.0113

Hawkes, K., O'Connell, J. F., & Blurton Jones, N. G. (1999). Comment to 'The raw and the stolen - Cooking and the ecology of human origins'. *Current Anthropology, 40*(5), 581–582. https://doi.org/10.1086/300083

Hawks, J. (2016). Still evolving (after all these years). *Scientific American, 25*(4), 100–105.

Hawks, J., Wang, E. T., Cochran, G. M., Harpending, H. C., & Moyzis, R. K. (2007). Recent acceleration of human adaptive evolution. *Proceedings of the National Academy of Sciences of the United States of America, 104*(52), 20753–20758. https://doi.org/10.1073/pnas.0707650104

Hay, O., Dar, G., Abbas, J., Stein, D., May, H., Masharawi, Y., Peled, N., & Hershkovitz, I. (2015). The Lumbar Lordosis in males and females, revisited. *PloS One, 10*(8), e0133685. https://doi.org/10.1371/journal.pone.0133685

Heaton, J. L., Pickering, T. R., Carlson, K. J., Crompton, R. H., Jashashvili, T., Beaudet, A., Bruxelles, L., Kuman, K., Heile, A. J., Stratford, D., & Clarke, R. J. (2019). The long limb bones of the StW 573 *Australopithecus* skeleton from Sterkfontein Member 2: Descriptions and proportions. *Journal of Human Evolution, 133*, 167–197. https://doi.org/10.1016/j.jhevol.2019.05.015

Heide, M., Haffner, C., Murayama, A., Kurotaki, Y., Shinohara, H., Okano, H., Sasaki, E., & Huttner, W. B. (2020). Human-specific ARHGAP11B increases size and folding of primate neocortex in the fetal marmoset. *Science (New York, N.Y.), 369*(6503), 546–550. https://doi.org/10.1126/science.abb2401

Heiss, C. N., & Olofsson, L. E. (2019). The role of the gut microbiota in development, function and disorders of the central nervous system and the enteric nervous system. *Journal of Neuroendocrinology, 31*(5), e12684. https://doi.org/10.1111/jne.12684

Helmuth, H. (1992). "Lucy's" body height and relative leg length: Human- or ape-like? *Zeitschrift für Morphologie und Anthropologie, 79*(1), 121–124.

Henry, A. G., Brooks, A. S., & Piperno, D. R. (2011). Microfossils in calculus demonstrate consumption of plants and cooked foods in Neanderthal diets (Shanidar III, Iraq; Spy I and II, Belgium). *Proceedings of the National Academy of Sciences of the United States of America, 108*(2), 486–491. https://doi.org/10.1073/pnas.1016868108

Henry, A. G., Brooks, A. S., & Piperno, D. R. (2014). Plant foods and the dietary ecology of Neanderthals and early modern humans. *Journal of Human Evolution, 69*, 44–54. https://doi.org/10.1016/j.jhevol.2013.12.014

Herculano-Houzel, S. (2011). Scaling of brain metabolism with a fixed energy budget per neuron: Implications for neuronal activity, plasticity and evolution. *PloS One, 6*(3), e17514. https://doi.org/10.1371/journal.pone.0017514

Herculano-Houzel, S. (2012). Neuronal scaling rules for primate brains: The primate advantage. In M. A. Hofman & D. Falk (Eds.), *Evolution of the primate brain: From neuron to behavior* (Vol. 195, pp. 325–340). Elsevier B.V. https://doi.org/10.1016/B978-0-444-53860-4.00015-5

Herculano-Houzel, S. (2016). *The human advantage: How our brains became remarkable*. MIT Press.

Herculano-Houzel, S. (2021). Remarkable but not extraordinary: The evolution of the human brain. In J. M. DeSilva (Ed.), *A most interesting problem. What Darwin's descent of man got right and wrong about human evolution* (pp. 46–62). Princeton University Press.

Herculano-Houzel, S., Catania, K., Manger, P. R., & Kaas, J. H. (2015). Mammalian brains are made of these: A dataset of the numbers and densities of neuronal and nonneuronal cells in the brain of glires, primates, scandentia, eulipotyphlans, afrotherians and artiodactyls, and their relationship with body mass. *Brain Behavior and Evolution, 86*(3-4), 145–163. https://doi.org/10.1159/000437413

Herculano-Houzel, S., Collins, C. E., Wong, P. Y., & Kaas, J. H. (2007). Cellular scaling rules for primate brains. *Proceedings of the National Academy of Sciences of the United States of America, 104*(9), 3562–3567. https://doi.org/10.1073/pnas.0611396104

Herculano-Houzel, S., & Kaas, J. H. (2011). Gorilla and orangutan brains conform to the primate cellular scaling rules: Implications for human evolution. *Brain Behav Evol, 77*(1), 33–44. https://doi.org/10.1159/000322729

Herculano-Houzel, S., Manger, P. R., & Kaas, J. H. (2014). Brain scaling in mammalian evolution as a consequence of concerted and mosaic changes in numbers of neurons and average neuronal cell size. *Frontiers in Neuroanatomy, 8*, 77. https://doi.org/10.3389/fnana.2014.00077

Herries, A. I. R., Curnoe, D., & Adams, J. W. (2009). A multi-disciplinary seriation of early *Homo* and *Paranthropus* bearing palaeocaves in southern Africa. *Quaternary International, 202*(1-2), 14–28. https://doi.org/10.1016/j.quaint.2008.05.017

Herries, A. I. R., Martin, J. M., Leece, A. B., Adams, J. W., Boschian, G., Joannes-Boyau, R., Edwards, T. R., Mallett, T., Massey, J., Murszewski, A., Neubauer, S., Pickering, R., Strait, D. S., Armstrong, B. J., Baker, S., Caruana, M. V., Denham, T., Hellstrom, J., Moggi-Cecchi, J., . . . & Menter, C. (2020). Contemporaneity of *Australopithecus, Paranthropus,* and early *Homo erectus* in South Africa. *Science (New York, N.Y.), 368*(6486), 1–19. https://doi.org/10.1126/science.aaw7293

Hershkovitz, I., Weber, G. W., Quam, R., Duval, M., Grun, R., Kinsley, L., Ayalon, A., Bar-Matthews, M., Valladas, H., Mercier, N., Arsuaga, J. L., Martinon-Torres, M., Bermudez de Castro, J. M., Fornai, C., Martin-Frances, L., Sarig, R., May, H., Krenn, V. A., Slon, V., . . . & Weinstein-Evron, M. (2018). The earliest modern humans outside Africa. *Science, 359*(6374), 456–459. https://doi.org/10.1126/science.aap8369

Hertler, C., Bruch, A., & Märker, M. (2013). The earliest stages of hominin dispersal in Africa and Eurasia. In *The encyclopedia of global human migration.* John Wiley & Sons. https://doi.org/10.1002/9781444351071.wbeghm802

Heyes, P., & MacDonald, K. (2015). Neandertal energetics: Uncertainty in body mass estimation limits comparisons with *Homo sapiens. Journal of Human Evolution, 85*, 193–197. https://doi.org/10.1016/j.jhevol.2015.04.007

Heyes, P. J., Anastasakis, K., de Jong, W., van Hoesel, A., Roebroeks, W., & Soressi, M. (2016). Selection and use of manganese dioxide by Neanderthals. *Scientific Reports, 6.* https://doi.org/10.1038/srep22159

Heymsfield, S. B., Gallagher, D., Mayer, L., Beetsch, J., & Pietrobelli, A. (2007). Scaling of human body composition to stature: New insights into body mass index. *American Journal of Clinical Nutrition, 86*(1), 82–91. https://doi.org/10.1093/ajcn/86.1.82

Higham, T., Douka, K., Wood, R., Ramsey, C. B., Brock, F., Basell, L., Camps, M., Arrizabalaga, A., Baena, J., Barroso-Ruiz, C., Bergman, C., Boitard, C., Boscato, P., Caparros, M., Conard, N. J., Draily, C., Froment, A., Galvan, B., Gambassini, P., . . . & Jacobi, R. (2014). The timing and spatiotemporal patterning of Neanderthal disappearance. *Nature, 512*(7514), 306–309. https://doi.org/10.1038/nature13621

Higuchi, S., Chaminade, T., Imamizu, H., & Kawato, M. (2009). Shared neural correlates for language and tool use in Broca's area. *Neuroreport, 20*(15), 1376–1381. https://doi.org/10.1097/WNR.0b013e3283315570

Hlubik, S., Berna, F., Feibel, C., Braun, D., & Harris, J. W. K. (2017). Researching the nature of fire at 1.5 Mya on the site of FxJj20 AB, Koobi Fora, Kenya, Using High-Resolution Spatial Analysis and FTIR Spectrometry. *Current Anthropology, 58*, S243–S257. https://doi.org/10.1086/692530

Hlubik, S., Cutts, R., Braun, D. R., Berna, F., Feibel, C. S., & Harris, J. W. K. (2019). Hominin fire use in the Okote member at Koobi Fora, Kenya: New evidence for the old debate. *Journal of Human Evolution, 133,* 214–229. https://doi.org/10.1016/j.jhevol.2019.01.010

Hlusko, L. J., Carlson, J. P., Chaplin, G., Elias, S. A., Hoffecker, J. F., Huffman, M., Jablonski, N. G., Monson, T. A., O'Rourke, D. H., Pilloud, M. A., & Scott, G. R. (2018). Environmental selection during the last ice age on the mother-to-infant transmission of vitamin D and fatty acids through breast milk. *Proceedings of the National Academy of Sciences of the United States of America, 115*(19), E4426–E4432. https://doi.org/10.1073/pnas.1711788115

Hobaiter, C., & Byrne, R. W. (2014). The meanings of chimpanzee gestures. *Current Biology: CB, 24*(14), 1596–1600. https://doi.org/10.1016/j.cub.2014.05.066

Hobaiter, C., Graham, K. E., & Byrne, R. W. (2022). Are ape gestures like words? Outstanding issues in detecting similarities and differences between human language and ape gesture. *Philosophical Transactions of the Royal Society of London. Series B, Biological Sciences, 377*(1860), 20210301. https://doi.org/10.1098/rstb.2021.0301

Hoberg, E. P., Alkire, N. L., de Queiroz, A., & Jones, A. (2001). Out of Africa: Origins of the *Taenia* tapeworms in humans. *Proceedings of the Royal Society B-Biological Sciences, 268*(1469), 781–787. https://doi.org/10.1098/rspb.2000.1579

Hogervorst, T., & Vereecke, E. E. (2015). Evolution of the human hip. Part 2: Muscling the double extension. *Journal of Hip Preservation Surgery, 2*(1), 3–14. https://doi.org/10.1093/jhps/hnu014

Holloway, R. L. (1970). Australopithecine endocast (Taung Specimen, 1924): A new volume determination. *Science, 168*(3934), 966. https://doi.org/10.1126/science.168.3934.966

Holloway, R. L. (1972). New australopithecine endocast, Sk 1585, from Swartkrans, South-Africa. *American Journal of Physical Anthropology, 37*(2), 173–185. https://doi.org/10.1002/ajpa.1330370203

Holloway, R. L. (2015). The evolution of the hominid brain. In W. Henke & I. Tattersall (Eds.), *Handbook of paleoanthropology* (Vol. 3, pp. 1961–1987). Springer-Verlag. https://doi.org/10.1007/978-3-642-39979-4_81

Holloway, R. L., Broadfield, D. C., & Carlson, K. J. (2014). New high-resolution computed tomography data of the Taung partial cranium and endocast and their bearing on metopism and hominin brain evolution. *Proceedings of the National Academy of Sciences of the United States of America, 111*(36), 13022–13027. https://doi.org/10.1073/pnas.1402905111

Holowka, N. B., O'Neill, M. C., Thompson, N. E., & Demes, B. (2017a). Chimpanzee and human midfoot motion during bipedal walking and the evolution of the longitudinal arch of the foot. *Journal of Human Evolution, 104,* 23–31. https://doi.org/10.1016/j.jhevol.2016.12.002

Holowka, N. B., O'Neill, M. C., Thompson, N. E., & Demes, B. (2017b). Chimpanzee ankle and foot joint kinematics: Arboreal versus terrestrial locomotion. *American Journal of Physical Anthropology, 164*(1), 131–147. https://doi.org/10.1002/ajpa.23262

Holowka, N. B., Richards, A., Sibson, B. E., & Lieberman, D. E. (2021). The human foot functions like a spring of adjustable stiffness during running. *The Journal of Experimental Biology, 224*(Pt 1). https://doi.org/10.1242/jeb.219667

Hopcroft, R. L. (2015). Sex differences in the relationship between status and number of offspring in the contemporary Uu.S. *Evolution and Human Behavior, 36*(2), 146–151. https://doi.org/10.1016/j.evolhumbehav.2014.10.003

Hora, M., Pontzer, H., & Sladek, V. (2019). Persistence hunting in Levant: Both Neandertals and modern humans could run down a horse *American Journal of Physical Anthropology*, *168*(S68), 107–107.

Hora, M., Pontzer, H., Struska, M., Entin, P., & Sladek, V. (2022). Comparing walking and running in persistence hunting. *Journal of Human Evolution*, *172*, 103247. https://doi.org/10.1016/j.jhevol.2022.103247

Hora, M., Pontzer, H., Wall-Scheffler, C. M., & Sladek, V. (2020). Dehydration and persistence hunting in *Homo erectus*. *Journal of Human Evolution*, *138*, 102682. https://doi.org/10.1016/j.jhevol.2019.102682

Hoyt, D. F., & Taylor, C. R. (1981). Gait and the energetics of locomotion in horses. *Nature*, *292*(5820), 239–240. https://doi.org/10.1038/292239a0

Hrvoj-Mihic, B., Bienvenu, T., Stefanacci, L., Muotri, A. R., & Semendeferi, K. (2013). Evolution, development, and plasticity of the human brain: From molecules to bones. *Frontiers in Human Neuroscience*, *7*, 1–18. https://doi.org/10.3389/fnhum.2013.00707

Hubisz, M. J., Williams, A. L., & Siepel, A. (2020). Mapping gene flow between ancient hominins through demography-aware inference of the ancestral recombination graph. *PLoS Genetics*, *16*(8), e1008895. https://doi.org/10.1371/journal.pgen.1008895

Hublin, J. J., Ben-Ncer, A., Bailey, S. E., Freidline, S. E., Neubauer, S., Skinner, M. M., Bergmann, I., Le Cabec, A., Benazzi, S., Harvati, K., & Gunz, P. (2017). New fossils from Jebel Irhoud, Morocco and the pan-African origin of *Homo sapiens*. *Nature*, *546*(7657), 289–292. https://doi.org/10.1038/nature22336

Hughes, J. F., Skaletsky, H., Pyntikova, T., Minx, P. J., Graves, T., Rozen, S., Wilson, R. K., & Page, D. C. (2005). Conservation of Y-linked genes during human evolution revealed by comparative sequencing in chimpanzee. *Nature*, *437*(7055), 100–103. https://doi.org/10.1038/nature04101

Huttenlocher, P. R., & Dabholkar, A. S. (1997). Regional differences in synaptogenesis in human cerebral cortex. *Journal of Comparative Neurology*, *387*(2), 167–178. https://doi.org/10.1002/(Sici)1096-9861(19971020)387:2<167::Aid-Cne1>3.0.Co;2-Z

Huxley, T. H. (1863). *Evidence as to man's place in nature*. D. Appleton and Company. https://archive.org/details/evidenceasto00huxl

Iasi, L. N. M., Chintalapati, M., Skov, L., Mesa, A. B., Hajdinjak, M., Peter, B. M., & Moorjani, P. (2024). Neanderthal ancestry through time: Insights from genomes of ancient and present-day humans. *Science*, *386*(6727), eadq3010. https://doi.org/10.1126/science.adq3010

Ilardo, M. A., Moltke, I., Korneliussen, T. S., Cheng, J., Stern, A. J., Racimo, F., de Barros Damgaard, P., Sikora, M., Seguin-Orlando, A., Rasmussen, S., van den Munckhof, I. C. L., Ter Horst, R., Joosten, L. A. B., Netea, M. G., Salingkat, S., Nielsen, R., & Willerslev, E. (2018). Physiological and genetic adaptations to diving in sea nomads. *Cell*, *173*(3), 569–580e515. https://doi.org/10.1016/j.cell.2018.03.054

Inchley, C. E., Larbey, C. D. A., Shwan, N. A. A., Pagani, L., Saag, L., Antao, T., Jacobs, G., Hudjashov, G., Metspalu, E., Mitt, M., Eichstaedt, C. A., Malyarchuk, B., Derenko, M., Wee, J., Abdullah, S., Ricaut, F. X., Mormina, M., Magi, R., Villems, R., Metspalu, M., Jones, M. K., Armour, J. A. L., & Kivisild, T. (2016). Selective sweep on human amylase genes postdates the split with Neanderthals. *Scientific Reports*, *6*, 37198. https://doi.org/10.1038/srep37198

Inouye, S. E. (1994). Ontogeny of knuckle-walking hand postures in African apes. *Journal of Human Evolution*, *26*(5-6), 459–485.

Ioannidou, M., Koufos, G. D., de Bonis, L., & Harvati, K. (2019). A new three-dimensional geometric morphometrics analysis of the *Ouranopithecus macedoniensis* cranium (Late Miocene, Central Macedonia, Greece). *American Journal of Physical Anthropology, 170*(2), 295–307. https://doi.org/10.1002/ajpa.23900

Irmak, M. K., Korkmaz, A., & Erogul, O. (2004). Selective brain cooling seems to be a mechanism leading to human craniofacial diversity observed in different geographical regions. *Medical Hypotheses, 63*(6), 974–979. https://doi.org/10.1016/j.mehy.2004.05.003

Ishikawa, M., Komi, P. V., Grey, M. J., Lepola, V., & Bruggemann, G. P. (2005). Muscle-tendon interaction and elastic energy usage in human walking. *Journal of Applied Physiology, 99*(2), 603–608.

Ito, A., Oishi, M., Endo, H., Hirasaki, E., & Ogihara, N. (2024). A cadaveric study of wrist-joint moments in chimpanzees and orangutans with implications for the evolution of knuckle-walking, *Journal of Human Evolution, 197*, 103600. https://doi.org/10.1016/j.jhevol.2024.103600

Izagirre, N., Garcia, I., Junquera, C., de la Rua, C., & Alonso, S. (2006). A scan for signatures of positive selection in candidate loci for skin pigmentation in humans. *Molecular Biology and Evolution, 23*(9), 1697–1706. https://doi.org/10.1093/molbev/msl030

Jablonski, N. G. (2004). The evolution of human skin and skin color. *Annual Review of Anthropology, 33*, 585–623. https://doi.org/10.1146/annurev.anthro.33.070203.143955

Jablonski, N. G. (2012). *Living colors. The biological and social meaning of skin color.* University of California Press.

Jablonski, N. G. (2017, 02/28/2017). The Real 'Skin in the Game': The History of Naked, Sweaty, and Colorful Skin in the Human Lineage. UC Berkeley. https://www.uctv.tv/shows/The-Real-Skin-in-the-Game-The-History-of-Naked-Sweaty-and-Colorful-Skin-in-the-Human-Lineage-32129?s=09

Jablonski, N. G. (2021). The evolution of human skin pigmentation involved the interactions of genetic, environmental, and cultural variables. *Pigment Cell & Melanoma Research, 34*(4), 707–729. https://doi.org/10.1111/pcmr.12976

Jablonski, N. G., & Chaplin, G. (2000). The evolution of human skin coloration. *Journal of Human Evolution, 39*, 57–106.

Jacobs, L. C., Wollstein, A., Lao, O., Hofman, A., Klaver, C. C., Uitterlinden, A. G., Nijsten, T., Kayser, M., & Liu, F. (2013). Comprehensive candidate gene study highlights *UGT1A* and *BNC2* as new genes determining continuous skin color variation in Europeans. *Human genetics, 132*(2), 147–158. https://doi.org/10.1007/s00439-012-1232-9

Jameson, N. M., Hou, Z. C., Sterner, K. N., Weckle, A., Goodman, M., Steiper, M. E., & Wildman, D. E. (2011). Genomic data reject the hypothesis of a prosimian primate clade. *Journal of Human Evolution, 61*(3), 295–305. https://doi.org/10.1016/j.jhevol.2011.04.004

Janvier, P. (2015). Facts and fancies about early fossil chordates and vertebrates. *Nature, 520*(7548), 483–489. https://doi.org/10.1038/nature14437

Jeffery, N., & Spoor, F. (2004). Prenatal growth and development of the modern human labyrinth. *Journal of Anatomy, 204*, 71–92.

Jessen, C. (1998). Brain cooling: An economy mode of temperature regulation in artiodactyls. *News in Physiological Sciences, 13*, 281–286.

Ji, Q., Wu, W., Ji, Y., Li, Q., & Ni, X. (2021). Late Middle Pleistocene Harbin cranium represents a new *Homo* species. *Innovation (New York, N.Y.), 2*(3), 100132. https://doi.org/10.1016/j.xinn.2021.100132

Johanson, D. C., Lovejoy, C. O., Kimbel, W. H., White, T. D., Ward, S. C., Bush, M. E., Latimer, B. M., & Coppens, Y. (1982). Morphology of the Pliocene partial hominid skeleton (Al 288-1) from the Hadar Formation, Ethiopia. *American Journal of Physical Anthropology*, *57*(4), 403–451. https://doi.org/10.1002/ajpa.1330570403

Johanson, D. C., & Taieb, M. (1976). Plio–Pleistocene hominid discoveries in Hadar, Ethiopia. *Nature*, *260*(5549), 293–297. https://www.ncbi.nlm.nih.gov/pubmed/815823

Johanson, D. C., & Wong, K. (2010). *Lucy's legacy. The quest for human origins*. Three Rivers Press.

Johnson, E. F., Mcclure, J., Herron, P., & Baskerville, K. A. (1993). Anatomical variation, human-diversity, and environmental adaptation. *Journal of the National Medical Association*, *85*(5), 337–338.

Jones, A. M. (2006). The physiology of the world record holder for the women's marathon. *Sport Science & Coaching*, *1*(2), 101–116.

Jones, S., Martin, R. D., & Pilbeam, D. (1994). *The Cambridge encyclopedia of human evolution*. Cambridge University Press.

Jordan, P. (1999). *Neanderthal*. Sutton Publishing Limited.

Jungers, W. L., Grabowski, M., Hatala, K. G., & Richmond, B. G. (2016). The evolution of body size and shape in the human career. *Philosophical Transactions of the Royal Society of London. Series B, Biological Sciences*, *371*(1698). https://doi.org/10.1098/rstb.2015.0247

Jurmain, R., Kilgore, L., Trevathan, W. R., Ciochon, R. L., & Bartelink, E. J. (2018). *Introduction to physical anthropology* (15 ed.). CENGAGE Learning.

Kaas, J. H. (2019). The origin and evolution of neocortex: From early mammals to modern humans. In M. A. Hofman (Ed.), *Progress in brain research* (2019/11/11 ed., Vol. 250, pp. 61–81). Elsevier B.V. https://doi.org/10.1016/bs.pbr.2019.03.017

Kabukcu, C., Hunt, C., Hill, E., Pomeroy, E., Reynolds, T., Barker, G., & Asouti, E. (2022). Cooking in caves: Palaeolithic carbonised plant food remains from Franchthi and Shanidar. *Antiquity*, *97*(391), 12–28. https://doi.org/10.15184/aqy.2022.143

Kaessmann, H., Wiebe, V., Weiss, G., & Pääbo, S. (2001). Great ape DNA sequences reveal a reduced diversity and an expansion in humans. *Nature Genetics*, *27*, 155–156.

Kahn, S., Ehrlich, P., Feldman, M., Sapolsky, R., & Wong, S. (2020). The jaw epidemic: Recognition, origins, cures, and prevention. *Bioscience*, *70*(9), 759–771. https://doi.org/10.1093/biosci/biaa073

Kamberov, Y. G., Guhan, S. M., DeMarchis, A., Jiang, J., Wright, S. S., Morgan, B. A., Sabeti, P. C., Tabin, C. J., & Lieberman, D. E. (2018). Comparative evidence for the independent evolution of hair and sweat gland traits in primates. *Journal of human evolution*, *125*, 99–105. https://doi.org/10.1016/j.jhevol.2018.10.008

Kaplan, H., Hill, K., Lancaster, J., & Hurtado, A. M. (2000). A theory of human life history evolution: Diet, intelligence, and longevity. *Evolutionary Anthropology*, *9*(4), 156–185. https://doi.org/10.1002/1520-6505(2000)9:4<156::AID-EVAN5>3.0.Co;2-7

Karakostis, F. A., Haeufle, D., Anastopoulou, I., Moraitis, K., Hotz, G., Tourloukis, V., & Harvati, K. (2021). Biomechanics of the human thumb and the evolution of dexterity. *Current Biology: CB*, *31*(6), 1317–1325.e1318. https://doi.org/10.1016/j.cub.2020.12.041

Kay, R. F. (2015). New world monkey origins. *Science*, *347*(6266), 1068–1069.

Keith, A. (1923). Hunterian lectures on man's posture: Its evolution and disorders. Lecture II. The evolution of the orthograde spine. *BMJ*, *1*, 499–502.

Kelly, A. P., Ocobock, C., Butaric, L. N., & Maddux, S. D. (2023). Metabolic demands and sexual dimorphism in human nasal morphology: A test of the respiratory-energetics hypothesis. *American Journal of Biological Anthropology, 180*(3), 453–471. https://doi.org/10.1002/ajpa.24692

Kelly, R. E. (2001). Tripedal knuckle-walking: A proposal for the evolution of human locomotion and handedness. *Journal of Theoretical Biology, 213*(3), 333–358. https://doi.org/10.1006/jtbi.2001.2421

Ker, R. F., Bennett, M. B., Bibby, S. R., Kester, R. C., & Alexander, R. M. (1987). The spring in the arch of the human foot. *Nature, 325*(7000), 147–149. https://doi.org/10.1038/325147a0

Key, A. J. M., Jarić, I., & Roberts, D. L. (2021). Modelling the end of the Acheulean at global and continental levels suggests widespread persistence into the Middle Palaeolithic. *Humanities and Social Sciences Communications, 8*(1). https://doi.org/10.1057/s41599-021-00735-8

Killgrove, K. (2021). A century of civilization intelligence, and (white) nationalism. In J. M. DeSilva (Ed.), *A most interesting problem. What Darwin's descent of man got right and wrong about human evolution* (pp. 103–124). Princeton University Press.

Kilroy, G. S., Tallman, S. D., & DiGangi, E. A. (2020). Secular change in morphological cranial and mandibular trait frequencies in European Americans born 1824–1987. *American Journal of Physical Anthropology, 173*(3), 589–605. https://doi.org/10.1002/ajpa.24115

Kim, J., Edge, M. D., Goldberg, A., & Rosenberg, N. A. (2021). Skin deep: The decoupling of genetic admixture levels from phenotypes that differed between source populations. *American Journal of Physical Anthropology, 175*(2), 406–421. https://doi.org/10.1002/ajpa.24261

Kimbel, W. H., & Villmoare, B. (2016). From *Australopithecus* to *Homo*: The transition that wasn't. *Philosophical Transactions of the Royal Society of London. Series B, Biological Sciences, 371*(1698). https://doi.org/10.1098/rstb.2015.0248

King, D. R., Bartlett, M. D., Gilman, C. A., Irschick, D. J., & Crosby, A. J. (2014). Creating gecko-like adhesives for "real world" surfaces. *Advanced Materials, 26*(25), 4345–4351. https://doi.org/10.1002/adma.201306259

King, W. (1864). The reputed fossil man of the Neanderthal. *Quarterly Journal of Science, 1*, 88–97.

Kingdon, J. (1997). *The Kingdon field guide to African mammals.* Academic Press, Harcourt Brace & Company.

Kinsbourne, M. (2013). Somatic twist: A model for the evolution of decussation. *Neuropsychology, 27*(5), 511–515. https://doi.org/10.1037/a0033662

Kitazawa, T., Takechi, M., Hirasawa, T., Adachi, N., Narboux-Neme, N., Kume, H., Maeda, K., Hirai, T., Miyagawa-Tomita, S., Kurihara, Y., Hitomi, J., Levi, G., Kuratani, S., & Kurihara, H. (2015). Developmental genetic bases behind the independent origin of the tympanic membrane in mammals and diapsids. *Nature Communications, 6*, 6853. https://doi.org/10.1038/ncomms7853

Kivell, T. L., Baraki, N., Lockwood, V., Williams-Hatala, E. M., & Wood, B. A. (2022). Form, function and evolution of the human hand. *American Journal of Biological Anthropology.* https://doi.org/10.1002/ajpa.24667

Kivell, T. L., Barros, A. P., & Smaers, J. B. (2013). Different evolutionary pathways underlie the morphology of wrist bones in hominoids. *BMC Evolutionary Biology, 13*, 1–12. https://doi.org/10.1186/1471-2148-13-229

Kivell, T. L., & Schmitt, D. (2009). Independent evolution of knuckle-walking in African apes shows that humans did not evolve from a knuckle-walking ancestor. *Proceedings of the National Academy of Sciences of the United States of America, 106*(34), 14241–14246. https://doi.org/10.1073/pnas.0901280106

Klein, J., & Takahata, N. (2002). *Where do we come from? The molecular evidence for human descent*. Springer-Verlag.

Knight, K. (2021). Humans ditched swivelling hips for shorter stride than chimps. *Journal of Experimental Biology, 224*(16). https://doi.org/10.1242/jeb.243185

Kolb, B., & Whishaw, I. Q. (2014). *An introduction to brain and behavior* (4th ed.). Worth Publishers.

Kolobova, K. A., Roberts, R. G., Chabai, V. P., Jacobs, Z., Krajcarz, M. T., Shalagina, A. V., Krivoshapkin, A. I., Li, B., Uthmeier, T., Markin, S. V., Morley, M. W., O'Gorman, K., Rudaya, N. A., Talamo, S., Viola, B., & Derevianko, A. P. (2020). Archaeological evidence for two separate dispersals of Neanderthals into southern Siberia. *Proceedings of the National Academy of Sciences of the United States of America, 117*(6), 2879–2885. https://doi.org/10.1073/pnas.1918047117

Komza, K., Viola, B., Netten, T., & Schroeder, L. (2022). Morphological integration in the hominid midfoot. *Journal of Human Evolution, 170*, 103231. https://doi.org/10.1016/j.jhevol.2022.103231

Kornack, D. R., & Rakic, P. (2001). Cell proliferation without neurogenesis in adult primate neocortex. *Science, 294*(5549), 2127–2130. https://doi.org/10.1126/science.1065467

Kozlov, M. (2021). Charts show how your brain expands and shrinks with age. *Nature, 604*, 230–231.

Kramer, P. A., & Eck, G. G. (2000). Locomotor energetics and leg length in hominid bipedality. *Journal of Human Evolution, 38*(5), 651–666. https://doi.org/10.1006/jhev.1999.0375

Krause, J., Fu, Q., Good, J. M., Viola, B., Shunkov, M. V., Derevianko, A. P., & Paabo, S. (2010). The complete mitochondrial DNA genome of an unknown hominin from southern Siberia. *Nature, 464*(7290), 894–897. https://doi.org/10.1038/nature08976

Krause, J., & Trappe, T. (2020). *De reis van onze genen* (R. Boley, Trans.). Nieuw Amsterdam.

Kriegstein, A., Noctor, S., & Martinez-Cerdeno, V. (2006). Patterns of neural stem and progenitor cell division may underlie evolutionary cortical expansion. *Nature Reviews Neuroscience, 7*(11), 883–890. https://doi.org/10.1038/nrn2008

Krings, M., Stone, A., Schmitz, R. W., Krainitzki, H., Stoneking, M., & Pääbo, S. (1997). Neanderthal DNA sequences and the origin of modern humans. *Cell, 90*, 19–30.

Kuderna, L. F. K., Gao, H., Janiak, M. C., Kuhlwilm, M., Orkin, J. D., Bataillon, T., Manu, S., Valenzuela, A., Bergman, J., Rousselle, M., Silva, F. E., Agueda, L., Blanc, J., Gut, M., de Vries, D., Goodhead, I., Harris, R. A., Raveendran, M., Jensen, A., Chuma, I. S.,. . . & Marques Bonet, T. (2023). A global catalog of whole-genome diversity from 233 primate species. *Science (New York, N.Y.), 380*(6648), 906–913. https://doi.org/10.1126/science.abn7829

Kudo, H., & Dunbar, R. I. M. (2001). Neocortex size and social network size in primates. *Animal Behaviour, 62*, 711–722. https://doi.org/10.1006/anbe.2001.1808

Kuhl, H. S., Kalan, A. K., Arandjelovic, M., Aubert, F., D'Auvergne, L., Goedmakers, A., Jones, S., Kehoe, L., Regnaut, S., Tickle, A., Ton, E., van Schijndel, J., Abwe, E. E., Angedakin, S., Agbor, A., Ayimisin, E. A., Bailey, E., Bessone, M., Bonnet, M.,. . . & Boesch, C. (2016). Chimpanzee accumulative stone throwing. *Sci Rep, 6*, 22219. https://doi.org/10.1038/srep22219

Kuhlwilm, M., & Boeckx, C. (2019). A catalog of single nucleotide changes distinguishing modern humans from archaic hominins. *Scientific Reports, 9*(1), 8463. https://doi.org/10.1038/s41598-019-44877-x

Kuhlwilm, M., Gronau, I., Hubisz, M. J., de Filippo, C., Prado-Martinez, J., Kircher, M., Fu, Q., Burbano, H. A., Lalueza-Fox, C., de la Rasilla, M., Rosas, A., Rudan, P., Brajkovic, D., Kucan, Ž., Gušic, I., Marques-Bonet, T., Andrés, A. M., Viola, B., Pääbo, S., Meyer, M., Siepel, A., & Castellano, S. (2016). Ancient gene flow from early modern humans into Eastern Neanderthals. *Nature, 530*(7591), 429–433. https://doi.org/10.1038/nature16544

Kun, E., & Narasimhan, V. M. (2023). Fast-evolving genomic regions underlie human brain development. *Nature.* https://doi.org/10.1038/d41586-023-00069-2

Kunze, J., Harvati, K., Hotz, G., & Karakostis, F. A. (2024). Humanlike manual activities in *Australopithecus. Journal of Human Evolution, 196,* 103591. https://doi.org/10.1016/j.jhevol.2024.103591

Kushlan, J. A. (1980). The evolution of hairlessness in man. *The American Naturalist, 116*(5), 727–729.

Kuzawa, C. W. (1998). Adipose tissue in human infancy and childhood: An evolutionary perspective. *American Journal of Biological Anthropology, 107*(S27), 177–209.

Kuzawa, C. W., Chugani, H. T., Grossman, L. I., Lipovich, L., Muzik, O., Hof, P. R., Wildman, D. E., Sherwood, C. C., Leonard, W. R., & Lange, N. (2014). Metabolic costs and evolutionary implications of human brain development. *Proceedings of the National Academy of Sciences of the United States of America, 111*(36), 13010–13015. https://doi.org/10.1073/pnas.1323099111

Laden, G., & Wrangham, R. (2005). The rise of the hominids as an adaptive shift in fallback foods: Plant underground storage organs (USOs) and australopith origins. *Journal of Human Evolution, 49*(4), 482–498. https://doi.org/10.1016/j.jhevol.2005.05.007

Lambrot, R., Xu, C., Saint-Phar, S., Chountalos, G., Cohen, T., Paquet, M., Suderman, M., Hallett, M., & Kimmins, S. (2013). Low paternal dietary folate alters the mouse sperm epigenome and is associated with negative pregnancy outcomes. *Nature Communications, 4,* 2889. https://doi.org/10.1038/ncomms3889

Landau, M. (1991). *Human evolution as narrative.* Yale University Press.

Landi, F., Profico, A., Veneziano, A., De Groote, I., & Manzi, G. (2020). Locomotion, posture, and the foramen magnum in primates: Reliability of indices and insights into hominin bipedalism. *American Journal of Primatology, 82*(9), e23170. https://doi.org/10.1002/ajp.23170

Langley, M. C., & Suddendorf, T. (2022). Archaeological evidence for thinking about possibilities in hominin evolution. *Philosophical Transactions of the Royal Society of London. Series B, Biological Sciences, 377*(1866), 20210350. https://doi.org/10.1098/rstb.2021.0350

Larbey, C., Mentzer, S. M., Ligouis, B., Wurz, S., & Jones, M. K. (2019). Cooked starchy food in hearths ca. 120 kya and 65 kya (MIS 5e and MIS 4) from Klasies River Cave, South Africa. *Journal of Human Evolution, 131,* 210–227. https://doi.org/10.1016/j.jhevol.2019.03.015

Larson, S. G. (1998). Parallel evolution in the hominoid trunk and forelimb. *Evolutionary Anthropology, 6*(3), 87–99. https://doi.org/10.1002/(Sici)1520-6505(1998)6:3<87::Aid-Evan3>3.0.Co;2-T

Lasisi, T., Smallcombe, J. W., Kenney, W. L., Shriver, M. D., Zydney, B., Jablonski, N. G., & Havenith, G. (2023). Human scalp hair as a thermoregulatory adaptation. *Proceedings of the National Academy of Sciences of the United States of America, 120*(24), e2301760120. https://doi.org/10.1073/pnas.2301760120

Laudicina, N. M., & Cartmill, M. (2022). Bony birth-canal dimensions and obstetric constraints in hominoids. *American Journal of Biological Anthropology*. https://doi.org/10.1002/ajpa.24659

Laudicina, N. M., Rodriguez, F., & DeSilva, J. M. (2019). Reconstructing birth in *Australopithecus sediba*. *PloS One, 14*(9). https://doi.org/10.1371/journal.pone.0221871

Lawler, A. (2011). Did modern humans travel out of Africa via Arabia? *Science, 331*(6016), 387–387. https://doi.org/10.1126/science.331.6016.387

Leakey, L. S. B. (1959). A new fossil skull from Olduvai. *Nature, 184*, 491–493.

Leakey, L. S. B., Tobias, P. V., & Napier, J. R. (1964). A new species of the genus *Homo* from Olduvai Gorge. *Nature*, 7–9.

Leakey, M. (1978). Pliocene footprints at Laetolil, Northern Tanzania. *Antiquity, 52*(205), 133. https://doi.org/10.1017/S0003598x00071969

Leakey, M. D., & Hay, R. L. (1979). Pliocene footprints in the Laetoli Beds at Laetoli, northern Tanzania. *Nature, 278*, 317–323.

Leakey, M. G., Feibel, C. S., McDougall, I., & Walker, A. (1995). New four-million-year-old hominid species from Kanapoi and Allia Bay, Kenya. *Nature, 376*(6541), 565–571. https://doi.org/10.1038/376565a0

Leakey, M. G., Feibel, C. S., McDougall, I., Ward, C., & Walker, A. (1998). New specimens and confirmation of an early age for *Australopithecus anamensis*. *Nature, 393*(6680), 62–66. https://doi.org/10.1038/29972

Le Cabec, A., Colard, T., Charabidze, D., Chaussain, C., Di Carlo, G., Gaudzinski-Windheuser, S., Hublin, J. J., Melis, R. T., Pioli, L., Ramirez-Rozzi, F., & Mussi, M. (2021). Insights into the palaeobiology of an early Homo infant: Multidisciplinary investigation of the GAR IVE hemi-mandible, Melka Kunture, Ethiopia. *Scientific Reports, 11*(1), 23087. https://doi.org/10.1038/s41598-021-02462-1

Lemelin, P., & Diogo, R. (2016). Anatomy, function, and evolution of the primate hand musculature. In L. Barrett (Ed.), *The evolution of the primate hand* (pp. 155–193). Sprinter.

Lemelin, P., & Schmitt, D. (2016). On primitiveness, prehensility, and opposability of the primate hand: The contributions of Frederic Wood Jones and John Russell Napier In L. Barrett (Ed.), *The evolution of the primate hand* (pp. 5–13). Sprinter.

Leonard, W. R., & Robertson, M. L. (1997). Comparative primate energetics and hominid evolution. *American Journal of Physical Anthropology, 102*(2), 265–281. https://doi.org/10.1002/(Sici)1096-8644(199702)102:2<265::Aid-Ajpa8>3.0.Co;2-X

Leonard, W. R., Stock, J. T., & Valeggia, C. R. (2010). Evolutionary perspectives on human diet and nutrition. *Evolutionary Anthropology, 19*(3), 85–86. https://doi.org/10.1002/evan.20250

Leutenegger, W. (1972). Newborn size and pelvic dimensions of *Australopithecus*. *Nature, 240*(5383), 568–569. https://doi.org/10.1038/240568a0

Leutenegger, W. (1973). Maternal-fetal weight relationships in primates. *Folia Primatologica, 20*(4), 280–293. https://doi.org/10.1159/000155580

Leutenegger, W. (1987). Neonatal brain size and neurocranial dimensions in pliocene hominids - Implications for obstetrics. *Journal of Human Evolution, 16*(3), 291–296. https://doi.org/10.1016/0047-2484(87)90004-2

Levine, R. V., & Norenzayan, A. (1999). The pace of life in 31 countries. *Journal of Cross-Cultural Psychology, 30*(2), 178–205. https://doi.org/10.1177/0022022199030002003

Levy, R. (2024). The prefrontal cortex: From monkey to man. *Brain, 147*(3), 794–815. https://doi.org/10.1093/brain/awad389

Lewis, J. E., Ward, C. V., Kimbel, W. H., Kidney, C. L., Brown, F. H., Quinn, R. L., Rowan, J., Lazagabaster, I. A., Sanders, W. J., Leakey, M. G., & Leakey, L. N. (2024). A 4.3-million-year-old *Australopithecus anamensis* mandible from Ileret, East Turkana, Kenya, and its paleoenvironmental context. *Journal of Human Evolution*, *194*, 103579. https://doi.org/10.1016/j.jhevol.2024.103579

Lieberman, D. (2013). *The story of the human body*. Penguin Books.

Lieberman, D. E. (2011). *The evolution of the human head*. The Belknap Press of Harvard University Press.

Lieberman, D. E. (2022). Standing up for the earliest bipedal hominins. *Nature*. https://doi.org/10.1038/d41586-022-02226-5

Lieberman, D. E., Pontzer, H., Cutright-Smith, E., & Raichlen, D. (2005). Why is the human gluteus so maximus? *American Journal of Physical Anthropology*, *126*(S40), 138–138.

Lieberman, D. E., Raichlen, D. A., Pontzer, H., Bramble, D. M., & Cutright-Smith, E. (2006). The human gluteus maximus and its role in running. *Journal of Experimental Biology*, *209*(11), 2143–2155. https://doi.org/10.1242/jeb.02255

Lieberman, P. (1993). On the Kebara KMH-2 hyoid and Neanderthal speech. *Current Anthropology*, *34*(2), 172–175. https://doi.org/10.1086/204155

Light, J. E., & Reed, D. L. (2009). Multigene analysis of phylogenetic relationships and divergence times of primate sucking lice (Phthiraptera: Anoplura). *Molecular Phylogenetics and Evolution*, *50*(2), 376–390. https://doi.org/10.1016/j.ympev.2008.10.023

Lindhout, F. W., Krienen, F. M., Pollard, K. S., & Lancaster, M. A. (2024). A molecular and cellular perspective on human brain evolution and tempo. *Nature*, *630*(8017), 596–608. https://doi.org/10.1038/s41586-024-07521-x

Linnaeus, C. (1758). *Systema naturae per regna tria naturae: secundum classes, ordines, genera, species, cum characteribus, differentiis, synonymis, locis* (10th ed.). Holmiae: Impensis Direct. Laurentii Salvii, 1758–1759. https://www.biodiversitylibrary.org/bibliography/559#/summary

Liu, W., Martinon-Torres, M., Cai, Y. J., Xing, S., Tong, H. W., Pei, S. W., Sier, M. J., Wu, X. H., Edwards, R. L., Cheng, H., Li, Y. Y., Yang, X. X., de Castro, J. M., & Wu, X. J. (2015). The earliest unequivocally modern humans in southern China. *Nature*, *526*(7575), 696–699. https://doi.org/10.1038/nature15696

Longren, L. L., Eigen, L., Shubitidze, A., Lieschnegg, O., Baum, D., Nyakatura, J. A., Hildebrandt, T., & Brecht, M. (2023). Dense reconstruction of elephant trunk musculature. *Current Biology: CB*, *33*(21), 4713–4720.e4713. https://doi.org/10.1016/j.cub.2023.09.007

Lopez-Rey, J. M., Garcia-Martinez, D., & Bastir, M. (2024). Shanidar 3 'rings the bell': Virtual ribcage reconstruction and its implications for understanding the Neanderthal bauplan. *Journal of Human Evolution*, *199*, 103629. https://doi.org/10.1016/j.jhevol.2024.103629

Lordkipanidze, D., Jashashvili, T., Vekua, A., Ponce de Leon, M. S., Zollikofer, C. P., Rightmire, G. P., Pontzer, H., Ferring, R., Oms, O., Tappen, M., Bukhsianidze, M., Agusti, J., Kahlke, R., Kiladze, G., Martinez-Navarro, B., Mouskhelishvili, A., Nioradze, M., & Rook, L. (2007). Postcranial evidence from early *Homo* from Dmanisi, Georgia. *Nature*, *449*(7160), 305–310. https://doi.org/10.1038/nature06134

Lorenzo, F. R., Huff, C., Myllymaki, M., Olenchock, B., Swierczek, S., Tashi, T., Gordeuk, V., Wuren, T., Ri-Li, G., McClain, D. A., Khan, T. M., Koul, P. A., Guchhait, P., Salama, M. E., Xing, J. C., Semenza, G. L., Liberzon, E., Wilson, A., Simonson, T. S., Jorde, L. B., Kaelin, W. G., Koivunen, P., & Prchal, J. T. (2014). A genetic mechanism for Tibetan high-altitude adaptation. *Nature Genetics*, *46*(9), 951–956. https://doi.org/10.1038/ng.3067

Lovejoy, C. O. (1981). The origin of man. *Science (New York, N.Y.), 211*(4480), 341–350. https://doi.org/10.1126/science.211.4480.341

Lovejoy, C. O. (2009). Reexamining human origins in light of *Ardipithecus ramidus. Science (New York, N.Y.), 326*(5949), 74e1–74e8. https://doi.org/10.1126/science.1175834

Lovejoy, C. O., Latimer, B., Suwa, G., Asfaw, B., & White, T. D. (2009). Combining prehension and propulsion: The foot of *Ardipithecus ramidus. Science (New York, N.Y.), 326*(5949), 72e1–72e8. https://doi.org/10.1126/science.1175832

Lovejoy, C. O., & McCollum, M. A. (2010). Spinopelvic pathways to bipedality: Why no hominids ever relied on a bent-hip-bent-knee gait. *Philosophical Transactions of the Royal Society of London. Series B, Biological Sciences, 365*(1556), 3289–3299. https://doi.org/10.1098/rstb.2010.0112

Lovejoy, C. O., Simpson, S. W., White, T. D., Asfaw, B., & Suwa, G. (2009). Careful climbing in the Miocene: The forelimbs of *Ardipithecus ramidus* and humans are primitive. *Science (New York, N.Y.), 326*(5949), 70e71-78. https://doi.org/10.1126/science.1175827

Lovejoy, C. O., Suwa, G., Simpson, S. W., Matternes, J. H., & White, T. D. (2009). The great divides: *Ardipithecus ramidus* reveals the postcrania of our last common ancestors with African apes. *Science, 326*(5949), 100–106. https://doi.org/10.1126/science.1175833

Lovejoy, C. O., Suwa, G., Spurlock, L., Asfaw, B., & White, T. D. (2009). The pelvis and femur of *Ardipithecus ramidus*: The emergence of upright walking. *Science (New York, N.Y.), 326*(5949), 71e71-76. https://doi.org/10.1126/science.1175831

Lowenstein, J. M., Molleson, T., & Washburn, S. L. (1982). Piltdown jaw confirmed as orang. *Nature, 299*(5881), 294–294. https://doi.org/10.1038/299294a0

Lowery, R. K., Uribe, G., Jimenez, E. B., Weiss, M. A., Herrera, K. J., Regueiro, M., & Herrera, R. J. (2013). Neanderthal and Denisova genetic affinities with contemporary humans: Introgression versus common ancestral polymorphisms. *Gene, 530*(1), 83–94. https://doi.org/10.1016/j.gene.2013.06.005

Lu, C. P., Polak, L., Keyes, B. E., & Fuchs, E. (2016). Spatiotemporal antagonism in mesenchymal-epithelial signaling in sweat versus hair fate decision. *Science, 354*(6319). https://doi.org/10.1126/science.aah6102

Ludecke, T., Leichliter, J. N., Stratford, D., Sigman, D. M., Vonhof, H., Haug, G. H., Bamford, M. K., & Martinez-Garcia, A. (2025). *Australopithecus* at Sterkfontein did not consume substantial mammalian meat. *Science, 387*(6731), 309–314. https://doi.org/10.1126/science.adq7315

Luef, E. M. (2018). Tracing the human brain's classical language areas in extant and extinct hominids. In E. M. Luef, & M. M. Marin (Eds.), *The talking species: Perspectives on the evolutionary, neuronal and cultural foundations of language* (pp. 29–56). Uni-Press Graz.

Luo, Y. (2020). Neanderthal DNA raises risk of severe COVID risk factors. *Nature, 587*(7835), 552–553. https://doi.org/10.1038/d41586-020-02957-3

Lynch, G., Hechtel, S., & Jacobs, D. (1983). Neonate size and evolution of brain size in the anthropoid primates. *Journal of Human Evolution, 12*(6), 519–522. https://doi.org/10.1016/S0047-2484(83)80031-1

Macchiarelli, R., Bergeret-Medina, A., Marchi, D., & Wood, B. (2020). Nature and relationships of *Sahelanthropus tchadensis. Journal of Human Evolution, 149*, 102898. https://doi.org/10.1016/j.jhevol.2020.102898

MacDonald, K., Martinón-Torres, M., Dennell, R. W., & Bermúdez de Castro, J. M. (2012). Discontinuity in the record for hominin occupation in south-western Europe: Implications for occupation of the middle latitudes of Europe. *Quaternary International, 271*, 84–97. https://doi.org/10.1016/j.quaint.2011.10.009

Machnicki, A. L., & Reno, P. L. (2020). Great apes and humans evolved from a long-backed ancestor. *Journal of Human Evolution, 144*, 102791. https://doi.org/10.1016/j.jhevol.2020.102791

MacLean, E. L., Barrickman, N. L., Johnson, E. M., & Wall, C. E. (2009). Sociality, ecology, and relative brain size in lemurs. *Journal of Human Evolution, 56*(5), 471–478. https://doi.org/10.1016/j.jhevol.2008.12.005

Maddison, D. R. (1991). African origin of human mitochondrial-DNA reexamined. *Systematic Zoology, 40*(3), 355–363. https://doi.org/10.2307/2992327

Madella, M., Jones, M. K., Goldberg, P., Goren, Y., & Hovers, E. (2002). The exploitation of plant resources by Neanderthals in Amud Cave (Israel): The evidence from Phytolith studies. *Journal of Archaeological Science, 29*(7), 703–719. https://doi.org/10.1006/jasc.2001.0743

Malone, M. A., MacLatchy, L. M., Mitani, J. C., Kityo, R., & Kingston, J. D. (2021). A chimpanzee enamel-diet delta(13)C enrichment factor and a refined enamel sampling strategy: Implications for dietary reconstructions. *Journal of Human Evolution, 159*, 103062. https://doi.org/10.1016/j.jhevol.2021.103062

Manaka, Y., & Sugita, Y. (2009). Insufficient visual information leads to spontaneous bipedal walking in Japanese monkeys. *Behavioural Processes, 80*(1), 104–106. https://doi.org/10.1016/j.beproc.2008.10.010

Mao, F., Hu, Y., Li, C., Wang, Y., Chase, M. H., Smith, A. K., & Meng, J. (2020). Integrated hearing and chewing modules decoupled in a Cretaceous stem therian mammal. *Science (New York, N.Y.), 367*(6475), 305–308. https://doi.org/10.1126/science.aay9220

Marchi, D., Rimoldi, A., García-Martínez, D., & Bastir, M. (2022). Morphological correlates of distal fibular morphology with locomotion in great apes, humans, and *Australopithecus afarensis. American Journal of Biological Anthropology*. https://doi.org/10.1002/ajpa.24507

Margaria, R., Cerretelli, P., Aghemo, P., & Sassi, G. (1963). Energy cost of running. *Journal of Applied Physiology, 18*(2), 367–370.

Marhounova, L., Kotrschal, A., Kverkova, K., Kolm, N., & Nemec, P. (2019). Artificial selection on brain size leads to matching changes in overall number of neurons. *Evolution, 73*(9), 2003–2012. https://doi.org/10.1111/evo.13805

Marino, F. E., Sibson, B. E., & Lieberman, D. E. (2022). The evolution of human fatigue resistance. *Journal of Comparative Physiology. B, Biochemical, Systemic, and Environmental Physiology*. https://doi.org/10.1007/s00360-022-01439-4

Martin, R. D. (1981). Relative brain size and basal metabolic rate in terrestrial vertebrates. *Nature, 293*(5827), 57–60. https://doi.org/10.1038/293057a0

Martin, R. D. (2007). The evolution of human reproduction: A primatological perspective. *American Journal of Biological Anthropology, 134*(S45), 59–84. https://doi.org/10.1002/ajpa.20734

Martinon-Torres, M., Dennell, R., & Bermudez de Castro, J. M. (2011). The Denisova hominin need not be an out of Africa story. *Journal of Human Evolution, 60*(2), 251–255. https://doi.org/10.1016/j.jhevol.2010.10.005

Marzke, M. W. (1997). Precision grips, hand morphology, and tools. *American Journal of Physical Anthropology, 102*(1), 91–110.

Masao, F. T., Ichumbaki, E. B., Cherin, M., Barili, A., Boschian, G., Iurino, D. A., Menconero, S., Moggi-Cecchi, J., & Manzi, G. (2016). New footprints from Laetoli (Tanzania) provide evidence for marked body size variation in early hominins. *eLife, 5*. https://doi.org/10.7554/eLife.19568

Masharawi, Y., Dar, G., Peleg, S., Steinberg, N., Medlej, B., May, H., Abbas, J., & Hershkovitz, I. (2010). A morphological adaptation of the thoracic and lumbar vertebrae to lumbar hyperlordosis in young and adult females. *European Spine Journal, 19*(5), 768–773. https://doi.org/10.1007/s00586-009-1256-6

Matarazzo, S. (2008). Knuckle walking signal in the manual digits of *Pan* and *Gorilla*. *American Journal of Physical Anthropology, 135*(1), 27–33.

Matsu'ura, S., Kondo, M., Danhara, T., Sakata, S., Iwano, H., Hirata, T., Kurniawan, I., Setiyabudi, E., Takeshita, Y., Hyodo, M., Kitaba, I., Sudo, M., Danhara, Y., & Aziz, F. (2020). Age control of the first appearance datum for Javanese *Homo erectus* in the Sangiran area. *Science (New York, N.Y.), 367*(6474), 210–214. https://doi.org/10.1126/science.aau8556

Matsuda, I., Takano, T., Shintaku, Y., & Clauss, M. (2022). Gastrointestinal morphology and ontogeny of foregut-fermenting primates. *American Journal of Biological Anthropology, 177*(4), 735–747. https://doi.org/10.1002/ajpa.24476

Matsui, A., Go, Y., & Niimura, Y. (2010). Degeneration of olfactory receptor gene repertories in primates: No direct link to full trichromatic vision. *Molecular Biology and Evolution, 27*(5), 1192–1200. https://doi.org/10.1093/molbev/msq003

Mayer, D. E. B. Y., Vandermeersch, B., & Bar-Yosef, O. (2009). Shells and ochre in Middle Paleolithic Qafzeh Cave, Israel: Indications for modern behavior. *Journal of Human Evolution, 56*(3), 307–314. https://doi.org/10.1016/j.jhevol.2008.10.005

Mayr, E. (1950). Taxonomic categories of fossil hominids. *Cold Spring Harbor Symp Quant Biol, 25*, 109–118.

McBrearty, S., & Jablonski, N. G. (2005). First fossil chimpanzee. *Nature, 437*(7055), 105–108. https://doi.org/10.1038/nature04008

McDonald, B., & McCoy, K. D. (2019). Maternal microbiota in pregnancy and early life. *Science, 365*(6457), 984–985.

McDougall, I., Brown, F. H., & Fleagle, J. G. (2005). Stratigraphic placement and age of modern humans from Kibish, Ethiopia. *Nature, 433*(7027), 733–736. https://doi.org/10.1038/nature03258

McGann, J. P. (2017). Poor human olfaction is a 19th-century myth. *Science (New York, N.Y.), 356*(6338). https://doi.org/10.1126/science.aam7263

McGrath, K., Limmer, L. S., Lockey, A. L., Guatelli-Steinberg, D., Reid, D. J., Witzel, C., Bocaege, E., McFarlin, S. C., & El Zaatari, S. (2021). 3D enamel profilometry reveals faster growth but similar stress severity in Neanderthal versus *Homo sapiens* teeth. *Scientific Reports, 11*(1), 522. https://doi.org/10.1038/s41598-020-80148-w

McGrosky, A., & Pontzer, H. (2023). The fire of evolution: Energy expenditure and ecology in primates and other endotherms. *The Journal of Experimental Biology, 226*(5). https://doi.org/10.1242/jeb.245272

Mchenry, H. M. (1992). Body size and proportions in early hominids. *American Journal of Physical Anthropology, 87*(4), 407–431. https://doi.org/10.1002/ajpa.1330870404

Mclean, M., Bisits, A., Davies, J., Woods, R., Lowry, P., & Smith, R. (1995). A placental clock controlling the length of human-pregnancy. *Nature Medicine, 1*(5), 460–463. https://doi.org/10.1038/nm0595-460

McLean, C. Y., Reno, P. L., Pollen, A. A., Bassan, A. I., Capellini, T. D., Guenther, C., Indjeian, V. B., Lim, X., Menke, D. B., Schaar, B. T., Wenger, A. M., Bejerano, G., & Kingsley, D.

M. (2011). Human-specific loss of regulatory DNA and the evolution of human-specific traits. *Nature, 471*(7337), 216–219. https://doi.org/10.1038/nature09774

McNutt, E. J., Hatala, K. G., Miller, C., Adams, J., Casana, J., Deane, A. S., Dominy, N. J., Fabian, K., Fannin, L. D., Gaughan, S., Gill, S. V., Gurtu, J., Gustafson, E., Hill, A. C., Johnson, C., Kallindo, S., Kilham, B., Kilham, P., Kim, E.,. . . & DeSilva, J. M. (2021). Footprint evidence of early hominin locomotor diversity at Laetoli, Tanzania. *Nature.* https://doi.org/10.1038/s41586-021-04187-7

McPherron, S. P., Alemseged, Z., Marean, C. W., Wynn, J. G., Reed, D., Geraads, D., Bobe, R., & Bearat, H. A. (2010). Evidence for stone-tool-assisted consumption of animal tissues before 3.39 million years ago at Dikika, Ethiopia. *Nature, 466*(7308), 857–860. https://doi.org/10.1038/nature09248

Meckel, L. A., McDaneld, C. P., & Wescott, D. J. (2018). White-tailed deer as a taphonomic agent: Photographic evidence of white-tailed deer gnawing on human bone. *Journal of Forensic Sciences, 63*(1), 292–294. https://doi.org/10.1111/1556-4029.13514

Melillo, S. M., Gibert, L., Saylor, B. Z., Deino, A., Alene, M., Ryan, T. M., & Haile-Selassie, Y. (2021). New Pliocene hominin remains from the Leado Dido'a area of Woranso-Mille, Ethiopia. *Journal of Human Evolution, 153*, 102956. https://doi.org/10.1016/j.jhevol.2021.102956

Mellars, P. (2006). A new radiocarbon revolution and the dispersal of modern humans in Eurasia. *Nature, 439*(7079), 931–935.

Mendez, F. L., Poznik, G. D., Castellano, S., & Bustamantel, C. D. (2016). The divergence of Neandertal and modern human Y chromosomes. *American Journal of Human Genetics, 98*(4), 728–734. https://doi.org/10.1016/j.ajhg.2016.02.023

Mercader, J., Akuku, P., Boivin, N., Bugumba, R., Bushozi, P., Camacho, A., Carter, T., Clarke, S., Cueva-Temprana, A., Durkin, P., Favreau, J., Fella, K., Haberle, S., Hubbard, S., Inwood, J., Itambu, M., Koromo, S., Lee, P., Mohammed, A.,. . . & Petraglia, M. (2021). Earliest Olduvai hominins exploited unstable environments ~ 2 million years ago. *Nature Communications, 12*(1), 3. https://doi.org/10.1038/s41467-020-20176-2

Merceron, G., Kaiser, T. M., Kostopoulos, D. S., & Schulz, E. (2010). Ruminant diets and the Miocene extinction of European great apes. *Proceedings. Biological Sciences, 277*(1697), 3105–3112. https://doi.org/10.1098/rspb.2010.0523

Meyer, M., Arsuaga, J. L., de Filippo, C., Nagel, S., Aximu-Petri, A., Nickel, B., Martinez, I., Gracia, A., de Castro, J. M. B., Carbonell, E., Viola, B., Kelso, J., Pruffer, K., & Paabo, S. (2016). Nuclear DNA sequences from the Middle Pleistocene Sima de los Huesos hominins. *Nature, 531*(7595), 504–507. https://doi.org/10.1038/nature17405

Meyer, M., Kircher, M., Gansauge, M. T., Li, H., Racimo, F., Mallick, S., Schraiber, J. G., Jay, F., Prufer, K., de Filippo, C., Sudmant, P. H., Alkan, C., Fu, Q., Do, R., Rohland, N., Tandon, A., Siebauer, M., Green, R. E., Bryc, K.,. . . & Paabo, S. (2012). A high-coverage genome sequence from an archaic Denisovan individual. *Science, 338*, 222–226. https://doi.org/10.1126/science.1224344

Meyer, M. R., Jung, J. P., Spear, J. K., Araiza, I. F., Galway-Witham, J., & Williams, S. A. (2023). Knuckle-walking in *Sahelanthropus*? Locomotor inferences from the ulnae of fossil hominins and other hominoids. *Journal of Human Evolution, 179*, 103355. https://doi.org/10.1016/j.jhevol.2023.103355

Meyer, M. R., Williams, S. A., Smith, M. P., & Sawyer, G. J. (2015). Lucy's back: Reassessment of fossils associated with the A.L. 288-1 vertebral column. *Journal of Human Evolution, 85*, 174–180. https://doi.org/10.1016/j.jhevol.2015.05.007

Michel, V., Valladas, H., Shen, G., Wang, W., Zhao, J. X., Shen, C. C., Valensi, P., & Bae, C. J. (2016). The earliest modern *Homo sapiens* in China? *Journal of Human Evolution*, *101*, 101–104, https://doi.org/10.1016/j.jhevol.2016.07.008

Michon, M., Lopez, V., & Aboitiz, F. (2019). Origin and evolution of human speech: Emergence from a trimodal auditory, visual and vocal network. In M. A. Hofman (Ed.), *Progress in brain research* (2019/11/11 ed., Vol. 250, pp. 345–371). Elsevier B.V. https://doi.org/10.1016/bs.pbr.2019.01.005

Milinski, M., Croy, I., Hummel, T., & Boehm, T. (2013). Major histocompatibility complex peptide ligands as olfactory cues in human body odour assessment. *Proceedings. Biological Sciences*, *280*(1755), 20122889. https://doi.org/10.1098/rspb.2012.2889

Milton, K. (1999a). Comment to 'The raw and the stolen: Cooking and the ecology of human origins'. *Current Anthropology*, *40*(5), 583–584. https://doi.org/10.1086/300083

Milton, K. (1999b). A hypothesis to explain the role of meat-eating in human evolution. *Evolutionary Anthropology*, *8*(1), 11–21. https://doi.org/10.1002/(Sici)1520-6505(1999)8:1<11::Aid-Evan6>3.0.Co;2-M

Milton, K. (2012). Diet and primate evolution. *Scientific American*, *12*, 22–29.

Mink, J. W., Blumenschine, R. J., & Adams, D. B. (1981). Ratio of Central nervous-system to body metabolism in vertebrates: Its constancy and functional basis. *American Journal of Physiology*, *241*(3), R203–R212.

Mitteroecker, P., & Fischer, B. (2024). Evolution of the human birth canal. *American Journal of Obstetrics and Gynecology*, *230*(3S), S841–S855. https://doi.org/10.1016/j.ajog.2022.09.010

Mitteroecker, P., Grunstra, N. D. S., Stansfield, E., Waltenberger, L., & Fischer, B. (2021). Did population differences in human pelvic form evolve by drift or selection? *PMSAP*, *33*(1), 11–26.

Moffett, E. A. (2021). Sexual dimorphism in the size and shape of the non-obstetric pelvis across anthropoids. *American Journal of Physical Anthropology*. https://doi.org/10.1002/ajpa.24398

Moncel, M. H., & Ashton, N. (2018). From 800 to 500 ka in Western Europe. The oldest evidence of Acheuleans in their technological, chronological, and geographical framework. In R. Gallotti, & M. Mussi (Eds.), *Emergence of the Acheulean in East Africa and beyond: Contributions in honor of Jean Chavaillon* (pp. 215–235). Springer International Publishing AG. https://doi.org/10.1007/978-3-319-75985-2_11

Moncel, M. H., Ashton, N., Arzarello, M., Fontana, F., Lamotte, A., Scott, B., Muttillo, B., Berruti, G., Nenzioni, G., Tuffreau, A., & Peretto, C. (2020). Early Levallois core technology between marine isotope stage 12 and 9 in Western Europe. *Journal of Human Evolution*, *139*. https://doi.org/10.1016/j.jhevol.2019.102735

Monson, T. A., Weitz, A. P., Brasil, M. F., & Hlusko, L. J. (2022). Teeth, prenatal growth rates, and the evolution of human-like pregnancy in later *Homo*. *Proceedings of the National Academy of Sciences of the United States of America*, *119*(41), e2200689119. https://doi.org/10.1073/pnas.2200689119

Montagna, W. (1972). The skin of nonhuman primates. *American Zoologist*, *12*, 109–124.

Moorjani, P., Amorim, C. E., Arndt, P. F., & Przeworski, M. (2016). Variation in the molecular clock of primates. *Proceedings of the National Academy of Sciences of the United States of America*, *113*(38), 10607–10612. https://doi.org/10.1073/pnas.1600374113

Morgan, M. H., & Carrier, D. R. (2012). Protective buttressing of the human fist and the evolution of hominin hands. *Journal of Experimental Biology, 216*(2), 236–244. https://doi.org/10.1242/jeb.075713

Mori, F., Kaneko, A., Matsuzawa, T., & Nishimura, T. (2021). Computational fluid dynamics simulation wall model predicting air temperature of the nasal passage for nonhuman primates. *American Journal of Physical Anthropology, 174*(4), 839–845. https://doi.org/10.1002/ajpa.24221

Mortensen, H. S., Pakkenberg, B., Dam, M., Dietz, R., Sonne, C., Mikkelsen, B., & Eriksen, N. (2014). Quantitative relationships in delphinid neocortex. *Frontiers in Neuroanatomy, 8,* 132. https://doi.org/10.3389/fnana.2014.00132

Mottram, D. S. (1998). Flavour formation in meat and meat products: a review. *Food Chemistry, 62*(4), 415–424. https://doi.org/10.1016/S0308-8146(98)00076-4

Mounier, A., Marchal, F., & Condemi, S. (2009). Is *Homo heidelbergensis* a distinct species? New insight on the Mauer mandible. *Journal of Human Evolution, 56*(3), 219–246. https://doi.org/10.1016/j.jhevol.2008.12.006

Muchlinski, M. N., Hemingway, H. W., Pastor, J., Omstead, K. M., & Burrows, A. M. (2018). How the brain May have shaped muscle anatomy and physiology: A preliminary study. *Anatomical Record (Hoboken), 301*(3), 528–537. https://doi.org/10.1002/ar.23746

Murray, N., Norton, H. L., & Parra, E. J. (2015). Distribution of two OCA2 polymorphisms associated with pigmentation in East-Asian populations. *Human Genome Variation, 2,* 15058. https://doi.org/10.1038/hgv.2015.58

Muthukrishna, M., Doebeli, M., Chudek, M., & Henrich, J. (2018). The cultural brain hypothesis: How culture drives brain expansion, sociality, and life history. *PLoS Computational Biology, 14*(11), e1006504. https://doi.org/10.1371/journal.pcbi.1006504

Myatt, J. P., Schilling, N., & Thorpe, S. K. S. (2011). Distribution patterns of fibre types in the triceps surae muscle group of chimpanzees and orangutans. *Journal of Anatomy, 218*(4), 402–412. https://doi.org/10.1111/j.1469-7580.2010.01338.x

Mylopotamitaki, D., Weiss, M., Fewlass, H., Zavala, E. I., Rougier, H., Sumer, A. P., Hajdinjak, M., Smith, G. M., Ruebens, K., Sinet-Mathiot, V., Pederzani, S., Essel, E., Harking, F. S., Xia, H., Hansen, J., Kirchner, A., Lauer, T., Stahlschmidt, M., Hein, M., Talamo, S., Wacker, L., Meller, H., Dietl, H., Orschiedt, J., Olsen, J. V., Zeberg, H., Prufer, K., Krause, J., Meyer, M., Welker, F., McPherron, S. P., Schuler, T., & Hublin, J. J. (2024). *Homo sapiens* reached the higher latitudes of Europe by 45,000 years ago. *Nature, 626*(7998), 341–346. https://doi.org/10.1038/s41586-023-06923-7

Nagano, A., Umberger, B. R., Marzke, M. W., & Gerritsen, K. G. (2005). Neuromusculoskeletal computer modeling and simulation of upright, straight-legged, bipedal locomotion of *Australopithecus afarensis* (A.L. 288-1). *American Journal of Physical Anthropology, 126*(1), 2–13. https://doi.org/10.1002/ajpa.10408

Nakatsukasa, M., Pickford, M., Egi, N., & Senut, B. (2007). Femur length, body mass, and stature estimates of *Orrorin tugenensis*, a 6 Ma hominid from Kenya. *Primates, 48*(3), 171–178. https://doi.org/10.1007/s10329-007-0040-7

Nakatsukasa, M., Tsujikawa, H., Shimizu, D., Takano, T., Kunimatsu, Y., Nakano, Y., & Ishida, H. (2003). Definitive evidence for tail loss in *Nacholapithecus*, an East African Miocene hominoid. *Journal of Human Evolution, 45*(2), 179–186. https://doi.org/10.1016/s0047-2484(03)00092-7

Nakatsukasa, M., Ward, C. V., Walker, A., Teaford, M. F., Kunimatsu, Y., & Ogihara, N. (2004). Tail loss in *Proconsul heseloni*. *Journal of Human Evolution*, *46*(6), 777–784. https://doi.org/10.1016/j.jhevol.2004.04.005

Negash, E. W., Alemseged, Z., Barr, W. A., Behrensmeyer, A. K., Blumenthal, S. A., Bobe, R., Carvalho, S., Cerling, T. E., Chritz, K. L., McGuire, E., Uno, K. T., Wood, B., & Wynn, J. G. (2024). Modern African ecosystems as landscape-scale analogues for reconstructing woody cover and early hominin environments. *Journal of Human Evolution*, *197*, 103604. https://doi.org/10.1016/j.jhevol.2024.103604

Nelissen, M. (2015). *De bril van Darwin. Op zoek naar de wortels van ons gedrag* (12e druk ed.). Lannoo NV.

Neubauer, S., Gunz, P., & Hublin, J. J. (2010). Endocranial shape changes during growth in chimpanzees and humans: A morphometric analysis of unique and shared aspects. *Journal of Human Evolution*, *59*(5), 555–566. https://doi.org/10.1016/j.jhevol.2010.06.011

Neubauer, S., Gunz, P., Weber, G. W., & Hublin, J. J. (2012). Endocranial volume of *Australopithecus africanus*: New CT-based estimates and the effects of missing data and small sample size. *Journal of Human Evolution*. https://doi.org/10.1016/j.jhevol.2012.01.005

Neubauer, S., Hublin, J. J., & Gunz, P. (2018). The evolution of modern human brain shape. *Science Advances*, *4*, eaao5961.

Neves, A. G. M., & Serva, M. (2012). Extremely rare interbreeding events can explain Neanderthal DNA in living humans. *PloS One*, *7*(10), e47076. https://doi.org/10.1371/journal.pone.0047076

Nevo, O., Valenta, K., Helman, A., Ganzhorn, J. U., & Ayasse, M. (2022). Fruit scent as an honest signal for fruit quality. *BMC Ecology and Evolution*, *22*(1), 139. https://doi.org/10.1186/s12862-022-02064-z

Nevo, O., Valenta, K., Razafimandimby, D., Melin, A. D., Ayasse, M., & Chapman, C. A. (2018). Frugivores and the evolution of fruit colour. *Biology Letters*, *14*(9), 20180377. https://doi.org/10.1098/rsbl.2018.0377

Ni, X., Ji, Q., Wu, W., Shao, Q., Ji, Y., Zhang, C., Liang, L., Ge, J., Guo, Z., Li, J., Li, Q., Grun, R., & Stringer, C. (2021). Massive cranium from Harbin in northeastern China establishes a new Middle Pleistocene human lineage. *Innovation (New York, N.Y.)*, *2*(3), 100130. https://doi.org/10.1016/j.xinn.2021.100130

Nicolaï, M. P. J., Shawkey, M. D., Porchetta, S., Claus, R., & D'Alba, L. (2020). Exposure to UV radiance predicts repeated evolution of concealed black skin in birds. *Nature Communications*, *11*(1). https://doi.org/10.1038/s41467-020-15894-6

Niespolo, E. M., Sharp, W. D., Avery, G., & Dawson, T. E. (2021). Early, intensive marine resource exploitation by Middle Stone Age humans at Ysterfontein 1 rockshelter, South Africa. *Proceedings of the National Academy of Sciences of the United States of America*, *118*(16). https://doi.org/10.1073/pnas.2020042118

Nilsson, G. E. (1996). Brain and body oxygen requirements of *Gnathonemus petersii*, a fish with an exceptionally large brain. *Journal of Experimental Biology*, *199*(3), 603–607.

Nimchinsky, E. A., Gilissen, E., Allman, J. M., Perl, D. P., Erwin, J. M., & Hof, P. R. (1999). A neuronal morphologic type unique to humans and great apes. *Proceedings of the National Academy of Sciences of the United States of America*, *96*(9), 5268–5273. https://doi.org/10.1073/pnas.96.9.5268

Norton, H. L., Kittles, R. A., Parra, E., McKeigue, P., Mao, X., Cheng, K., Canfield, V. A., Bradley, D. G., McEvoy, B., & Shriver, M. D. (2007). Genetic evidence for the convergent evolution of light skin in Europeans and East Asians. *Molecular Biology and Evolution, 24*(3), 710–722. https://doi.org/10.1093/molbev/msl203

O'Brien, K., Hebdon, N., & Faith, J. T. (2023). Paleoecological evidence for environmental specialization in *Paranthropus boisei* compared to early *Homo. Journal of Human Evolution, 177*, 103325. https://doi.org/10.1016/j.jhevol.2023.103325

O'Connell, J. F., Allen, J., Williams, M. A. J., Williams, A. N., Turney, C. S. M., Spooner, N. A., Kamminga, J., Brown, G., & Cooper, A. (2018). When did *Homo sapiens* first reach Southeast Asia and Sahul? *Proceedings of the National Academy of Sciences of the United States of America, 115*(34), 8482–8490. https://doi.org/10.1073/pnas.1808385115

O'Connell, J. F., Hawkes, K., & Jones, N. B. (1988). Hadza Scavenging - Implications for Plio Pleistocene hominid subsistence. *Current Anthropology, 29*(2), 356–363. https://doi.org/10.1086/203648

O'Neill, M. C., Demes, B., Thompson, N. E., & Umberger, B. R. (2018). Three-dimensional kinematics and the origin of the hominin walking stride. *J R Soc Interface, 15*(145). https://doi.org/10.1098/rsif.2018.0205

O'Neill, M. C., Demes, B., Thompson, N. E., Larson, S. G., Stern, J. T., Jr., & Umberger, B. R. (2022). Adaptations for bipedal walking: Musculoskeletal structure and three-dimensional joint mechanics of humans and bipedal chimpanzees (*Pan troglodytes*). *Journal of Human Evolution, 168*, 103195. https://doi.org/10.1016/j.jhevol.2022.103195

O'Neill, M. C., Umberger, B. R., Holowka, N. B., Larson, S. G., & Reiser, P. J. (2017). Chimpanzee super strength and human skeletal muscle evolution. *Proceedings of the National Academy of Sciences of the United States of America, 114*(28), 7343–7348. https://doi.org/10.1073/pnas.1619071114

Ochando, J., Jiménez-Espejo, F. J., Giles-Guzmán, F., de Carvalho, C. N., Carrión, J. S., Muñiz, F., Rubiales, J. M., Cura, P., Belo, J., Finlayson, S., Martrat, B., van Drooge, B. L., Jiménez-Moreno, G., García-Alix, A., Rodríguez, J. A. L., Albert, R. M., Ohkouchi, N., Ogawa, N., Suga, H., Camuera, J., Martínez-Ruiz, F., Villanueva, J., Teruel, O., Davtian, N., Moreira, N., Belaústegui, Z., Rodríguez-Vidal, J., Munuera, M., Menez, A., Finlayson, G., & Finlayson, C. (2024). A Neanderthal's specialised burning structure compatible with tar obtention. *Quaternary Science Reviews, 346*. https://doi.org/10.1016/j.quascirev.2024.109025

Okerblom, J., Fletes, W., Patel, H. H., Schenk, S., Varki, A., & Breen, E. C. (2018). Human-like Cmah inactivation in mice increases running endurance and decreases muscle fatigability: Implications for human evolution. *Proceedings of the Royal Society B: Biological Sciences, 285*(1886). https://doi.org/10.1098/rspb.2018.1656

Olkowicz, S., Kocourek, M., Lucan, R. K., Portes, M., Fitch, W. T., Herculano-Houzel, S., & Nemec, P. (2016). Birds have primate-like numbers of neurons in the forebrain. *Proceedings of the National Academy of Sciences of the United States of America, 113*(26), 7255–7260. https://doi.org/10.1073/pnas.1517131113

Organ, C., Nunn, C. L., Machanda, Z., & Wrangham, R. W. (2011). Phylogenetic rate shifts in feeding time during the evolution of *Homo. Proceedings of the National Academy of Sciences of the United States of America, 108*(35), 14555–14559. https://doi.org/10.1073/pnas.1107806108

Organ, J. M., Muchlinski, M. N., & Deane, A. S. (2011). Mechanoreceptivity of prehensile tail skin varies between ateline and cebine primates. *Anatomical Record-Advances in Integrative Anatomy and Evolutionary Biology, 294*(12), 2064–2072. https://doi.org/10.1002/ar.21505

Orlando, L., Darlu, P., Toussaint, M., Bonjean, D., Otte, M., & Hanni, C. (2006). Revisiting Neandertal diversity with a 100,000 year old mtDNA sequence. *Current Biology*, *16*(11), R400–R402. https://doi.org/10.1016/j.cub.2006.05.019

Orr, C. M. (2005). Knuckle-walking anteater: A convergence test of adaptation for purported knuckle-walking features of African Hominidae. *American Journal of Physical Anthropology*, *128*(3), 639–658. https://doi.org/10.1002/ajpa.20192

Orr, C. M. (2018). Kinematics of the anthropoid os centrale and the functional consequences of scaphoid-centrale fusion in African apes and hominins. *Journal of Human Evolution*, *114*, 102–117. https://doi.org/10.1016/j.jhevol.2017.10.002

Orwant, R. (2005). Lessons from our closest cousins. *NewScientist*, *3 Sept*(2515), 6–7.

Pääbo, S. (2014). *Neanderthal man: In search of lost genomes*. Basic Books.

Pablos, A. (2015). The foot in the *Homo* fossil record. *Mitteilungen der Gesellschaft für Urgeschichte*, *24*, 11–28.

Pagano, A. S., Smith, C. M., Balzeau, A., Marquez, S., & Laitman, J. T. (2022). Nasopharyngeal morphology contributes to understanding the "muddle in the middle" of the Pleistocene hominin fossil record. *Anatomical Record (Hoboken, N.J.: 2007)*. https://doi.org/10.1002/ar.24913

Pakkenberg, B., & Gundersen, H. J. G. (1997). Neocortical neuron number in humans: Effect of sex and age. *The Journal of Comparative Neurology*, *384*(2), 312–320. https://doi.org/10.1002/(sici)1096-9861(19970728)384:2<312::Aid-cne10>3.0.Co;2-k

Palomero-Gallagher, N., & Zilles, K. (2019). Differences in cytoarchitecture of Broca's region between human, ape and macaque brains, *Cortex*, *118*, 132–153. https://doi.org/10.1016/j.cortex.2018.09.008

Pante, M., Torre, I., d'Errico, F., Njau, J., & Blumenschine, R. (2020). Bone tools from Beds II-IV, Olduvai Gorge, Tanzania, and implications for the origins and evolution of bone technology. *Journal of Human Evolution*, *148*, 102885. https://doi.org/10.1016/j.jhevol.2020.102885

Parker, S. T. (2015). Re-evaluating the extractive foraging hypothesis. *New Ideas in Psychology*, *37*, 1–12. https://doi.org/10.1016/j.newideapsych.2014.11.001

Parkinson, J. A., Plummer, T. W., Oliver, J. S., & Bishop, L. C. (2022). Meat on the menu: GIS spatial distribution analysis of bone surface damage indicates that Oldowan hominins at Kanjera South, Kenya had early access to carcasses. *Quaternary Science Reviews*, *277*. https://doi.org/10.1016/j.quascirev.2021.107314

Parra, F. C., Amado, R. C., Lambertucci, J. R., Rocha, J., Antunes, C. M., & Pena, S. D. J. (2003). Color and genomic ancestry in Brazilians. *Proceedings of the National Academy of Sciences of the United States of America*, *100*(1), 177–182. https://doi.org/10.1073/pnas.0126614100

Patterson, D. B., Du, A., Faith, J. T., Rowan, J., Uno, K., Behrensmeyer, A. K., Braun, D. R., & Wood, B. A. (2022). Did vegetation change drive the extinction of *Paranthropus boisei*? *Journal of Human Evolution*, *103154*. https://doi.org/10.1016/j.jhevol.2022.103154

Patterson, F. G. P., & Gordon, W. (2001). Twenty-seven years of project Koko and Michael. In B. M. F. Galdikas, N. E. Briggs, L. K. Sheeran, G. L. Shapiro, & J. Goodall (Eds.), *All apes great and small* (Vol. 1, pp. 165–176). Springer. https://doi.org/10.1007/b107546

Pearce, E., Stringer, C., & Dunbar, R. I. M. (2013). New insights into differences in brain organization between Neanderthals and anatomically modern humans. *Proceedings of the Royal Society B: Biological Sciences*, *280*(1758), 20130168. https://doi.org/10.1098/rspb.2013.0168

Pease, A., & Pease, B. (1999). Waarom mannen niet luisteren en vrouwen niet kunnen kaartlezen (L. C. bvan Twisk, Trans.). *Het Spectrum*. (Why man don't listen & Women can't read maps).

Pennisi, E. (2021). Genomes arising. *Science, 371*(6529), 556–559. https://doi.org/10.1126/science.371.6529.556

Perry, P. (2018, 28/02/2018). Are humans domesticated animals? *BigThink*. Retrieved 22/04/2021 from https://bigthink.com/philip-perry/humans-were-the-first-domesticated-species-hypothesis-states

Peterson, M. R., Cherukuri, V., Paulson, J. N., Ssentongo, P., Kulkarni, A. V., Warf, B. C., Monga, V., & Schiff, S. J. (2021). Normal childhood brain growth and a universal sex and anthropomorphic relationship to cerebrospinal fluid. *Journal of Neurosurgery. Pediatrics, 28*(4), 458–468. https://doi.org/10.3171/2021.2.PEDS201006

Petr, M., Hajdinjak, M., Fu, Q., Essel, E., Rougier, H., Crevecoeur, I., Semal, P., Golovanova, L. V., Doronichev, V. B., Lalueza-Fox, C., de la Rasilla, M., Rosas, A., Shunkov, M. V., Kozlikin, M. B., Derevianko, A. P., Vernot, B., Meyer, M., & Kelso, J. (2020). The evolutionary history of Neanderthal and Denisovan Y chromosomes. *Science (New York, N.Y.), 369*(6511), 1653–1656. https://doi.org/10.1126/science.abb6460

Peyregne, S., Boyle, M. J., Dannemann, M., & Prufer, K. (2017). Detecting ancient positive selection in humans using extended lineage sorting. *Genome Research, 27*(9), 1563–1572. https://doi.org/10.1101/gr.219493.116

Picin, A., Peresani, M., Falgueres, C., Gruppioni, G., & Bahain, J. J. (2013). San Bernardino Cave (Italy) and the appearance of Levallois technology in Europe: Results of a radiometric and technological reassessment. *PloS One, 8*(10), e76182. https://doi.org/10.1371/journal.pone.0076182

Pilbeam, D. (2004). The anthropoid postcranial axial skeleton: Comments on development, variation, and evolution. *Journal of Experimental Zoology Part B-Molecular and Developmental Evolution, 302b*(3), 241–267. https://doi.org/10.1002/jez.b.22

Pinson, A., Xing, L., Namba, T., Kalebic, N., Peters, J., Oegema, C. E., Traikov, S., Reppe, K., Riesenberg, S., Maricic, T., Derihaci, R., Wimberger, P., Paabo, S., & Huttner, W. B. (2022). Human TKTL1 implies greater neurogenesis in frontal neocortex of modern humans than Neanderthals. *Science (New York, N.Y.), 377*(6611), eabl6422. https://doi.org/10.1126/science.abl6422

Plummer, T. W., Oliver, J. S., Finestone, E. M., Ditchfield, P. W., Bishop, L. C., Blumenthal, S. A., Lemorini, C., Caricola, I., Bailey, S. E., Herries, A. I. R., Parkinson, J. A., Whitfield, E., Hertel, F., Kinyanjui, R. N., Vincent, T. H., Li, Y., Louys, J., Frost, S. R., Braun, D. R., Reeves, J. S., Early, E. D. G., Onyango, B., Lamela-Lopez, R., Forrest, F. L., He, H., Lane, T. P., Frouin, M., Nomade, S., Wilson, E. P., Bartilol, S. K., Rotich, N. K., & Potts, R. (2023). Expanded geographic distribution and dietary strategies of the earliest Oldowan hominins and *Paranthropus*. *Science (New York, N.Y.), 379*(6632), 561–566. https://doi.org/10.1126/science.abo7452

Polk, J. D. (2004). Influences of limb proportions and body size on locomotor kinematics in terrestrial primates and fossil hominins. *Journal of Human Evolution, 47*(4), 237–252. https://doi.org/10.1016/j.jhevol.2004.07.003

Ponce de Leon, M. S., Golovanova, L., Doronichev, V., Romanova, G., Akazawa, T., Kondo, O., Ishida, H., & Zollikofer, C. P. (2008). Neanderthal brain size at birth provides insights into the evolution of human life history. *Proceedings of the National Academy of Sciences* of the United States of America, *105*(37), 13764–13768. https://doi.org/10.1073/pnas. 0803917105

Ponce de León, M. S., Bienvenu, T., Akazawa, T., & Zollikofer, C. P. (2016). Brain development is similar in Neanderthals and modern humans. *Current biology: CB*, *26*(14), R665–R666. https://doi.org/10.1016/j.cub.2016.06.022

Ponce de León, M. S., Bienvenu, T., Marom, A., Engel, S., Tafforeau, P., Alatorre Warren, J. L., Lordkipanidze, D., Kurniawan, I., Murti, D. B., Suriyanto, R. A., Koesbardiati, T., & Zollikofer, C. P. E. (2021). The primitive brain of early *Homo*. *Science (New York, N.Y.)*, *372*(6538), 165–171. https://doi.org/10.1126/science.aaz0032

Pontzer, H. (2007). Predicting the energy cost of terrestrial locomotion: A test of the LiMb model in humans and quadrupeds. *Journal of Experimental Biology, 210*(3), 484–494. https://doi.org/10.1242/jeb.02662

Pontzer, H., Brown, M. H., Wood, B. M., Raichlen, D. A., Mabulla, A. Z. P., Harris, J. A., Dunsworth, H., Hare, B., Walker, K., Luke, A., Dugas, L. R., Schoeller, D., Plange-Rhule, J., Bovet, P., Forrester, T. E., Thompson, M. E., Shumaker, R. W., Rothman, J. M., Vogel, E., Sulistyo, F., Alavi, S., Prasetyo, D., Urlacher, S. S., & Ross, S. R. (2021). Evolution of water conservation in humans. *Current Biology: CB, 31*(8), 1804–1810.e1805. https://doi. org/10.1016/j.cub.2021.02.045

Pontzer, H., Rolian, C., Rightmire, G. P., Jashashvili, T., de Leon, M. S. P., Lordkipanidze, D., & Zollikofer, C. P. E. (2010). Locomotor anatomy and biomechanics of the Dmanisi hominins. *Journal of Human Evolution, 58*(6), 492–504. https://doi.org/10.1016/j.jhevol.2010.03.006

Porter, A. M. W. (1995). The body-weight of Al-288-1 (Lucy): A new approach using estimates of skeletal length and the body-mass index. *International Journal of Osteoarchaeology, 5*(3), 203–212. https://doi.org/10.1002/oa.1390050302

Potts, R. (2012). Environmental and behavioral evidence pertaining to the evolution of early *Homo. Current Anthropology, 53*(S6), S299–S317. https://doi.org/10.1086/667704

Potts, R. (2013). Hominin evolution in settings of strong environmental variability. *Quaternary Science Reviews, 73*, 1–13. https://doi.org/10.1016/j.quascirev.2013.04.003

Potts, R. (2020). Microevolution in our megadont relative. *Nature Ecology & Evolution, 5*(1), 14–16. https://doi.org/10.1038/s41559-020-01339-2

Pouydebat, E., Berge, C., Gorce, P., & Coppens, Y. (2006). Apr-May). Grasping among Primates: Precision, tools and evolutionary implications. *Comptes Rendus Palevol, 5*(3-4), 597–602.

Power, R. C., Salazar-Garcia, D. C., Rubini, M., Darlas, A., Harvati, K., Walker, M., Hublin, J. J., & Henry, A. G. (2018). Dental calculus indicates widespread plant use within the stable Neanderthal dietary niche. *Journal of Human Evolution, 119*, 27–41. https://doi. org/10.1016/j.jhevol.2018.02.009

Prabhakar, S., Visel, A., Akiyama, J. A., Shoukry, M., Lewis, K. D., Holt, A., Plajzer-Frick, I., Morrison, H., FitzPatrick, D. R., Afzal, V., Pennacchio, L. A., Rubin, E. M., & Noonan, J. P. (2008). Human-specific gain of function in a developmental enhancer. *Science, 321*(5894), 1346–1350. https://doi.org/10.1126/science.1159974

Prado-Martinez, J., Sudmant, P. H., Kidd, J. M., Li, H., Kelley, J. L., Lorente-Galdos, B., Veeramah, K. R., Woerner, A. E., O'Connor, T. D., Santpere, G., Cagan, A., Theunert, C., Casals, F., Laayouni, H., Munch, K., Hobolth, A., Halager, A. E., Malig, M., Hernandez-Rodriguez, J., Hernando-Herraez, I., Prufer, K., Pybus, M., Johnstone, L., Lachmann, M., Alkan, C., Twigg, D., Petit, N., Baker, C., Hormozdiari, F., Fernandez-Callejo, M., Dabad, M., Wilson, M. L., Stevison, L., Camprubi, C., Carvalho, T., Ruiz-Herrera, A., Vives, L., Mele, M.,

Abello, T., Kondova, I., Bontrop, R. E., Pusey, A., Lankester, F., Kiyang, J. A., Bergl, R. A., Lonsdorf, E., Myers, S., Ventura, M., Gagneux, P., Comas, D., Siegismund, H., Blanc, J., Agueda-Calpena, L., Gut, M., Fulton, L., Tishkoff, S. A., Mullikin, J. C., Wilson, R. K., Gut, I. G., Gonder, M. K., Ryder, O. A., Hahn, B. H., Navarro, A., Akey, J. M., Bertranpetit, J., Reich, D., Mailund, T., Schierup, M. H., Hvilsom, C., Andres, A. M., Wall, J. D., Bustamante, C. D., Hammer, M. F., Eichler, E. E., & Marques-Bonet, T. (2013). Great ape genetic diversity and population history. *Nature, 499*(7459), 471–475. https://doi.org/10.1038/nature12228

Praet, D. (2013). Biologie in het antieke Griekenland: van de eerste natuurwetenschappers tot de school van Aristoteles. In L. Van Speybroeck, & J. Braeckman (Eds.), *Fascinerend leven. Markante figuren en ideeën uit de geschiedenis van de biologie* (pp. 25–71). Academia Press.

Prang, T. C. (2019). The African ape-like foot of *Ardipithecus ramidus* and its implications for the origin of bipedalism. *eLife, 8*, 1–17. https://doi.org/10.7554/eLife.44433

Prat, S. (2022). Emergence of the genus *Homo*: From concept to taxonomy. *Anthropologie, 126*(4). https://doi.org/10.1016/j.anthro.2022.103068

Pritchard, J. K. (2013). How we are evolving? *Scientific American, 22*(1), 99–105.

Pritchard, V. E., Malone, S. A., Burgoyne, K., Heron-Delaney, M., Bishop, D. V. M., & Hulme, C. (2019). Stage 1 registered report: The relationship between handedness and language ability in children. *Wellcome Open Research, 4*, 30. https://doi.org/10.12688/wellcomeopenres.15077.1

Profousova, I., Mihalikova, K., Laho, T., Varadyova, Z., Petrzelkova, K. J., Modry, D., & Kisidayova, S. (2011). The ciliate, *Troglodytella abrassarti*, contributes to polysaccharide hydrolytic activities in the chimpanzee colon. *Folia Microbiologica, 56*(4), 339–343. https://doi.org/10.1007/s12223-011-0053-x

Prüfer, K., de Filippo, C., Grote, S., Mafessoni, F., Korlevic, P., Hajdinjak, M., Vernot, B., Skov, L., Hsieh, P. S., Peyregne, S., Reher, D., Hopfe, C., Nagel, S., Maricic, T., Fu, Q. M., Theunert, C., Rogers, R., Skoglund, P., Chintalapati, M., Dannemann, M., Nelson, B. J., Key, F. M., Rudan, P., Kucan, Z., Gusic, I., Golovanova, L. V., Doronichev, V. B., Patterson, N., Reich, D., Eichler, E. E., Slatkin, M., Schierup, M. H., Andres, A. M., Kelso, J., Meyer, M., & Paabo, S. (2017). A high-coverage Neandertal genome from Vindija Cave in Croatia. *Science, 358*(6363), 655–658. https://doi.org/10.1126/science.aao1887

Prüfer, K., Posth, C., Yu, H., Stoessel, A., Spyrou, M. A., Deviese, T., Mattonai, M., Ribechini, E., Higham, T., Velemínský, P., Brůžek, J., & Krause, J. (2021). A genome sequence from a modern human skull over 45,000 years old from Zlatý kůň in Czechia. *Nature ecology & evolution, 5*(6), 820–825. https://doi.org/10.1038/s41559-021-01443-x

Prüfer, K., Racimo, F., Patterson, N., Jay, F., Sankararaman, S., Sawyer, S., Heinze, A., Renaud, G., Sudmant, P. H., de Filippo, C., Li, H., Mallick, S., Dannemann, M., Fu, Q., Kircher, M., Kuhlwilm, M., Lachmann, M., Meyer, M., Ongyerth, M., Siebauer, M., Theunert, C., Tandon, A., Moorjani, P., Pickrell, J., Mullikin, J. C., Vohr, S. H., Green, R. E., Hellmann, I., Johnson, P. L., Blanche, H., Cann, H., Kitzman, J. O., Shendure, J., Eichler, E. E., Lein, E. S., Bakken, T. E., Golovanova, L. V., Doronichev, V. B., Shunkov, M. V., Derevianko, A. P., Viola, B., Slatkin, M., Reich, D., Kelso, J., & Paabo, S. (2014, Jan 2). The complete genome sequence of a Neanderthal from the Altai Mountains. *Nature, 505*(7481), 43–49. https://doi.org/10.1038/nature12886

Pugh, K. D. (2022). Phylogenetic analysis of Middle-Late Miocene apes. *Journal of Human Evolution, 165*, 103140. https://doi.org/10.1016/j.jhevol.2021.103140

Pujol, A., Rissech, C., Ventura, J., Badosa, J., & Turbón, D. (2014). Ontogeny of the female femur: Geometric morphometric analysis applied on current living individuals of a Spanish population. *Journal of Anatomy, 225*(3), 346–357. https://doi.org/10.1111/joa.12209

Purves, D., & Williams, M. (2001). *Neuroscience* (2nd ed.). Sinauer Associates.

Püschel, T. A., Nicholson, S. L., Baker, J., Barton, R. A., & Venditti, C. (2024). Hominin brain size increase has emerged from within-species encephalization. *Proceedings of the National Academy of Sciences of the United States of America, 121*(49), e2409542121. https://doi.org/10.1073/pnas.2409542121

Quillen, E. E., Norton, H. L., Parra, E. J., Lona-Durazo, F., Ang, K. C., Mircea Illiescu, F., Pearson, L. N., Shriver, M. D., Lasisi, T., Gokcumen, O., Starr, I., Lin, Y.-L., Martin, A. R., & Jablonski, N. G. (2018). Shades of complexity: New perspectives on the evolution and genetic architecture of human skin. *American Journal of Physical Anthropology, 168*(Suppl 67), 1–23. https://doi.org/10.1002/ajpa.23737

Radivojević, M., Rehren, T., Pernicka, E., Šljivar, D., Brauns, M., & Borić, D. (2010). On the origins of extractive metallurgy: New evidence from Europe. *Journal of Archaeological Science, 37*(11), 2775–2787. https://doi.org/10.1016/j.jas.2010.06.012

Radovcic, D., Srsen, A. O., Radovcic, J., & Frayer, D. W. (2015). Evidence for Neandertal jewelry: Modified white-tailed eagle claws at Krapina. *PloS One, 10*(3), e0119802. https://doi.org/10.1371/journal.pone.0119802

Raff, J. (2021). Journey into the Americas. *Scientific American, 324*(5), 26–33.

Raghav, S., Uppal, V., & Gupta, A. (2022). Comparative study on distribution of sebaceous and sweat glands in skin of different domestic animals. *Indian Journal of Animal Research, 56*(11), 1356–1360. https://doi.org/10.18805/Ijar.B-4228

Ragsdale, A. P., Weaver, T. D., Atkinson, E. G., Hoal, E. G., Moller, M., Henn, B. M., & Gravel, S. (2023). A weakly structured stem for human origins in Africa. *Nature.* https://doi.org/10.1038/s41586-023-06055-y

Raichlen, D. A., & Gordon, A. D. (2017). Interpretation of footprints from Site S confirms human-like bipedal biomechanics in Laetoli hominins. *Journal of Human Evolution, 107,* 134–138. https://doi.org/10.1016/j.jhevol.2017.04.002

Raichlen, D. A., Gordon, A. D., Harcourt-Smith, W. E. H., Foster, A. D., & Haas, W. R. (2010). Laetoli footprints preserve earliest direct evidence of human-like bipedal biomechanics. *PloS One, 5*(3). https://doi.org/10.1371/journal.pone.0009769

Raichlen, D. A., & Polk, J. D. (2013). Linking brains and brawn: Exercise and the evolution of human neurobiology. *Proceedings of the Royal Society B-Biological Sciences, 280*(1750). https://doi.org/10.1098/rspb.2012.2250

Raichlen, D. A., Pontzer, H., & Sockol, M. D. (2008). The Laetoli footprints and early hominin locomotor kinematics. *Journal of Human Evolution, 54*(1), 112–117. https://doi.org/10.1016/j.jhevol.2007.07.005

Railsback, L. B., Gibbard, P. L., Head, M. J., Voarintsoa, N. R. G., & Toucanne, S. (2015). An optimized scheme of lettered marine isotope substages for the last 1.0 million years, and the climatostratigraphic nature of isotope stages and substages. *Quaternary Science Reviews, 111,* 94–106. https://doi.org/10.1016/j.quascirev.2015.01.012

Rak, Y. (1991). Lucy's pelvic anatomy: Its role in bipedal gait. *Journal of Human Evolution, 20,* 283–290.

Rakovac, M. (2021). On evolution and development of human gait. In *Measurement and Analysis of Human Locomotion* (pp. 39–59). https://doi.org/10.1007/978-3-030-79685-3_3

Ramirez Rozzi, F. V., & Bermudez De Castro, J. M. (2004). Surprisingly rapid growth in Neanderthals. *Nature, 428*(6986), 936–939. https://doi.org/10.1038/nature02428

Rasmussen, M., Guo, X. S., Wang, Y., Lohmueller, K. E., Rasmussen, S., Albrechtsen, A., Skotte, L., Lindgreen, S., Metspalu, M., Jombart, T., Kivisild, T., Zhai, W. W., Eriksson, A., Manica, A., Orlando, L., De La Vega, F. M., Tridico, S., Metspalu, E., Nielsen, K., Avila-Arcos, M. C., Moreno-Mayar, J. V., Muller, C., Dortch, J., Gilbert, M. T. P., Lund, O., Wesolowska, A., Karmin, M., Weinert, L. A., Wang, B., Li, J., Tai, S. S., Xiao, F., Hanihara, T., van Driem, G., Jha, A.R., Ricaut, F.X., de Knijff, P., Migliano, A.B., Romero, I.G., Kristiansen, K., Lambert, D.M., Brunak, S., Forster, P., Brinkmann, B., Nehlich, O., Bunce, M., Richards, M., Gupta, R., Bustamante, C. D., Krogh, A., Foley, R. A., Lahr, M. M., Balloux, F., Sicheritz-Ponten, T., Villems, R., Nielsen, R., Wang, J., & Willerslev, E. (2011). An Aboriginal Australian genome reveals separate human dispersals into Asia. *Science, 334*(6052), 94–98. https://doi.org/10.1126/science.1211177

Rathkey, J. K., & Wall-Scheffler, C. M. (2017). People choose to run at their optimal speed. *American Journal of Physical Anthropology, 163*(1), 85–93. https://doi.org/10.1002/ajpa.23187

Raviv, L., Jacobson, S. L., Plotnik, J. M., Bowman, J., Lynch, V., & Benitez-Burraco, A. (2023). Elephants as an animal model for self-domestication. *Proceedings of the National Academy of Sciences of the United States of America, 120*(15), e2208607120. https://doi.org/10.1073/pnas.2208607120

Reader, S. M., & Laland, K. N. (2002). Social intelligence, innovation, and enhanced brain size in primates. *Proceedings of the National Academy of Sciences of the United States of America, 99*(7), 4436–4441. https://doi.org/10.1073/pnas.062041299

Reed, D. N., Raney, E., Johnson, J., Jackson, H., Virabalin, N., & Mbonu, N. (2023). Hominin nomenclature and the importance of information systems for managing complexity in paleoanthropology. *Journal of Human Evolution, 175*, 103308. https://doi.org/10.1016/j.jhevol.2022.103308

Reich, D., Green, R. E., Kircher, M., Krause, J., Patterson, N., Durand, E. Y., Viola, B., Briggs, A. W., Stenzel, U., Johnson, P. L., Maricic, T., Good, J. M., Marques-Bonet, T., Alkan, C., Fu, Q., Mallick, S., Li, H., Meyer, M., Eichler, E. E., Stoneking, M., Richards, M., Talamo, S., Shunkov, M. V., Derevianko, A. P., Hublin, J. J., Kelso, J., Slatkin, M., & Paabo, S. (2010). Genetic history of an archaic hominin group from Denisova Cave in Siberia. *Nature, 468*(7327), 1053–1060. https://doi.org/10.1038/nature09710

Rein, T. R., Harrison, T., Carlson, K. J., & Harvati, K. (2017). Adaptation to suspensory locomotion in *Australopithecus sediba*. *Journal of Human Evolution, 104*, 1–12. https://doi.org/10.1016/j.jhevol.2016.12.005

Rey-Rodriguez, I., Lopez-Garcia, J. M., Bennasar, M., Banuls-Cardona, S., Blain, H. A., Blanco-Lapaz, A., Rodriguez-Alvarez, X. P., de Lombera-Hermida, A., Diaz-Rodriguez, M., Ameijenda-Iglesias, A., Agusti, J., & Fabregas-Valcarce, R. (2016). Last Neanderthals and first anatomically modern humans in the NW Iberian Peninsula: Climatic and environmental conditions inferred from the Cova Eiros small-vertebrate assemblage during MIS 3. *Quaternary Science Reviews, 151*, 185–197. https://doi.org/10.1016/j.quascirev.2016.08.030

Ribeiro, P. F., Ventura-Antunes, L., Gabi, M., Mota, B., Grinberg, L. T., Farfel, J. M., Ferretti-Rebustini, R. E., Leite, R. E., Filho, W. J., & Herculano-Houzel, S. (2013). The human cerebral cortex is neither one nor many: Neuronal distribution reveals two quantitatively different zones in the gray matter, three in the white matter, and explains local variations in cortical folding. *Frontiers in Neuroanatomy, 7,* 28. https://doi.org/10.3389/fnana.2013. 00028

Ribot Trafi, F., Garcia Bartual, M., Garcia-Nos, E., Altamirano Enciso, A. J., Nevgloski, A. J., & Wang, Q. (2020). Another interpretation of *Homo antecessor. J Anthropol Sci, 98,* 161–170. https://doi.org/10.4436/JASS.98016

Richards, G. D., Jabbour, R. S., Guipert, G., & Defleur, A. (2024). Early Neanderthal mandibular remains from Baume Moula-Guercy (Soyons, Ardeche). *Anatomical Record (Hoboken).* https://doi.org/10.1002/ar.25550

Richerson, P., Gavrilets, S., & de Waal, F. B. M. (2021). Modern theories of human evolution foreshadowed by Darwin's Descent of Man. *Science, 372*(eaba3776), 1–10. https://doi. org/10.1126/science.aba3776

Richmond, B. G., & Jungers, W. L. (2008). *Orrorin tugenensis* femoral morphology and the evolution of hominin bipedalism. *Science, 319*(5870), 1662–1665. https://doi.org/10.1126/ science.1154197

Richmond, B. G., & Strait, D. S. (2000). Evidence that humans evolved from a knuckle-walking ancestor. *Nature, 404*(6776), 382–385. https://doi.org/10.1038/35006045

Richter, C., Behringer, V., Manig, F., Henle, T., Hohmann, G., & Zierau, O. (2022). Traces of dietary patterns in saliva of hominids: Profiling salivary amino acid fingerprints in great apes and humans. *Journal of Human Evolution, 175,* 103305. https://doi.org/10.1016/j. jhevol.2022.103305

Ritchie, S. J., Cox, S. R., Shen, X., Lombardo, M. V., Reus, L. M., Alloza, C., Harris, M. A., Alderson, H. L., Hunter, S., Neilson, E., Liewald, D. C. M., Auyeung, B., Whalley, H. C., Lawrie, S. M., Gale, C. R., Bastin, M. E., McIntosh, A. M., & Deary, I. J. (2018). Sex differences in the adult human brain: Evidence from 5216 UK biobank participants. *Cerebral cortex (New York, N.Y.: 1991), 28*(8), 2959–2975. https://doi.org/10.1093/cercor/ bhy109

Rizal, Y., Westaway, K. E., Zaim, Y., van den Bergh, G. D., Bettis, E. A., 3rd, Morwood, M. J., Huffman, O. F., Grun, R., Joannes-Boyau, R., Bailey, R. M., Sidarto, Westaway, M. C., Kurniawan, I., Moore, M. W., Storey, M., Aziz, F., Suminto, Zhao, J. X., Aswan, Sipola, M. E., Larick, R., Zonneveld, J. P., Scott, R., Putt, S., & Ciochon, R. L. (2020). Last appearance of *Homo erectus* at Ngandong, Java, 117,000-108,000 years ago. *Nature, 577*(7790), 381–385. https://doi.org/10.1038/s41586-019-1863-2

Roberts, A. (2021). The fetus, the fish heart and the fruit fly. In J. M. DeSilva (Ed.), *A most interesting problem. What Darwin's descent of man got right and wrong about human evolution* (pp. 24–45). Princeton University Press.

Rodman, P. S., & McHenry, H. M. (1980). Bioenergetics and the origin of hominid bipedalism. *American Journal of Physical Anthropology, 52*(1), 103–106. https://doi.org/10.1002/ ajpa.1330520113

Roebroeks, W. (2006). The human colonisation of Europe: Where are we? *Journal of Quaternary Science, 21*(5), 425–435. https://doi.org/10.1002/jqs.1044

Roebroeks, W., Sier, M. J., Nielsen, T. K., De Loecker, D., Pares, J. M., Arps, C. E., & Mucher, H. J. (2012). Use of red ochre by early Neandertals. *Proceedings of the National Academy of Sciences of the United States of America, 109*(6), 1889–1894. https://doi.org/10.1073/ pnas.1112261109

Roebroeks, W., & Villa, P. (2011). On the earliest evidence for habitual use of fire in Europe. *Proceedings of the National Academy of Sciences, 108*(13), 5209–5214. https://doi. org/10.1073/pnas.1018116108

Rogers, A. R., Harris, N. S., & Achenbach, A. A. (2020). Neanderthal-Denisovan ancestors inter-bred with a distantly-related hominin. *Science Advances, 6*(eaay5483), 1–7. https://doi.org/10.1101/657247

Rogers, R. L. (2015). Chromosomal rearrangements as barriers to genetic homogeniza-tion between archaic and modern humans. *Molecular Biology and Evolution, 32*(12), 3064–3078. https://doi.org/10.1093/molbev/msv204

Rolian, C. (2009). Integration and evolvability in primate hands and feet. *Evolutionary Biology, 36*(1), 100–117. https://doi.org/10.1007/s11692-009-9049-8

Rolian, C. (2014). Genes, development, and evolvability in primate evolution. *Evolutionary Anthropology, 23*(3), 93–104. https://doi.org/10.1002/evan.21409

Rolian, C., Liberman, D., & Hallgrimsson, B. (2010). Did human fingers and toes coevolve? *Integrative and Comparative Biology, 50,* E289.

Rolian, C., Lieberman, D. E., & Hallgrimsson, B. (2010). The coevolution of human hands and feet. *Evolution, 64*(6), 1558–1568. https://doi.org/10.1111/j.1558-5646.2010.00944.x

Rolian, C., Lieberman, D. E., & Zermeno, J. P. (2011). Hand biomechanics during simulated stone tool use. *Journal of Human Evolution, 61*(1), 26–41. https://doi.org/10.1016/j.jhevol.2011.01.008

Rose, K. D., Chester, S. G., Dunn, R. H., Boyer, D. M., & Bloch, J. I. (2011). New fossils of the oldest North American euprimate *Teilhardina brandti* (Omomyidae) from the paleocene-eocene thermal maximum. *American Journal of Physical Anthropology, 146*(2), 281–305. https://doi.org/10.1002/ajpa.21579

Rosenberg, K., & Trevathan, W. (2002). Birth, obstetrics and human evolution. *BJOG-an International Journal of Obstetrics and Gynaecology, 109*(11), 1199–1206. https://doi.org/10.1046/j.1471-0528.2002.00010.x

Rosenberger, A. L. (2010). The skull of *Tarsius*: Functional morphology, eyeballs, and the non-pursuit predatory lifestyle. *International Journal of Primatology, 31*(6), 1032–1054. https://doi.org/10.1007/s10764-010-9447-x

Rosenberg-Yefet, T., Shemer, M., & Barkai, R. (2021). Acheulian shortcuts: Cumulative culture and the use of handaxes as cores for the production of predetermined blanks. *Journal of Archaeological Science: Reports, 36.* https://doi.org/10.1016/j.jasrep.2021.102822

Ross, C. F., Washington, R. L., Eckhardt, A., Reed, D. A., Vogel, E. R., Dominy, N. J., & Machanda, Z. P. (2009). Ecological consequences of scaling of chew cycle duration and daily feeding time in Primates. *Journal of Human Evolution, 56*(6), 570–585.

Rote, N. S., Chakrabarti, S., & Stetzer, B. P. (2004). The role of human endogenous retroviruses in trophoblast differentiation and placental development. *Placenta, 25*(8-9), 673–683. https://doi.org/10.1016/j.placenta.2004.02.008

Roth, G., & Dicke, U. (2019). Origin and evolution of human cognition. In M. A. Hofman (Ed.), *Progress in brain research* (2019/11/11 ed., Vol. 250, pp. 285–316). Elsevier B.V. https://doi.org/10.1016/bs.pbr.2019.02.004

Rotival, M., Cossart, P., & Quintana-Murci, L. (2021). Reconstructing 50,000 years of human history from our DNA: Lessons from modern genomics. *Comptes Rendus Biologies, 344*(2), 177–187. https://doi.org/10.5802/crbiol.55

Rouault, M., & Koechlin, E. (2018). Prefrontal function and cognitive control: From action to language. *Current Opinion in Behavioral Sciences, 21,* 106–111. https://doi.org/10.1016/j.cobeha.2018.03.008

Roura, E. C. (1999). Comment to 'The raw and the stolen. Cooking and the ecology of human origins'. *Current Anthropology, 40*(5), 580–581. https://doi.org/10.1086/300083

Rowlett, R. M. (1999). Comment to 'The raw and the stolen. Cooking and the ecology of human origins'. *Current Anthropology, 40*(5), 584–585. https://doi.org/10.1086/300083

Rowley-Conwy, P. (2024). Hunter-gatherers and earliest farmers in western Europe. *Proceedings of the National Academy of Sciences of the United States of America, 121*(10), e2322683121. https://doi.org/10.1073/pnas.2322683121

Rucklin, M., King, B., Cunningham, J. A., Johanson, Z., Marone, F., & Donoghue, P. C. J. (2021). Acanthodian dental development and the origin of gnathostome dentitions. *Nature ecology & evolution, 5*(7), 919–926. https://doi.org/10.1038/s41559-021-01458-4

Ruff, C. B. (1991). Climate and body shape in hominid evolution. *Journal of Human Evolution, 21*(2), 81–105. https://doi.org/10.1016/0047-2484(91)90001-C

Ruff, C. B. (1994). Morphological adaptation to climate in modern and fossil hominids. *Yearbook of Physical Anthropology, 37*, 65–107.

Ruff, C. B. (1995). Biomechanics of the hip and birth in early *Homo. American Journal of Biological Anthropology, 98*(4), 527–574. https://doi.org/10.1002/ajpa.1330980412

Ruff, C. B., Puymerail, L., Macchiarelli, R., Sipla, J., & Ciochon, R. L. (2015). Structure and composition of the Trinil femora: Functional and taxonomic implications. *Journal of Human Evolution, 80*, 147–158. https://doi.org/10.1016/j.jhevol.2014.12.004

Ruff, C. B., Trinkaus, E., & Holliday, T. W. (1997). Body mass and encephalization in Pleistocene *Homo. Nature, 387*(6629), 173–176. https://doi.org/10.1038/387173a0

Russo, G. A., & Kirk, E. C. (2017). Another look at the foramen magnum in bipedal mammals. *Journal of Human Evolution, 105*, 24–40. https://doi.org/10.1016/j.jhevol.2017.01.018

Rutherford, A. (2019). *Het boek over de mensheid* (F. van Delft & H. Moerdijk, Trans.). Luitingh-Sijthoff.

Sabater Pi, J., Vea, J. J., & Serrallonga, J. (1997). Did the first hominids build nests? *Current Anthropology, 38*(5), 914–916. https://doi.org/10.1086/204682

Sabeti, P. C., Varilly, P., Fry, B., Lohmueller, J., Hostetter, E., Cotsapas, C., Xie, X. H., Byrne, E. H., McCarroll, S. A., Gaudet, R., Schaffner, S. F., Lander, E. S., & Consortium, I. H. (2007). Genome-wide detection and characterization of positive selection in human populations. *Nature, 449*(7164), 913-U912. https://doi.org/10.1038/nature06250

Saibene, F., & Minetti, A. E. (2003). Biomechanical and physiological aspects of legged locomotion in humans. *Eur J Appl Physiol, 88*(4-5), 297–316. https://doi.org/10.1007/s00421-002-0654-9

Saladin, K. S. (2004). *Anatomy and physiology: The unity of form and function*. McGraw-Hill.

Sanada, K., Nakajima, S., Kurokawa, S., Barcelo-Soler, A., Ikuse, D., Hirata, A., Yoshizawa, A., Tomizawa, Y., Salas-Valero, M., Noda, Y., Mimura, M., Iwanami, A., & Kishimoto, T. (2020). Gut microbiota and major depressive disorder: A systematic review and meta-analysis. *Journal of Affective Disorders, 266*, 1–13. https://doi.org/10.1016/j.jad.2020.01.102

Sankararaman, S., Mallick, S., Dannemann, M., Prufer, K., Kelso, J., Paabo, S., Patterson, N., & Reich, D. (2014). The genomic landscape of Neanderthal ancestry in present-day humans. *Nature, 507*(7492), 354–357. https://doi.org/10.1038/nature12961

Sano, K., Beyene, Y., Katoh, S., Koyabu, D., Endo, H., Sasaki, T., Asfaw, B., & Suwa, G. (2020). A 1.4-million-year-old bone handaxe from Konso, Ethiopia, shows advanced tool technology in the early Acheulean. *Proceedings of the National Academy of Sciences of the United States of America, 117*(31), 18393–18400. https://doi.org/10.1073/pnas.2006370117

Sargis, E. J., Boyer, D. M., Bloch, J. I., & Silcox, M. T. (2007). Evolution of pedal grasping in Primates. *Journal of Human Evolution, 53*(1), 103–107.

Sarringhaus, L., Mitani, J. C., & MacLatchy, L. M. (2022). The ontogeny of knuckle-walking and dorsal metacarpal ridge prominence in chimpanzees. *American Journal of Biological Anthropology*. https://doi.org/10.1002/ajpa.24477

Sawyer, G. J., & Maley, B. (2005). Neanderthal reconstructed. *Anatomical Record. Part B, New Anatomist, 283*(1), 23–31. https://doi.org/10.1002/ar.b.20057 ((2006))

Scerri, E. M. L., Thomas, M. G., Manica, A., Gunz, P., Stock, J. T., Stringer, C., Grove, M., Groucutt, H. S., Timmermann, A., Rightmire, G. P., d'Errico, F., Tryon, C. A., Drake, N. A., Brooks, A. S., Dennell, R. W., Durbin, R., Henn, B. M., Lee-Thorp, J., deMenocal, P., Petraglia, M. D., Thompson, J. C., Scally, A., & Chikhi, L. (2018). Did our species evolve in subdivided populations across Africa, and why does it matter? *Trends Ecol Evol, 33*(8), 582–594. https://doi.org/10.1016/j.tree.2018.05.005

Schenker, N. M., Buxhoeveden, D. P., Blackmon, W. L., Amunts, K., Zilles, K., & Semendeferi, K. (2008). A comparative quantitative analysis of cytoarchitecture and minicolumnar organization in Broca's area in humans and great apes. *Journal of Comparative Neurology, 510*(1), 117–128. https://doi.org/10.1002/cne.21792

Schmidt, P., Blessing, M., Rageot, M., Iovita, R., Pfleging, J., Nickel, K. G., Righetti, L., & Tennie, C. (2019). Birch tar production does not prove Neanderthal behavioral complexity. *Proceedings of the National Academy of Sciences, 116*(36), 17707–17711. https://doi.org/10.1073/pnas.1911137116

Schmidt, P., Charrie-Duhaut, A., February, E., & Wadley, L. (2024). Adhesive technology based on biomass tar documents engineering capabilities in the African Middle Stone Age. *Journal of Human Evolution, 194*, 103578. https://doi.org/10.1016/j.jhevol.2024.103578

Schmitt, D. (2003). Insights into the evolution of human bipedalism from experimental studies of humans and other primates. *Journal of Experimental Biology, 206*(9), 1437–1448. https://doi.org/10.1242/jeb.00279

Schmitt, D., & Lemelin, P. (2002). Origins of primate locomotion: Gait mechanics of the Woolly Opossum. *American Journal of Physical Anthropology, 118*(3), 231–238.

Schoetensack, O. (1908). *Der Unterkiefer des Homo heidelbergensis, aus den Sanden von Mauer bei Heidelberg*. Verlag von Wilhelm Engelmann. https://gdz.sub.uni-goettingen.de/id/PPN361493010?tify={%22pages%22:[6],%22view%22:%22export%22}

Scholz, M., Bachmann, L., Nicholson, G. J., Bachmann, J., Giddings, I., Rüschoff-Thale, B., Czarnetzki, A., & Pusch, C. M. (2000). Genomic differentiation of Neanderthals and anatomically modern man allows a fossil-DNA-based classification of morphologically indistinguishable hominid bones. *The American Journal of Human Genetics, 66*, 1927–1932.

Scholz, M. N., D'Aout, K., Bobbert, M. F., & Aerts, P. (2006). Vertical jumping performance of bonobo (*Pan Paniscus*) suggests superior muscle properties. *Proceedings of the Royal Society B-Biological Sciences, 273*(1598), 2177–2184. https://doi.org/10.1098/rspb.2006.3568

Schrago, C. G. (2007). On the time scale of New World primate diversification. *American Journal of Physical Anthropology, 132*(3), 344–354. https://doi.org/10.1002/ajpa.20459

Schultz, A. H., & Straus, W. L., Jr. (1945). The numbers of vertebrae in primates. *Proceedings of the American Philosophical Society, 89*(4), 601–626.

Schultz, J. A. (2020). Eat and listen—how chewing and hearing evolved mammalian middle ear bones separated from the jaw of vertebrate ancestors. *Science, 367*(6475), 244–246.

Schwartz, G. G., & Rosenblum, L. A. (1981). Allometry of primate hair density and the evolution of human hairlessness. *American Journal of Physical Anthropology, 55*(1), 9–12. https://doi.org/10.1002/ajpa.1330550103

Scott, J. E. (2018). Reevaluating cases of trait-dependent diversification in primates. *American Journal of Physical Anthropology, 167*(2), 244–256. https://doi.org/10.1002/ajpa.23621

Segers, V., Aerts, P., Lenoir, M., & De Clercq, D. (2007). Dynamics of the body centre of mass during actual acceleration across transition speed. *The Journal of Experimental Biology, 210*(Pt 4), 578–585. https://doi.org/10.1242/jeb.02693

Seghier, M. L. (2013). The angular gyrus: Multiple functions and multiple subdivisions. *The Neuroscientist, 19*(1), 43–61. https://doi.org/10.1177/1073858412440596

Segurel, L., & Bon, C. (2017). On the evolution of Lactase Persistence in humans. *Annu Rev Genomics Hum Genet, 18*, 297–319. https://doi.org/10.1146/annurev-genom-091416-035340

Sehner, S., Fichtel, C., & Kappeler, P. M. (2018). Primate tails: Ancestral state reconstruction and determinants of interspecific variation in primate tail length. *American Journal of Physical Anthropology, 167*(4), 750–759. https://doi.org/10.1002/ajpa.23703

Sekhavati, Y., & Strait, D. (2024). Estimating ancestral ranges and biogeographical processes in early hominins. *Journal of Human Evolution, 191*, 103547. https://doi.org/10.1016/j.jhevol.2024.103547

Sellers, W. I., Cain, G. M., Wang, W. J., & Crompton, R. H. (2005). Stride lengths, speed and energy costs in walking of *shebel*: Using evolutionary robotics to predict locomotion of early human ancestors. *Journal of the Royal Society Interface, 2*(5), 431–441. https://doi.org/10.1098/rsif.2005.0060

Semaw, S., Renne, P., Harris, J. W., Feibel, C. S., Bernor, R. L., Fesseha, N., & Mowbray, K. (1997). 2.5-million-year-old stone tools from Gona, Ethiopia. *Nature, 385*(6614), 333–336. https://doi.org/10.1038/385333a0

Semaw, S., Rogers, M. J., Simpson, S. W., Levin, N. E., Quade, J., Dunbar, N., McIntosh, W. C., Caceres, I., Stinchcomb, G. E., Holloway, R. L., Brown, F. H., Butler, R. F., Stout, D., & Everett, M. (2020). Co-occurrence of Acheulian and Oldowan artifacts with Homo erectus cranial fossils from Gona, Afar, Ethiopia. *Science Advances, 6*(10), eaaw4694. https://doi.org/10.1126/sciadv.aaw4694

Semendeferi, K., Armstrong, E., Schleicher, A., Zilles, K., & Van Hoesen, G. W. (2001). Prefrontal cortex in humans and apes: A comparative study of area 10. *American Journal of Physical Anthropology, 114*(3), 224–241. https://doi.org/10.1002/1096-8644(200103)114:3<224::Aid-Ajpa1022>3.0.Co;2-I

Semendeferi, K., Lu, A., Schenker, N., & Damasio, H. (2002). Humans and great apes share a large frontal cortex. *Nature Neuroscience, 5*(3), 272–276. https://doi.org/10.1038/nn814

Semendeferi, K., Teffer, K., Buxhoeveden, D. P., Park, M. S., Bludau, S., Amunts, K., Travis, K., & Buckwalter, J. (2011). Spatial organization of neurons in the frontal Pole sets humans apart from great apes. *Cerebral Cortex, 21*(7), 1485–1497. https://doi.org/10.1093/cercor/bhq191

Senut, B. (2020). *Orrorin tugenensis* and the origins of man: A synthesis. *Bulletin De L Academie Nationale De Medecine, 204*(3), 258–267. https://doi.org/10.1016/j.banm.2019.12.018

Senut, B., Pickford, M., Gommery, D., Mein, P., Cheboi, K., & Coppens, Y. (2001, 2001). First hominid from the Miocene (Lukeino Formation, Kenya). *C.R. Academie des Sciences, Paris, Sciences de la Terre et des planètes, 332*, 137–144.

Setchell, J. M., & Dixson, A. F. (2001). Changes in the secondary sexual adornments of male mandrills (*Mandrillus sphinx*) are associated with gain and loss of alpha status. *Hormones and Behavior, 39*(3), 177–184. https://doi.org/10.1006/hbeh.2000.1628

Seymour, R. S., Bosiocic, V., & Snelling, E. P. (2016). Fossil skulls reveal that blood flow rate to the brain increased faster than brain volume during human evolution. *Royal Society Open Science, 3*(8), 160305. https://doi.org/10.1098/rsos.160305

Shaw, C. N., & Stock, J. T. (2013). Extreme mobility in the late pleistocene? Comparing limb biomechanics among fossil *Homo*, varsity athletes and Holocene foragers. *Journal of Human Evolution, 64*(4), 242–249. https://doi.org/10.1016/j.jhevol.2013.01.004

Shea, J. J. (2010). Complex projectile technology and *Homo sapiens* dispersal into Western Eurasia. *PaleoAnthropology, 100-122*. https://doi.org/10.4207/pa.2010.Art36

Shefelbine, S. J., Tardieu, C., & Carter, D. R. (2002). Development of the femoral bicondylar angle in hominid bipedalism. *Bone, 30*(5), 765–770. https://doi.org/10.1016/S8756-3282(02)00700-7

Shen, G., Gao, X., Gao, B., & Granger, D. E. (2009). Age of Zhoukoudian *Homo erectus* determined with (26)Al/(10)Be burial dating. *Nature, 458*(7235), 198–200. https://doi.org/10.1038/nature07741

Sherwood, L., Klandorf, H., & Yancey, P. H. (2005). *Animal physiology. From genes to organisms*. Thomson Brooks/Cole.

Shipton, C. (2022). Predetermined refinement: The earliest levallois of the Kapthurin Formation. *Journal of Paleolithic Archaeology, 5*(1). https://doi.org/10.1007/s41982-021-00109-1

Shipton, C., Blinkhorn, J., Breeze, P. S., Cuthbertson, P., Drake, N., Groucutt, H. S., Jennings, R. P., Parton, A., Scerri, E. M. L., Alsharekh, A., & Petraglia, M. D. (2018). Acheulean technology and landscape use at Dawadmi, central Arabia. *PloS one, 13*(7), e0200497. https://doi.org/10.1371/journal.pone.0200497

Shubin, N. H. (2008). *Your inner fish*. Pantheon Books.

Shubin, N. H., Daeschler, E. B., & Jenkins, F. A. (2006). The pectoral fin of *Tiktaalik roseae* and the origin of the tetrapod limb. *Nature, 440*(7085), 764–771.

Shunkov, M. V., Fedorchenko, A. Y., Kozlikin, M. B., & Derevianko, A. P. (2020. Initial upper palaeolithic ornaments and formal bone tools from the East chamber of Denisova Cave in the Russian Altai. *Quaternary International, 559*, 47–67. https://doi.org/10.1016/j.quaint.2020.07.027

Silcox, M. T., Bloch, J. I., Boyer, D. M., Chester, S. G. B., & Lopez-Torres, S. (2017). The evolutionary radiation of plesiadapiforms. *Evolutionary Anthropology, 26*(2), 74–94. https://doi.org/10.1002/evan.21526

Silvestro, D., Bacon, C. D., Ding, W., Zhang, Q., Donoghue, P. C. J., Antonelli, A., & Xing, Y. (2021). Fossil data support a pre-Cretaceous origin of flowering plants. *Nature ecology & evolution, 5*(4), 449–457. https://doi.org/10.1038/s41559-020-01387-8

Simmen, B., & Hladik, C. M. (1998). Sweet and Bitter Taste Discrimination in Primates: Scaling Effects across Species. *Folia Primatologica, 69*, 129–138.

Simmen, B., Pasquet, P., Masi, S., Koppert, G. J. A., Wells, J. C. K., & Hladik, C. M. (2017). Primate energy input and the evolutionary transition to energy-dense diets in humans. *Proceedings of the Royal Society B-Biological Sciences, 284*(1856). https://doi.org/10.1098/rspb.2017.0577

Simmons, R. E., & Scheepers, L. (1996). Winning by a neck: Sexual selection in the evolution of giraffe. *The American Naturalist, 148*(5), 771–786.

Simoes, L. G., Peyroteo-Stjerna, R., Marchand, G., Bernhardsson, C., Vialet, A., Chetty, D., Alacamli, E., Edlund, H., Bouquin, D., Dina, C., Garmond, N., Gunther, T., & Jakobsson, M. (2024). Genomic ancestry and social dynamics of the last hunter-gatherers of atlantic France. *Proceedings of the National Academy of Sciences of the United States of America, 121*(10), e2310545121. https://doi.org/10.1073/pnas.2310545121

Simpson, S. W., Latimer, B., & Lovejoy, C. O. (2018). Why do knuckle-walking African apes Knuckle-Walk? *Anatomical Record (Hoboken), 301*(3), 496–514. https://doi.org/10.1002/ar.23743

Simpson, S. W., Quade, J., Levin, N. E., Butler, R., Dupont-Nivet, G., Everett, M. A., & Semaw, S. (2009). A female *Homo erectus* pelvis from Gona, Ethiopia. *American Journal of Physical Anthropology*, 241–241. https://www.webofscience.com/wos/woscc/full-record/WOS:000263442701299

Skov, L., Coll Macià, M., Sveinbjörnsson, G., Mafessoni, F., Lucotte, E. A., Einarsdóttir, M. S., Jonsson, H., Halldorsson, B., Gudbjartsson, D. F., Helgason, A., Schierup, M. H., & Stefansson, K. (2020). The nature of Neanderthal introgression revealed by 27,566 Icelandic genomes. *Nature*. https://doi.org/10.1038/s41586-020-2225-9

Slimak, L., Zanolli, C., Higham, T., Frouin, M., Schwenninger, J. L., Arnold, L. J., Demuro, M., Douka, K., Mercier, N., Guerin, G., Valladas, H., Yvorra, P., Giraud, Y., Seguin-Orlando, A., Orlando, L., Lewis, J. E., Muth, X., Camus, H., Vandevelde, S., Buckley, M., Mallol, C., Stringer, C., & Metz, L. (2022). Modern human incursion into Neanderthal territories 54,000 years ago at Mandrin, France. *Science Advances, 8*(6), 1–16. https://doi.org/10.1126/sciadv.abj9496

Slon, V., Mafessoni, F., Vernot, B., de Filippo, C., Grote, S., Viola, B., Hajdinjak, M., Peyrégne, S., Nagel, S., Brown, S., Douka, K., Higham, T., Kozlikin, M. B., Shunkov, M. V., Derevianko, A. P., Kelso, J., Meyer, M., Prüfer, K., & Pääbo, S. (2018). The genome of the offspring of a Neanderthal mother and a Denisovan father. *Nature*. https://doi.org/10.1038/s41586-018-0455-x

Smaers, J. B., Dechmann, D. K. N., Goswami, A., Soligo, C., & Safi, K. (2012). Comparative analyses of evolutionary rates reveal different pathways to encephalization in bats, carnivorans, and primates. *Proceedings of the National Academy of Sciences of the United States of America, 109*(44), 18006–18011. https://doi.org/10.1073/pnas.1212181109

Smil, V. (2002). Eating meat: Evolution, patterns, and consequences. *Population and Development Review, 28*(4), 599–639.

Smith, B. H. (1999). Comment to 'The raw and the stolen. Cooking and the ecology of human origins. *Current Anthropology, 40*(5), 585–586. https://doi.org/10.1086/300083

Smith, F. A. (2022). The road to a larger brain. *Science, 376*(6588), 27–28.

Smith, M. M., & Coates, M. I. (1998). Evolutionary origins of the vertebrate dentition: Phylogenetic patterns and developmental evolution. *European Journal of Oral Sciences, 106*(suppl 1), 482–500.

Smith, T., Rose, K. D., & Gingerich, P. D. (2006). Rapid Asia-Europe-North America geographic dispersal of earliest eocene primate teilhardina during the paleocene-eocene thermal maximum. *Proceedings of the National Academy of Sciences of the United States of America, 103*(30), 11223–11227. https://doi.org/10.1073/pnas.0511296103

Snir, A., Nadel, D., Groman-Yaroslavski, I., Melamed, Y., Sternberg, M., Bar-Yosef, O., & Weiss, E. (2015). The origin of cultivation and proto-weeds, long Before neolithic farming. *PloS One, 10*(7), e0131422. https://doi.org/10.1371/journal.pone.0131422

Sockol, M. D., Raichlen, D. A., & Pontzer, H. (2007). Chimpanzee locomotor energetics and the origin of human bipedalism. *Proceedings of the National Academy of Sciences of the United States of America, 104*(30), 12265–12269. https://doi.org/10.1073/pnas.0703267104

Song, Y., Boncompagni, A. C., Kim, S. S., Gochnauer, H. R., Zhang, Y., Loots, G. G., Wu, D., Li, Y., Xu, M., & Millar, S. E. (2018). Regional control of hairless versus hair-bearing skin by Dkk2. *Cell Reports, 25*(11), 2981–2991 e2983. https://doi.org/10.1016/j.celrep.2018.11.017

Sorensen, M. V., & Leonard, W. R. (2001). Neandertal energetics and foraging efficiency. *Journal of Human Evolution, 40*(6), 483–495. https://doi.org/10.1006/jhev.2001.0472

Sorrentino, R., Stephens, N. B., Marchi, D., DeMars, L. J. D., Figus, C., Bortolini, E., Badino, F., Saers, J. P. P., Bettuzzi, M., Boschin, F., Capecchi, G., Feletti, F., Guarnieri, T., May, H., Morigi, M. P., Parr, W., Ricci, S., Ronchitelli, A., Stock, J. T., Carlson, K. J., Ryan, T. M., Belcastro, M. G., & Benazzi, S. (2021). Unique foot posture in Neanderthals reflects their body mass and high mechanical stress. *Journal of Human Evolution, 161*, 103093. https://doi.org/10.1016/j.jhevol.2021.103093

Spear, J. K. (2025). Phylogenetic comparative analysis of suspensory adaptations in primates. *Journal of Human Evolution, 198*, 103616. https://doi.org/10.1016/j.jhevol.2024.103616

Sperber, J. (1983). A probable disease model for hominid evolution and site location. *Journal of Human Evolution, 12*(8), 695–695. https://doi.org/10.1016/S0047-2484(83)80062-1 10.1016/S0047-2484(83)80062-1

Spiteri, E., Konopka, G., Coppola, G., Bomar, J., Oldham, M., Ou, J., Vernes, S. C., Fisher, S. E., Ren, B., & Geschwind, D. H. (2007). Identification of the transcriptional targets of *FOXP2*, a gene linked to speech and language, in developing human brain. *American Journal of Human Genetics, 81*(6), 1144–1157. https://doi.org/10.1086/522237

Spocter, M. A., Hopkins, W. D., Garrison, A. R., Bauernfeind, A. L., Stimpson, C. D., Hof, P. R., & Sherwood, C. C. (2010). Wernicke's area homologue in chimpanzees (Pan troglodytes) and its relation to the appearance of modern human language. *Proceedings of the Royal Society B-Biological Sciences, 277*(1691), 2165–2174. https://doi.org/10.1098/rspb.2010.0011

Sponheimer, M., Alemseged, Z., Cerling, T. E., Grine, F. E., Kimbel, W. H., Leakey, M. G., Lee-Thorp, J. A., Manthi, F. K., Reed, K. E., Wood, B. A., & Wynn, J. G. (2013). Isotopic evidence of early hominin diets. *Proceedings of the National Academy of Sciences of the United States of America, 110*(26), 10513–10518. https://doi.org/10.1073/pnas.1222579110

Sponheimer, M., Daegling, D. J., Ungar, P. S., Bobe, R., & Paine, O. C. C. (2022). Problems with *paranthropus*. *Quaternary International, 650*, 40–51. https://doi.org/10.1016/j.quaint.2022.03.024

Spoor, F., Gunz, P., Neubauer, S., Stelzer, S., Scott, N., Kwekason, A., & Dean, M. C. (2015). Reconstructed *Homo habilis* type OH 7 suggests deep-rooted species diversity in early *homo*. *Nature, 519*(7541), 83–86. https://doi.org/10.1038/nature14224

Spoor, F., Hublin, J.-J., Braun, M., & Zonneveld, F. (2003). The bony labyrinth of Neanderthals. *Journal of Human Evolution, 44*(2), 141–165. https://doi.org/10.1016/s0047-2484(02)00166-5

Stahl, A. B. (1984). Hominid dietary selection before fire. *Current Anthropology, 25*(2), 151–168. https://doi.org/10.1086/203106

Standen, E. M., Du, T. Y., & Larsson, H. C. (2014). Developmental plasticity and the origin of tetrapods. *Nature, 513*, 54–58. https://doi.org/10.1038/nature13708

Standring, S. (2020). *Gray's anatomy: The anatomical basis of clinical practice* (S. Standring, Ed. 42 ed.). Elsevier LTD.

Stanhewicz, A. E., & Kenney, W. L. (2017). Role of folic acid in nitric oxide bioavailability and vascular endothelial function. *Nutrition Reviews, 75*(1), 61–70. https://doi.org/10.1093/nutrit/nuw053

Stansfield, E., Fischer, B., Grunstra, N. D. S., Pouca, M. V., & Mitteroecker, P. (2021). The evolution of pelvic canal shape and rotational birth in humans. *BMC Biol, 19*(1), 224. https://doi.org/10.1186/s12915-021-01150-w

Stansfield, E., Kumar, K., Mitteroecker, P., & Grunstra, N. D. S. (2021). Biomechanical trade-offs in the pelvic floor constrain the evolution of the human birth canal. *Proceedings of the National Academy of Sciences, 118*(16), e2022159118. https://doi.org/10.1073/pnas.2022159118

Stedman, H. H., Kozyak, B. W., Nelson, A., Thesier, D. M., Su, L. T., Low, D. W., Bridges, C. R., Shrager, J. B., Minugh-Purvis, N., & Mitchell, M. A. (2004). Myosin gene mutation correlates with anatomical changes in the human lineage. *Nature, 428*(6981), 415–418. https://doi.org/10.1038/nature02358

Stephan, H., Frahm, H., & Baron, G. (1981). New and revised data on volumes of brain structures in insectivores and primates. *Folia Primatologica, 35*(1), 1–29. https://doi.org/10.1159/000155963

Stern, J. T., & Susman, R. L. (1983). The locomotor anatomy of *Australopithecus afarensis*. *American Journal of Physical Anthropology, 60*(3), 279–317. https://doi.org/10.1002/ajpa.1330600302

Steudel-Numbers, K. L. (2006). Energetics in *Homo erectus* and other early hominins: The consequences of increased lower-limb length. *Journal of Human Evolution, 51*(5), 445–453. https://doi.org/10.1016/j.jhevol.2006.05.001

Steudel-Numbers, K. L., & Tilkens, M. J. (2004). The effect of lower limb length on the energetic cost of locomotion: Implications for fossil hominins. *Journal of Human Evolution, 47*(1-2), 95–109. https://doi.org/10.1016/j.jhevol.2004.06.002

Stewart, T. A., Lemberg, J. B., Taft, N. K., Yoo, I., Daeschler, E. B., & Shubin, N. H. (2019). Fin ray patterns at the fin-to-limb transition. *Proceedings of the National Academy of Sciences of the United States of America, 117*(3) 1612–1620. https://doi.org/10.1073/pnas.1915983117

Stiassny, M. L. J. (2003). Atavism. In B. K. Hall, & W. Olsen (Eds.), *Keywords and concepts in evolutionary developmental biology* (pp. 476). Harvard University Press.

Stiles, J., & Jernigan, T. L. (2010). The basics of brain development. *Neuropsychology Review, 20*(4), 327–348. https://doi.org/10.1007/s11065-010-9148-4

Stock, J. T., Pomeroy, E., Ruff, C. B., Brown, M., Gasperetti, M. A., Li, F. J., Maher, L., Malone, C., Mushrif-Tripathy, V., Parkinson, E., Rivera, M., Siew, Y. Y., Stefanovic, S., Stoddart, S., Zarina, G., & Wells, J. C. K. (2023). Long-term trends in human body size track regional variation in subsistence transitions and growth acceleration linked to dairying. *Proceedings of the National Academy of Sciences of the United States of America, 120*(4), e2209482119. https://doi.org/10.1073/pnas.2209482119

Stokstad, E. (2017). New ape found, sparking fears for its survival. *Science, 358*(6363), 572–573.

Stout, D., & Chaminade, T. (2012). Stone tools, language and the brain in human evolution. *Philosophical Transactions of the Royal Society B-Biological Sciences, 367*(1585), 75–87. https://doi.org/10.1098/rstb.2011.0099

Straus, C., Vasilakos, K., Wilson, R. J., Oshima, T., Zelter, M., Derenne, J. P., Similowski, T., & Whitelaw, W. A. (2003). A phylogenetic hypothesis for the origin of hiccough. *Bioessays, 25*(2), 182–188. https://doi.org/10.1002/bies.10224

Stringer, C. (2003). Human evolution: Out of Ethiopia. *Nature, 423*(6941), 692–693, 695. https://doi.org/10.1038/423692a

Stringer, C. (2011). The changing landscapes of the earliest human occupation of Britain and Europe. In N. Ashton, S. G. Lewis, & C. Stringer *(Eds.), The Ancient Human Occupation of Britain* (pp. 1–10). https://doi.org/10.1016/b978-0-444-53597-9.00001-7

Stringer, C. (2021, November, 2-4th 2021). *The Middle Pleistocene cranium from Harbin establishes a new human lineage in China.* Human Evolution - From Fossils to Ancient and Modern Genomes, Virtual Conference; Welcome Connecting Science. https://coursesandconferences. wellcomeconnectingscience.org/event/human-evolution-from-fossils-to-ancient-and-modern-genomes-20250428/

Stringer, C. B. (1996). Current issues in modern human origins. In W. E. Meikle, F. C. Howell, & N. G. Jablonski (Eds.), *Contemporary issues in human evolution* (Vol. 21, pp. 115–134). California Academy of Sciences Memoir.

Struska, M., Jaresova, P. A., Hora, M., Wall-Scheffler, C., Michalek, T., & Sladek, V. (2024). Impact of relative lower-limb length on heat loss and body temperature during running. *American Journal of Biological Anthropology, 185*(4), e25036. https://doi.org/10.1002/ajpa.25036

Stuart-Fox, D., Newton, E., & Clusella-Trullas, S. (2017). Thermal consequences of colour and near-infrared reflectance. *Philosophical Transactions of the Royal Society of London. Series B, Biological Sciences, 372*(1724). https://doi.org/10.1098/rstb.2016.0345

Sturm, R. A., Box, N. F., & Ramsay, M. (1998). Human pigmentation genetics: The difference is only skin deep. *Bioessays, 20*(9), 712–721. https://doi.org/10.1002/(Sici)1521-1878(199809)20:9<712::Aid-Bies4>3.0.Co;2-I

Suddendorf, T., Kirkland, K., Bulley, A., Redshaw, J., & Langley, M. C. (2020). It's in the bag: Mobile containers in human evolution and child development. *Evolutionary Human Sciences, 2*, e48. https://doi.org/10.1017/ehs.2020.47

Sümer, A. P., Rougier, H., Villalba-Mouco, V., Huang, Y., Iasi, L. N. M., Essel, E., Mesa, A. B., Furtwaengler, A., Peyregne, S., de Filippo, C., Rohrlach, A. B., Pierini, F., Mafessoni, F., Fewlass, H., Zavala, E. I., Mylopotamitaki, D., Bianco, R. A., Schmidt, A., Zorn, J.,. . . & Krause, J. (2024). Earliest modern human genomes constrain timing of Neanderthal admixture. *Nature*. https://doi.org/10.1038/s41586-024-08420-x

Sun, G., Dilcher, D. L., Zheng, S., & Zhou, Z. (1998). In search of the first flower: A Jurrasic angiosperm *Archaefructus*, from Northeast China. *Science, 282*, 1692–1695.

Sun, J., Ma, P. C., Cheng, H. Z., Wang, C. Z., Li, Y. L., Cui, Y. Q., Yao, H. B., Wen, S. Q., & Wei, L. H. (2021). Post-last glacial maximum expansion of Y-chromosome haplogroup C2a-L1373 in northern Asia and its implications for the origin of Native Americans. *American Journal of Physical Anthropology, 174*(2), 363–374. https://doi.org/10.1002/ajpa.24173

Suwa, G. (1984). Could *Australopithecus afarensis* have made the Laetoli footprints. *American Journal of Physical Anthropology, 63*(2), 224–225.

Suwa, G., Asfaw, B., Kono, R. T., Kubo, D., Lovejoy, C. O., & White, T. D. (2009). The *Ardipithecus ramidus* skull and its implications for hominid origins. *Science, 326*(5949), 68e1–68e7. https://doi.org/10.1126/science.1175825

Suwa, G., Kono, R. T., Katoh, S., Asfaw, B., & Beyene, Y. (2007). A new species of great ape from the late Miocene epoch in Ethiopia. *Nature, 448*(7156), 921–924. https://doi.org/10.1038/nature06113

Suwa, G., Kono, R. T., Simpson, S. W., Asfaw, B., Lovejoy, C. O., & White, T. D. (2009). Paleobiological implications of the *Ardipithecus ramidus* dentition. *Science (New York, N.Y.), 326*(5949), 94–99. https://doi.org/10.1126/science.1175824

Suwa, G., Sasaki, T., Semaw, S., Rogers, M. J., Simpson, S. W., Kunimatsu, Y., Nakatsukasa, M., Kono, R. T., Zhang, Y., Beyene, Y., Asfaw, B., & White, T. D. (2021). Canine sexual dimorphism in *Ardipithecus ramidus* was nearly human-like. *Proceedings of the National Academy of Sciences of the United States of America, 118*(49), e2116630118. https://doi.org/10.1073/pnas.2116630118

Swindler, D. R. (2002). *Primate dentition. An introduction to the teeth of non-human primates.* Cambridge University Press. https://www.dropbox.com/home/Evomorph/Evomorph%20E-books/Human%20anatomy%20and%20evolution?preview=Primate+dentition.pdf

Sylvester, A. D., Lautzenheiser, S. G., & Kramer, P. A. (2021). A review of musculoskeletal modelling of human locomotion. *Interface Focus, 11*(5). https://doi.org/10.1098/rsfs.2020.0060

Szamado, S., & Szathmary, E. (2006). Selective scenarios for the emergence of natural language. *Trends Ecol Evol, 21*(10), 555–561. https://doi.org/10.1016/j.tree.2006.06.021

Tagore, D., & Akey, J. M. (2024). Archaic hominin admixture and its consequences for modern humans. *Current Opinion in Genetics & Development, 90*, 102280. https://doi.org/10.1016/j.gde.2024.102280

Tague, R. G. (2009). High assimilation of the sacrum in a sample of American skeletons: Prevalence, pelvic size, and obstetrical and evolutionary implications. *American Journal of Physical Anthropology, 138*(4), 429–438. https://doi.org/10.1002/ajpa.20958

Tague, R. G. (2011). Fusion of coccyx to sacrum in humans: Prevalence, correlates, and effect on pelvic size, with obstetrical and evolutionary implications. *American Journal of Physical Anthropology, 145*(3), 426–437. https://doi.org/10.1002/ajpa.21518

Tague, R. G. (2012). Small anatomical variant has profound implications for evolution of human birth and brain development. *Proceedings of the National Academy of Sciences of the United States of America, 109*(22), 8360–8361. https://doi.org/10.1073/pnas.1205763109

Tague, R. G., & Lovejoy, C. O. (1998). AL 288-1–Lucy or Lucifer: Gender confusion in the Pliocene. *Journal of Human Evolution, 35*(1), 75–94. https://doi.org/10.1006/jhev.1998.0223

Tanaka, I., Tokida, E., Takefushi, H., & Hagiware, T. (2007). Tube test in free-ranging Japanese macaques: Use of sticks and stones to obtain fruit from a transparent pipe. In T. Matsuzawa (Ed.), *Primate origins of human cognition and behavior* (pp. 509–518). Springer.

Tardieu, C. (1998). Short adolescence in early hominids: Infantile and adolescent growth of the human femur. *American Journal of Biological Anthropology, 107*(2), 163–178. https://doi.org/10.1002/(sici)1096-8644(199810)107:2<163::Aid-ajpa3>3.0.Co;2-w

Tardieu, C. (2010). Development of the human hind limb and its importance for the evolution of bipedalism. *Evolutionary Anthropology: Issues, News, and Reviews, 19*(5), 174–186. https://doi.org/10.1002/evan.20276

Tardieu, C., & Trinkaus, E. (1994). Early ontogeny of the human femoral bicondylar angle. *American Journal of Biological Anthropology, 95*(2), 183–195. https://doi.org/10.1002/ajpa.1330950206

Teixeira, J. C., Jacobs, G. S., Stringer, C., Tuke, J., Hudjashov, G., Purnomo, G. A., Sudoyo, H., Cox, M. P., Tobler, R., Turney, C. S. M., Cooper, A., & Helgen, K. M. (2021). Widespread Denisovan ancestry in Island Southeast Asia but no evidence of substantial super-archaic hominin admixture. *Nature Ecology & Evolution, 5*(5), 616–624. https://doi.org/10.1038/s41559-021-01408-0

Templeton, A. R. (2002). Out of Africa again and again. *Nature, 416*(6876), 45–51. https://doi.org/10.1038/416045a

Thibault, S., Py, R., Gervasi, A. M., Salemme, R., Koun, E., Lövden, M., Boulenger, V., Roy, A. C., & Brozzoli, C. (2021). Tool use and language share syntactic processes and neural patterns in the basal ganglia. *Science, 374*(6569). https://doi.org/10.1126/science.abe0874

Thomas, D. Q., Fernhall, B., & Granat, H. (1999). Changes in running economy during a 5-km run in trained men and women runners. *Journal of Strength and Conditioning Research, 13*(2), 162–167.

Thompson, N. E., & Almecija, S. (2017). The evolution of vertebral formulae in Hominoidea. *Journal of Human Evolution, 110*, 18–36. https://doi.org/10.1016/j.jhevol.2017.05.012

Thompson, N. E., Rubinstein, D., Parrella-O'Donnell, W., Brett, M. A., Demes, B., Larson, S. G., & O'Neill, M. C. (2021). The loss of the 'pelvic step' in human evolution. *The Journal of Experimental Biology, 224*(16), jeb240440. https://doi.org/10.1242/jeb.240440

Thompson, N. W., Mockford, B. J., & Cran, G. W. (2001). Absence of palmaris longus muscle: A population study. *The Ulster Medical Journal, 70*(1), 22–24.

Thorne, A. G., & Wolpoff, M. H. (1992). The multiregional evolution of humans. *Scientific American, 266*(4), 76–83. https://www.jstor.org/stable/pdf/24939019.pdf?refreqid=excelsior%3Acb7db58556e7a3c65fd6eb3f343ba020&ab_segments=&origin=&initiator=&acceptTC=1

Thorpe, S. K. S., Crompton, R. H., Gunther, M. M., Ker, R. F., & Alexander, R. M. (1999). Dimensions and moment arms of the hind- and forelimb muscles of common chimpanzees (*Pan troglodytes*). *American Journal of Physical Anthropology, 110*(2), 179–199. https://doi.org/10.1002/(Sici)1096-8644(199910)110:2<179::Aid-Ajpa5>3.0.Co;2-Z

Thorpe, S. K. S., Holder, R. L., & Crompton, R. H. (2007). Origin of human bipedalism as an adaptation for locomotion on flexible branches. *Science, 316*(5829), 1328–1331. https://doi.org/10.1126/science.1140799

Timmermann, A. (2020). Quantifying the potential causes of Neanderthal extinction: Abrupt climate change versus competition and interbreeding. *Quaternary Science Reviews, 238*, 106331. https://doi.org/10.1016/j.quascirev.2020.106331

Timmermann, A., Yun, K. S., Raia, P., Ruan, J., Mondanaro, A., Zeller, E., Zollikofer, C., Ponce de Leon, M., Lemmon, D., Willeit, M., & Ganopolski, A. (2022). Climate effects on archaic human habitats and species successions. *Nature, 604*, 495–501. https://doi.org/10.1038/s41586-022-04600-9

Tobler, R., Souilmi, Y., Huber, C. D., Bean, N., Turney, C. S. M., Grey, S. T., & Cooper, A. (2023). The role of genetic selection and climatic factors in the dispersal of anatomically modern humans out of Africa. *Proceedings of the National Academy of Sciences of the United States of America, 120*(22), e2213061120. https://doi.org/10.1073/pnas.2213061120

Tocheri, M. W., Orr, C. M., Jacofsky, M. C., & Marzke, M. W. (2008). The evolutionary history of the hominin hand since the last common ancestor of *Pan* and *Homo*. *Journal of Anatomy, 212*(4), 544–562. https://doi.org/10.1111/j.1469-7580.2008.00865.x

Torti, V. (2021). From whispers to howling cries: Sound production and perception in nonhuman primates. *In Neuroendocrine Regulation of Animal Vocalization* (pp. 133–147). https://doi.org/10.1016/b978-0-12-815160-0.00016-5

Toth, N., & Schick, K. (2019). Why did the Acheulean happen? Experimental studies into the manufacture and function of Acheulean artifacts. *Anthropologie, 123*(4-5), 724–768. https://doi.org/10.1016/j.anthro.2017.10.008

Toussaint, S., Llamosi, A., Morino, L., & Youlatos, D. (2020). The Central role of small vertical substrates for the origin of grasping in early primates. *Current Biology, 30*(9), 1600–1613.e3.

Towle, I., Irish, J. D., & Loch, C. (2021). *Paranthropus robustus* tooth chipping patterns do not support regular hard food mastication. *Journal of Human Evolution, 158*, 103044. https://doi.org/10.1016/j.jhevol.2021.103044

Towle, I., MacIntosh, A. J. J., Hirata, K., Kubo, M. O., & Loch, C. (2022). Atypical tooth wear found in fossil hominins also present in a Japanese macaque population. *American Journal of Biological Anthropology*. https://doi.org/10.1002/ajpa.24500

Trevathan, W. (2015). Primate pelvic anatomy and implications for birth. *Philosophical Transactions of the Royal Society of London. Series B, Biological Sciences, 370*(1663), 20140065. https://doi.org/10.1098/rstb.2014.0065

Trevathan, W., & Rosenberg, K. (2000). The shoulders follow the head: Postcranial constraints on human childbirth. *Journal of Human Evolution, 39*(6), 583–586. https://doi.org/10.1006/jhev.2000.0434

Trinkaus, E. (1985). Cannibalism and burial at Krapina. *Journal of Human Evolution, 14*(2), 203–216. https://doi.org/10.1016/S0047-2484(85)80007-5

Trinkaus, E., Moldovan, O., Milota, T., Bilgar, A., Sarcina, L., Athreya, S., Bailey, S. E., Rodrigo, R., Mircea, G., Higham, T., Ramsey, C. B., & van der Plicht, J. (2003). An early modern human from the Pestera cu Oase, Romania. *Proceedings of the National Academy of Sciences of the United States of America, 100*(20), 11231–11236. https://doi.org/10.1073/pnas.2035108100 ((2006))

Trinkaus, E., & Rhoads, M. L. (1999). Neandertal knees: Power lifters in the pleistocene? *Journal of Human Evolution, 37*(6), 833–859. https://doi.org/10.1006/jhev.1999.0317

Trulsson, M. (2006). Sensory-motor function of human periodontal mechanoreceptors. *Journal of oral Rehabilitation, 33*(4), 262–273. https://doi.org/10.1111/j.1365-2842.2006.01629.x

Tucker, V. A. (1975). Energetic cost of moving about. *American Scientist, 63*(4), 413–419.

Turk, M., Turk, I., Dimkaroski, L., Blackwell, B. A. B., Horusitzky, F. Z., Otte, M., Bastiani, G., & Korat, L. (2018). The Mousterian musical instrument from the Divje Babe I cave (Slovenia): Arguments on the material evidence for Neanderthal musical behaviour. *Anthropologie, 122*(4), 679–706. https://doi.org/10.1016/j.anthro.2018.10.001

Turk, M., Turk, I., & Otte, M. (2020). The Neanderthal musical instrument from Divje Babe I Cave (Slovenia): A critical review of the discussion. *Applied Sciences-Basel, 10*(4), 1226. https://doi.org/10.3390/app10041226

Uchida, T. K., & Delp, S. L. (2020). *Biomechanics of movements. The science of sports, robotics, and rehabilitation.* The MIT Press.

Urciuoli, A., Zanolli, C., Beaudet, A., Dumoncel, J., Santos, F., Moya-Sola, S., & Alba, D. M. (2020). The evolution of the vestibular apparatus in apes and humans. *eLife, 9*(e51261), 1–33. https://doi.org/10.7554/eLife.51261

Valamoti, S. M., Petridou, C., Berihuete-Azorin, M., Stika, H. P., Papadopoulou, L., & Mimi, I. (2021). Deciphering ancient 'recipes' from charred cereal fragments: An integrated methodological approach using experimental, ethnographic and archaeological evidence. *Journal of Archaeological Science, 128*, 1–18. https://doi.org/10.1016/j.jas.2021.105347

Valleau, J. C., & Sullivan, E. L. (2014). The impact of leptin on perinatal development and psychopathology. *Journal of Chemical Neuroanatomy, 61-62*, 221–232. https://doi.org/10.1016/j.jchemneu.2014.05.001

Vallender, E. J. (2019). Genetics of human brain evolution. In M. A. Hofman (Ed.), *Progress in brain research* (2019/11/11 ed., Vol. 250, pp. 3–39). Elsevier B.V. https://doi.org/10.1016/bs.pbr.2019.01.003

Vallini, L., Marciani, G., Aneli, S., Bortolini, E., Benazzi, S., Pievani, T., & Pagani, L. (2022). Genetics and material culture support repeated expansions into Paleolithic Eurasia from a population hub out of Africa. *Genome Biology and Evolution, 14*(4). https://doi.org/10.1093/gbe/evac045

van 't Hoog, A. (2005). Op zoek naar het verschil. *Bionieuws, 2*(15), 8–9. ((2006))

van Beesel, J., Hutchinson, J. R., Hublin, J. J., & Melillo, S. (2022). Comparison of the arm-lowering performance between *Gorilla* and *Homo* through musculoskeletal modeling. *American Journal of Biological Anthropology*. https://doi.org/10.1002/ajpa.24511

van Casteren, A., Codd, J. R., Kupczik, K., Plasqui, G., Sellers, W. I., & Henry, A. G. (2022). The cost of chewing: The energetics and evolutionary significance of mastication in humans. *Science Advances, 8*(eabn8351), 1–8.

Van Daele, P. A. A. G., Faulkes, E., Verheyen, E., & Adriaens, D. (2007). African mole-rats (Bathyergidae): A complex radiation in tropical soils. In S. Begall, H. Burda, & C. E. Schleich (Eds.), *Subterranean rodents: News from underground* (pp. 357–373). Springer-Verlag.

van Heteren, A. H. (2013). Is LB1 diseased or modern? A review of the proposed pathologies. *Gene, 528*(1), 12–20. https://doi.org/10.1016/j.gene.2013.06.010

van Praag, H., Kempermann, G., & Gage, F. H. (1999). Running increases cell proliferation and neurogenesis in the adult mouse dentate gyrus. *Nature Neuroscience, 2*(3), 266–270. https://doi.org/10.1038/6368

Van Schaik, C. (2006). Why are some animals so smart? *Scientific American, 294*(4), 48–55.

VanSickle, C., Cofran, Z., & Hunt, D. (2020). Did Neanderthals have large brains? Factors affecting endocranial volume comparisons. *American Journal of Physical Anthropology, 173*(4), 1–8. https://doi.org/10.1002/ajpa.24124

van Straalen, N. M., & Roelofs, D. (2017). *Evolueren wij nog? Alles wat je wilt weten over ontwikkeling en evolutie van ons lichaam.* Amsterdam University Press B.V.

Vereecke, E., D'Aout, K., De Clercq, D., Van Elsacker, L., & Aerts, P. (2003). Dynamic plantar pressure distribution during terrestrial locomotion of bonobos (*Pan Paniscus*). *American Journal of Physical Anthropology, 120*(4), 373–383.

Vereecke, E. E., D'Aout, K., & Aerts, P. (2006). Locomotor versatility in the white-handed gibbon (Hylobates lar): A spatiotemporal analysis of the bipedal, tripedal, and quadrupedal gaits. *Journal of Human Evolution, 50*(5), 552–567. https://doi.org/10.1016/j.jhevol.2005.12.011

Vereecke, E. E., & Wunderlich, R. E. (2016). Experimental research on hand use in primates. In L. Barrett (Ed.), *The evolution of the primate hand* (pp. 259–284). Sprinter.

Verendeev, A., Thomas, C., McFarlin, S. C., Hopkins, W. D., Phillips, K. A., & Sherwood, C. C. (2015). Comparative analysis of Meissner's corpuscles in the fingertips of primates. *Journal of Anatomy, 227*(1), 72–80.

Vermassen, G. (2020, 02/03/20). *De evolutionaire oorsprong van het patriarchaat.* Humanistisch Verbond. Retrieved 22/06/21 from https://humanistischverbond.be/blog/227/de-evolutionaire-oorsprong-van-het-patriarchaat/

Vernes, S. C., Spiteri, E., Nicod, J., Groszer, M., Taylor, J. M., Davies, K. E., Geschwind, D. H., & Fisher, S. E. (2007). High-throughput analysis of promoter occupancy reveals direct neural targets of *FOXP2*, a gene mutated in speech and language disorders. *American Journal of Human Genetics, 81*(6), 1232–1250. https://doi.org/10.1086/522238

Vernot, B., & Akey, J. M. (2014). Resurrecting surviving neandertal lineages from modern human genomes. *Science, 343*(6174), 1017–1021. https://doi.org/10.1126/science.1245938

Vernot, B., Tucci, S., Kelso, J., Schraiber, J. G., Wolf, A. B., Gittelman, R. M., Dannemann, M., Grote, S., McCoy, R. C., Norton, H., Scheinfeldt, L. B., Merriwether, D. A., Koki, G., Friedlaender, J. S., Wakefield, J., Paabo, S., & Akey, J. M. (2016). Excavating Neandertal and Denisovan DNA from the genomes of Melanesian individuals. *Science (New York, N.Y.), 352*(6282), 235–239. https://doi.org/10.1126/science.aad9416

Vernot, B., Zavala, E. I., Gomez-Olivencia, A., Jacobs, Z., Slon, V., Mafessoni, F., Romagne, F., Pearson, A., Petr, M., Sala, N., Pablos, A., Aranburu, A., de Castro, J. M. B., Carbonell, E., Li, B., Krajcarz, M. T., Krivoshapkin, A. I., Kolobova, K. A., Kozlikin, M. B., Shunkov, M. V., Derevianko, A. P., Viola, B., Grote, S., Essel, E., Herraez, D. L., Nagel, S., Nickel, B., Richter, J., Schmidt, A., Peter, B., Kelso, J., Roberts, R. G., Arsuaga, J. L., & Meyer, M. (2021). Unearthing Neanderthal population history using nuclear and mitochondrial DNA from cave sediments. *Science (New York, N.Y.)*. https://doi.org/10.1126/science.abf1667

Vickery, S., Patil, K. R., Dahnke, R., Hopkins, W. D., Sherwood, C. C., Caspers, S., Eickhoff, S. B., & Hoffstaedter, F. (2024, Aug 30). The uniqueness of human vulnerability to brain aging in great ape evolution. *Sci Adv, 10*(35), eado2733. https://doi.org/10.1126/sciadv.ado2733

Vidal, C. M., Lane, C. S., Asrat, A., Barfod, D. N., Mark, D. F., Tomlinson, E. L., Tadesse, A. Z., Yirgu, G., Deino, A., Hutchison, W., Mounier, A., & Oppenheimer, C. (2022). Age of the oldest known *Homo sapiens* from eastern Africa. *Nature*. https://doi.org/10.1038/s41586-021-04275-8

Videan, E. N., & McGrew, W. C. (2001). Are bonobos (*Pan Paniscus*) really more bipedal than chimpanzees (*Pan troglodytes*)? *American Journal of Primatology, 54*(4), 233–239. https://doi.org/10.1002/ajp.1033

Videan, E. N., & McGrew, W. C. (2002). Bipedality in chimpanzee (*Pan troglodytes*) and Bonobo (*Pan Paniscus*): Testing hypotheses on the evolution of bipedalism. *American Journal of Physical Anthropology, 118*(2), 184–190.

Villa, P., Boschian, G., Pollarolo, L., Sacca, D., Marra, F., Nomade, S., & Pereira, A. (2021). Elephant bones for the Middle Pleistocene toolmaker. *PloS One, 16*(8), e0256090. https://doi.org/10.1371/journal.pone.0256090

Villamil, C. I., & Middleton, E. R. (2024). Conserved patterns and locomotor-related evolutionary constraints in the hominoid vertebral column. *Journal of Human Evolution, 190*, 103528. https://doi.org/10.1016/j.jhevol.2024.103528

Villmoare, B., & Grabowski, M. (2022). Did the transition to complex societies in the Holocene drive a reduction in brain size? A reassessment of the DeSilva et al. (2021) hypothesis. *Frontiers in Ecology and Evolution, 10*. https://doi.org/10.3389/fevo.2022.963568

Villmoare, B., Kimbel, W. H., Seyoum, C., Campisano, C. J., DiMaggio, E., Rowan, J., Braun, D. R., Arrowsmith, J. R., & Reed, K. E. (2015). Early *Homo* at 2.8 Ma from Ledi-Geraru, Afar, Ethiopia. *Sciencexpress, 1343*, 1–7.

Vinyard, C. J., Ravosa, M. J., Williams, S. H., Wall, C. E., Johnson, K. R., & Hylander, W. L. (2007). Jaw-muscle function and the origin of primates. In M. J. Ravosa, & M. Dagosto (Eds.), *Primate origins: Adaptations and evolution* (pp. 179–231). Springer.

Vrba, E. S. (1998). Multiphasic growth models and the evolution of prolonged growth exemplified by human brain evolution. *Journal of Theoretical Biology, 190*(3), 227–239. https://doi.org/10.1006/jtbi.1997.0549

Vyas, D. N., Al-Meeri, A., & Mulligan, C. J. (2017). Testing support for the northern and southern dispersal routes out of Africa: An analysis of Levantine and southern Arabian populations. *American Journal of Physical Anthropology, 164*(4), 736–749. https://doi.org/10.1002/ajpa.23312

Wadley, L., Esteban, I., de la Pena, P., Wojcieszak, M., Stratford, D., Lennox, S., d'Errico, F., Rosso, D. E., Orange, F., Backwell, L., & Sievers, C. (2020). Fire and grass-bedding construction 200 thousand years ago at Border Cave, South Africa. *Science, 369*(6505), 863–866. https://doi.org/10.1126/science.abc7239

Waelkens, S. (2009). *Samenstelling van melk.* KULeuven. Retrieved 19/05/2023 from https://wet.kuleuven.be/wetenschapinbreedbeeld/lesmateriaal_biochemie/zure_melk/melk_samenstelling#:~:text=Het%20bevat%20eiwitten%2C%20lipiden%20(vetten,in%20lage%20concentratie%20aanwezig%20zijn.

Wagner, G. A., Krbetschek, M., Degering, D., Bahain, J. J., Shao, Q., Falgueres, C., Voinchet, P., Dolo, J. M., Garcia, T., & Rightmire, G. P. (2010). Radiometric dating of the type-site for *Homo heidelbergensis* at Mauer, Germany. *Proceedings of the National Academy of Sciences of the United States of America, 107*(46), 19726–19730. https://doi.org/10.1073/pnas.1012722107

Walker, R., Burger, O., Wagner, J., & Von Rueden, C. R. (2006). Evolution of brain size and juvenile periods in primates. *Journal of Human Evolution, 51*(5), 480–489. https://doi.org/10.1016/j.jhevol.2006.06.002

Wallace, A. R. (1858). On the tendency of varieties to depart indefinitely from the original type. *Journal of the Proceedings of the Linnean Society: Zoology, 3*(9), 53–62.

Wang, J., Jiang, L., & Sun, H. (2021). Early evidence for beer drinking in a 9000-year-old platform mound in southern China. *PloS One, 16*(8), e0255833. https://doi.org/10.1371/journal.pone.0255833

Wang, W. J., Crompton, R. H., Li, Y., & Gunther, M. M. (2003). Energy transformation during erect and 'bent-hip, bent-knee' walking by humans with implications for the evolution of bipedalism. *Journal of Human Evolution, 44*(5), 563–579. https://doi.org/10.1016/s0047-2484(03)00045-9

Ward, C. V., Kimbel, W. H., Harmon, E. H., & Johanson, D. C. (2012). New postcranial fossils of *Australopithecus afarensis* from Hadar, Ethiopia (1990–2007). *Journal of Human Evolution, 63*(1), 1–51. https://doi.org/10.1016/j.jhevol.2011.11.012

Ward, C. V., Leakey, M. G., & Walker, A. (2001). Morphology of *Australopithecus anamensis* from Kanapoi and Allia Bay, Kenya. *Journal of Human Evolution, 41*(4), 255–368. https://doi.org/10.1006/jhev.2001.0507

Ward, P. (2013). What may become of *Homo sapiens*? *Scientific American, 22*(1), 107–111.

Warrener, A. G., Lewton, K. L., Pontzer, H., & Lieberman, D. E. (2015). A wider pelvis does not increase locomotor cost in humans, with implications for the evolution of childbirth. *PloS One, 10*(3). https://doi.org/10.1371/journal.pone.0118903

Watts, D. P. (2020). Meat eating by nonhuman primates: A review and synthesis. *Journal of Human Evolution, 149.* https://doi.org/10.1016/j.jhevol.2020.102882

Weaver, A. H. (2005). Reciprocal evolution of the cerebellum and neocortex in fossil humans. *Proceedings of the National Academy of Sciences of the United States of America, 102*(10), 3576–3580. https://doi.org/10.1073/pnas.0500692102

Weaver, T. D. (2009). The meaning of Neandertal skeletal morphology. *Proceedings of the National Academy of Sciences of the United States of America, 106*(38), 16028–16033. https://doi.org/10.1073/pnas.0903864106

Weaver, T. D., & Hublin, J. J. (2009). Neandertal birth canal shape and the evolution of human childbirth. *Proceedings of the National Academy of Sciences of the United States of America, 106*(20), 8151–8156. https://doi.org/10.1073/pnas.0812554106

Welker, F., Ramos-Madrigal, J., Kuhlwilm, M., Liao, W., Gutenbrunner, P., de Manuel, M., Samodova, D., Mackie, M., Allentoft, M. E., Bacon, A.-M., Collins, M. J., Cox, J., Lalueza-Fox, C., Olsen, J. V., Demeter, F., Wang, W., Marques-Bonet, T., & Cappellini, E. (2019). Enamel proteome shows that *Gigantopithecus* was an early diverging pongine. *Nature.* https://doi.org/10.1038/s41586-019-1728-8

Wellik, D. M. (2007). Hox patterning of the vertebrate axial skeleton. *Developmental Dynamics, 236*(9), 2454–2463. https://doi.org/10.1002/dvdy.21286

Westergaard, G. C., Kuhn, H. E., & Suomi, S. J. (1998). Bipedal posture and hand preference in humans and other primates. *Journal of Comparative Psychology, 112*(1), 55–64. https://doi.org/10.1037//0735-7036.112.1.55

Westerman, F. (2018). *Wij, de mens.* Querido Fosfor.

Wheeler, P. E. (1984). The evolution of bipedality and loss of functional body hair in hominids. *Journal of Human Evolution, 13*(1), 91–98. https://doi.org/10.1016/S0047-2484(84)80079-2

Wheeler, P. E. (1985). The loss of functional body hair in man: The influence of thermal environment, body form and bipedality. *Journal of Human Evolution, 14*(1), 23–28. https://doi.org/10.1016/S0047-2484(85)80091-9

Wheeler, P. E. (1991). The thermoregulatory advantages of hominid bipedalism in open equatorial environments: The contribution of increased convective heat-loss and cutaneous evaporative cooling. *Journal of Human Evolution, 21*(2), 107–115. https://doi.org/10.1016/0047-2484(91)90002-D

Wheeler, P. E. (1992a). The influence of the loss of functional body hair on the water budgets of early hominids. *Journal of Human Evolution, 23*(5), 379–388. https://doi.org/10.1016/0047-2484(92)90086-O

Wheeler, P. E. (1992b). The thermoregulatory advantages of large body size for hominids foraging in savannah environments. *Journal of Human Evolution, 23*(4), 351–362. https://doi.org/10.1016/0047-2484(92)90071-g

Wheeler, P. E. (1994). The thermoregulatory advantages of heat-storage and shade-seeking behavior to hominids foraging in equatorial Savanna environments. *Journal of Human Evolution, 26*(4), 339–350. https://doi.org/10.1006/jhev.1994.1021

Whitcome, K. K. (2012). Functional implications of variation in lumbar vertebral count among hominins. *Journal of Human Evolution, 62*(4), 486–497. https://doi.org/10.1016/j.jhevol.2012.01.008

Whitcome, K. K., Miller, E. E., & Burns, J. L. (2017). Pelvic rotation effect on human stride length: Releasing the constraint of obstetric selection. *Anatomical Record (Hoboken), 300*(4), 752–763. https://doi.org/10.1002/ar.23551

Whitcome, K. K., Shapiro, L. J., & Lieberman, D. E. (2007). Fetal load and the evolution of lumbar lordosis in bipedal hominins. *Nature, 450*(7172), 1075–1078. https://doi.org/10.1038/nature06342

White, T. D., Asfaw, B., Beyene, Y., Haile-Selassie, Y., Lovejoy, C. O., Suwa, G., & WoldeGabriel, G. (2009). *Ardipithecus ramidus* and the paleobiology of early hominids. *Science (New York, N.Y.), 326*(5949), 75–86. https://doi.org/10.1126/science.1175802

White, T. D., Asfaw, B., DeGusta, D., Gilbert, H., Richards, G. D., & Howell, F. C. (2003). Pleistocene *Homo sapiens* from Middle Awash, Ethiopia. *Nature, 423*(6941), 742–747.

White, T. D., Lovejoy, C. O., Asfaw, B., Carlson, J. P., & Suwa, G. (2015). Neither chimpanzee nor human, *Ardipithecus* reveals the surprising ancestry of both. *Proceedings of the National Academy of Sciences of the United States of America, 112*(16), 4877–4884. https://doi.org/10.1073/pnas.1403659111

White, T. D., Suwa, G., & Asfaw, B. (1994). *Australopithecus ramidus*, a new species of early hominid from Aramis, Ethiopia. *Nature, 371*(6495), 306–312. https://doi.org/10.1038/371306a0

White, T. D., Suwa, G., & Asfaw, B. (1995). Corrigendum: *Australopithecus ramidus*, a new species of early hominid from Aramis, Ethiopië. *Nature, 375*(6526), 88.

Whiten, A., Goodall, J., McGrew, W. C., Nishida, T., Reynolds, V., Sugiyama, Y., Tutin, C. E., Wrangham, R. W., & Boesch, C. (1999). Cultures in chimpanzees. *Nature, 399*(6737), 682–685. https://doi.org/10.1038/21415

WHO/FAO/UNU Expert Consultation. (2007). *Protein and amino acid requirements in human nutrition* (WHO Technical Report Series, Issue. WHO. https://apps.who.int/iris/bitstream/handle/10665/43411/WHO_TRS_935_eng.pdf

Wible, B. (2021). Complicated legacies: The human genome at 20. *Science, 371*(6529), 564–569.

Wibowo, M. C., Yang, Z., Borry, M., Hubner, A., Huang, K. D., Tierney, B. T., Zimmerman, S., Barajas-Olmos, F., Contreras-Cubas, C., Garcia-Ortiz, H., Martinez-Hernandez, A., Luber, J. M., Kirstahler, P., Blohm, T., Smiley, F. E., Arnold, R., Ballal, S. A., Pamp, S. J., Russ, J., Maixner, F., Rota-Stabelli, O., Segata, N., Reinhard, K., Orozco, L., Warinner, C., Snow, M., LeBlanc, S., & Kostic, A. D. (2021). Reconstruction of ancient microbial genomes from the human gut. *Nature.* https://doi.org/10.1038/s41586-021-03532-0

Wikipedia-bijdragers. (2022, 09/03/2022). *Bibliografische gegevens voor Geschiedenis van de auto (1885-1904)*. Wikipedia, de vrije encyclopedie. Retrieved 25/03/2022 from https://nl.wikipedia.org/w/index.php?title=Geschiedenis_van_de_auto_(1885-1904)&oldid=61270391

Wikipedia Contributors. (2022, 19/03/2022). *Herpes simplex virus*. Wikipedia, The Free Encyclopedia. Retrieved 30/03/2022 from https://en.wikipedia.org/wiki/Herpes_simplex_virus

Wilkins, J., Schoville, B. J., Brown, K. S., & Chazan, M. (2012). Evidence for early hafted hunting technology. *Science, 338*(6109), 942–946. https://doi.org/10.1126/science.1227608

Wilkins, J., Schoville, B. J., Pickering, R., Gliganic, L., Collins, B., Brown, K. S., von der Meden, J., Khumalo, W., Meyer, M. C., Maape, S., Blackwood, A. F., & Hatton, A. (2021). Innovative *Homo sapiens* behaviours 105,000 years ago in a wetter Kalahari. *Nature.* https://doi.org/10.1038/s41586-021-03419-0

Williams, S. A. (2012). Variation in anthropoid vertebral formulae: Implications for homology and homoplasy in hominoid evolution. *Journal of Experimental Zoology. Part B, Molecular and Developmental Evolution, 318*(2), 134–147. https://doi.org/10.1002/jezb.21451

Williams, S. A., Ostrofsky, K. R., Frater, N., Churchill, S. E., Schmid, P., & Berger, L. R. (2013). The vertebral column of *Australopithecus sediba*. *Science (New York, N.Y.)*, *340*(6129), 1232996. https://doi.org/10.1126/science.1232996

Williams, S. A., & Pilbeam, D. (2021). Homeotic change in segment identity derives the human vertebral formula from a chimpanzee-like one. *American Journal of Physical Anthropology*, *176*, 283–294. https://doi.org/10.1002/ajpa.24356

Williams, S. A., Prang, T. C., Meyer, M. R., Russo, G. A., & Shapiro, L. J. (2020). Reevaluating bipedalism in *Danuvius*. *Nature*, *586*, E1–E3. https://doi.org/10.1038/s41586-020-2736-4

Williams, S. A., Prang, T. C., Meyer, M. R., Nalley, T. K., Van Der Merwe, R., Yelverton, C., Garcia-Martinez, D., Russo, G. A., Ostrofsky, K. R., Spear, J., Eyre, J., Grabowski, M., Nalla, S., Bastir, M., Schmid, P., Churchill, S. E., & Berger, L. R. (2021). New fossils of *Australopithecus sediba* reveal a nearly complete lower back. *eLife*, *10*(e70447), 1–25. https://doi.org/10.7554/eLife.70447

Williams, S. A., & Russo, G. A. (2015). Evolution of the hominoid vertebral column: The long and the short of it. *Evolutionary Anthropology*, *24*(1), 15–32. https://doi.org/10.1002/evan.21437

Williams, S. A., Zeng, I., Guerra, J. S., Nalla, S., Elliott, M. C., Hawks, J., Berger, L. R., & Meyer, M. R. (2022). *Homo naledi* lumbar vertebrae and a new 3D method to quantify vertebral wedging. *American Journal of Biological Anthropology*. https://doi.org/10.1002/ajpa.24621

Williams-Hatala, E. M., Hatala, K. G., Gordon, M., Key, A., Kasper, M., & Kivell, T. L. (2018). The manual pressures of stone tool behaviors and their implications for the evolution of the human hand. *Journal of Human Evolution*, *119*, 14–26. https://doi.org/10.1016/j.jhevol.2018.02.008

Williams-Hatala, E. M., Hatala, K. G., Key, A., Dunmore, C. J., Kasper, M., Gordon, M., & Kivell, T. L. (2020). Kinetics of stone tool production among novice and expert tool makers. *American Journal of Physical Anthropology*, e24159. https://doi.org/10.1002/ajpa.24159

Willoughby, P. R. (2021). Early humans far from the South African coast collected unusual objects. *Nature*. https://doi.org/10.1038/d41586-021-00795-5

Wilson Mantilla, G. P., Chester, S. G. B., Clemens, W. A., Moore, J. R., Sprain, C. J., Hovatter, B. T., Mitchell, W. S., Mans, W. W., Mundil, R., & Renne, P. R. (2021). Earliest Palaeocene purgatoriids and the initial radiation of stem primates. *Royal Society Open Science*, *8*(2), 210050. https://doi.org/10.1098/rsos.210050

Wiseman, A. L. A. (2023). Three-dimensional volumetric muscle reconstruction of the *Australopithecus afarensis* pelvis and limb, with estimations of limb leverage. *Royal Society Open Science*, *10*(6). https://doi.org/10.1098/rsos.230356

Wißing, C., Rougier, H., Baumann, C., Comeyne, A., Crevecoeur, I., Drucker, D. G., Gaudzinski-Windheuser, S., Germonpré, M., Gómez-Olivencia, A., Krause, J., Matthies, T., Naito, Y. I., Posth, C., Semal, P., Street, M., & Bocherens, H. (2019). Stable isotopes reveal patterns of diet and mobility in the last Neandertals and first modern humans in Europe. *Scientific Reports*, *9*(1). https://doi.org/10.1038/s41598-019-41033-3

Wolfe, S. W., Crisco, J. J., Orr, C. M., & Marzke, M. W. (2006). The dart-throwing motion of the wrist: Is it unique to humans? *J Hand Surg Am*, *31*(9), 1429–1437. https://doi.org/10.1016/j.jhsa.2006.08.010

Wolpoff, M. H., Senut, B., Pickford, M., & Hawks, J. (2002). *Sahelanthropus* or 'Sahelpithecus'? *Nature*, *419*(6907), 581–582. https://doi.org/10.1038/419581a

Wong, K. (2010). Swept away. *Scientific American*, *302*(4), 12.

Wong, K. (2015). Neandertal minds. *Scientific American, 312*(2), 36–43.

Wood, B. (2014). Fifty years after *Homo habilis. Nature, 508*(7494), 31–33. https://doi.org/10.1038/508031a

Wood, B., & Boyle, E. K. (2016). Hominin taxic diversity: Fact or fantasy? *American Journal of Physical Anthropology, 159*, 37–78. https://doi.org/10.1002/ajpa.22902

Wood, B., & Collard, M. (1999). The human genus. *Science, 284*, 65–71.

Wood, B., & Eve, K. B. (2016). Hominin taxic diversity: Fact or fantasy? *American Journal of Physical Anthropology, 159*(Suppl 61), S37–78. https://doi.org/10.1002/ajpa.22902

Wood, B., & Richmond, B. G. (2000). Human evolution: Taxonomy and paleobiology. *J Anat, 197*(Pt 1), 19–60. https://www.ncbi.nlm.nih.gov/pubmed/10999270

Wood, T. (2020). An expanded character set for evaluating the phylogenetic position of *Homo floresiensis. American Journal of Physical Anthropology, 171*, 312–312.

Wragg Sykes, R. (2020). *Kindred Neanderthal life, love, death and art.* Bloomsbury Sigma.

Wragg Sykes, R. M. (2017). Neanderthals in the Outermost West: Technological adaptation in the Late Middle Palaeolithic (re)-colonization of Britain, Marine Isotope Stage 4/3. *Quaternary International, 433*, 4–32. https://doi.org/10.1016/j.quaint.2015.12.087

Wrangham, R., & Carmody, R. (2010). Human adaptation to the control of fire. *Evolutionary Anthropology, 19*(5), 187–199. https://doi.org/10.1002/evan.20275

Wrangham, R. W., Jones, J. H., Laden, G., Pilbeam, D., & Conklin-Brittain, N. (1999a). The raw and the stolen: Cooking and the ecology of human origins. *Current Anthropology, 40*(5), 567–594. https://doi.org/10.1086/300083

Wrangham, R. W., Jones, J. H., Laden, G., Pilbeam, D., & Conklin-Brittain, N. (1999b). Reply to comments on The raw and The stolen: Cooking and The ecology of human origins. *Current Anthropology, 40*(5), 586–587. https://doi.org/10.1086/300083

Wright, E., Grawunder, S., Ndayishimiye, E., Galbany, J., McFarlin, S. C., Stoinski, T. S., & Robbins, M. M. (2021). Chest beats as an honest signal of body size in male mountain gorillas (*Gorilla beringei beringei*). *Scientific Reports, 11*(1). https://doi.org/10.1038/s41598-021-86261-8

Wroe, S., Parr, W. C. H., Ledogar, J. A., Bourke, J., Evans, S. P., Fiorenza, L., Benazzi, S., Hublin, J. J., Stringer, C., Kullmer, O., Curry, M., Rae, T. C., & Yokley, T. R. (2018). Computer simulations show that Neanderthal facial morphology represents adaptation to cold and high energy demands, but not heavy biting. *Proceedings of the Royal Society B. Biological Sciences, 285*(1876). https://doi.org/10.1098/rspb.2018.0085

Wu, X., & Bae, C. J. (2024). Xujiayao *Homo*: A new form of large brained hominin in Eastern Asia. *PaleoAnthropology, 202*. https://paleoanthropology.org/ojs/index.php/paleo/libraryFiles/downloadPublic/18

Wu, X. H., Zhang, C., Goldberg, P., Cohen, D., Pan, Y., Arpin, T., & Bar-Yosef, O. (2012). Early pottery at 20,000 years ago in Xianrendong Cave, China. *Science, 336*(6089), 1696–1700. https://doi.org/10.1126/science.1218643

Wu, X. J., Bae, C. J., Friess, M., Xing, S., Athreya, S., & Liu, W. (2022). Evolution of cranial capacity revisited: A view from the late Middle Pleistocene cranium from Xujiayao, China. *Journal of Human Evolution, 163*, 103119. https://doi.org/10.1016/j.jhevol.2021.103119

Wu, Y., Chen, K., Xing, C., Huang, M., Zhao, K., & Zhou, W. (2024). Human olfactory perception embeds fine temporal resolution within a single sniff. *Nat Hum Behav, 8*(11), 2168–2178. https://doi.org/10.1038/s41562-024-01984-8

Wu, Y., Wang, H., Wang, H., & Hadly, E. A. (2017). Rethinking the origin of primates by reconstructing their diel activity patterns using genetics and morphology. *Scientific Reports*, 7(1), 11837. https://doi.org/10.1038/s41598-017-12090-3

Wynn, J. G., Alemseged, Z., Bobe, R., Grine, F. E., Negash, E. W., & Sponheimer, M. (2020). Isotopic evidence for the timing of the dietary shift toward C4 foods in eastern African *Paranthropus*. *Proceedings of the National Academy of Sciences of the United States of America*. https://doi.org/10.1073/pnas.2006221117

Xia, B., Zhang, W., Wudzinska, A., Huang, E., Brosh, R., Pour, M., Miller, A., Dasen, J. S., Maurano, M. T., Kim, S. Y., Boeke, J. D., & Yanai, I. (2021). The genetic basis of tail-loss evolution in humans and apes. *bioRxiv*. https://doi.org/10.1101/2021.09.14.460388

Xia, B., Zhang, W., Zhao, G., Zhang, X., Bai, J., Brosh, R., Wudzinska, A., Huang, E., Ashe, H., Ellis, G., Pour, M., Zhao, Y., Coelho, C., Zhu, Y., Miller, A., Dasen, J. S., Maurano, M. T., Kim, S. Y., Boeke, J. D., & Yanai, I. (2024). On the genetic basis of tail-loss evolution in humans and apes. *Nature*, 626(8001), 1042–1048. https://doi.org/10.1038/s41586-024-07095-8

Xiang, L., Crow, T. J., Hopkins, W. D., & Roberts, N. (2020). Comparison of surface area and cortical thickness asymmetry in the human and chimpanzee brain. *Cerebral cortex (New York, N.Y.: 1991)*. https://doi.org/10.1093/cercor/bhaa202

Xing, S., O'Hara, M., Guatelli-Steinberg, D., Ge, J., & Liu, W. (2017). Dental scratches and handedness in East Asian early pleistocene hominins. *International Journal of Osteoarchaeology*, 27(6), 937–946. https://doi.org/10.1002/oa.2601

Xu, S. H., Li, S. L., Yang, Y. J., Tan, J. Z., Lou, H. Y., Jin, W. F., Yang, L., Pan, X. D., Wang, J. C., Shen, Y. P., Wu, B. L., Wang, H. Y., & Jin, L. (2011). A genome-wide search for signals of high-altitude adaptation in Tibetans. *Molecular Biology and Evolution*, 28(2), 1003–1011. https://doi.org/10.1093/molbev/msq277

Yang, Z., Bai, C., Pu, Y., Kong, Q., Guo, Y., Ouzhuluobu, Gengdeng, Liu, X., Zhao, Q., Qiu, Z., Zheng, W., He, Y., Lin, Y., Deng, L., Zhang, C., Xu, S., Peng, Y., Xiang, K., Zhang, X., Baimayangji, Cirenyangji, Cui, C., Baimakangzhuo, Gonggalanzi, Bianba, Pan, Y., Xin, J., Wang, Y., Liu, S., Wang, L., Guo, H., Feng, Z., Wang, S., Shi, H., Jiang, B., Wu, T., Qi, X., . . . Su, B. (2022). Genetic adaptation of skin pigmentation in highland Tibetans. *Proceedings of the National Academy of Sciences of the United States of America*, 119(40), e2200421119. https://doi.org/10.1073/pnas.2200421119

Yang, Z. H., Shi, H., Ma, P. C., Zhao, S. L., Kong, Q. H., Bian, T. H., Gong, C., Zhao, Q., Liu, Y., Qi, X. B., Zhang, X. M., Han, Y. L., Liu, J. W., Li, Q. W., Chen, H., & Su, B. (2018). Darwinian positive selection on the pleiotropic effects of *KITLG* explain skin pigmentation and winter temperature adaptation in eurasians. *Molecular Biology and Evolution*, 35(9), 2272–2283. https://doi.org/10.1093/molbev/msy136

Yao, Z., van Velthoven, C. T. J., Kunst, M., Zhang, M., McMillen, D., Lee, C., Jung, W., Goldy, J., Abdelhak, A., Aitken, M., Baker, K., Baker, P., Barkan, E., Bertagnolli, D., Bhandiwad, A., Bielstein, C., Bishwakarma, P., Campos, J., Carey, D., . . . & Zeng, H. (2023). A high-resolution transcriptomic and spatial atlas of cell types in the whole mouse brain. *Nature*, 624(7991), 317–332. https://doi.org/10.1038/s41586-023-06812-z

Ye, S., Sun, J., Craig, S. R., Di Rienzo, A., Witonsky, D., Yu, J. J., Moya, E. A., Simonson, T. S., Powell, F. L., Basnyat, B., Strohl, K. P., Hoit, B. D., & Beall, C. M. (2024). Higher oxygen content and transport characterize high-altitude ethnic Tibetan women with the highest lifetime reproductive success. *Proceedings of the National Academy of Sciences of the United States of America*, 121(45), e2403309121. https://doi.org/10.1073/pnas.2403309121

Yegian, A. K., Heymsfield, S. B., Castillo, E. R., Muller, M. J., Redman, L. M., & Lieberman, D. E. (2024). Metabolic scaling, energy allocation tradeoffs, and the evolution of humans' unique metabolism. *Proceedings of the National Academy of Sciences of the United States of America, 121*(48), e2409674121. https://doi.org/10.1073/pnas.2409674121

Yegian, A. K., Tucker, Y., Bramble, D. M., & Lieberman, D. E. (2021). Neuromechanical linkage between the head and forearm during running. *American Journal of Physical Anthropology.* https://doi.org/10.1002/ajpa.24234

Yegian, A. K., Tucker, Y., Gillinov, S., & Lieberman, D. E. (2021). Shorter distal forelimbs benefit bipedal walking and running mechanics: Implications for hominin forelimb evolution. *American Journal of Physical Anthropology.* https://doi.org/10.1002/ajpa.24274

Young, M. M. I., Winters, S., Young, C., Weiß, B. M., Troscianko, J., Ganswindt, A., Barrett, L., Henzi, S. P., Higham, J. P., & Widdig, A. (2020). Male characteristics as predictors of genital color and display variation in vervet monkeys. *Behavioral Ecology and Sociobiology, 74*(2). https://doi.org/10.1007/s00265-019-2787-4

Yum, S. M., Baek, I. K., Hong, D., Kim, J., Jung, K., Kim, S., Eom, K., Jang, J., Kim, S., Sattorov, M., Lee, M. G., Kim, S., Adams, M. J., & Park, G. S. (2020). Fingerprint ridges allow primates to regulate grip. *Proceedings of the National Academy of Sciences of the United States of America.* https://doi.org/10.1073/pnas.2001055117

Zachos, J., Pagani, M., Sloan, L., Thomas, E., & Billups, K. (2001). Trends, rhythms, and aberrations in global climate 65 Ma to present. *Science, 292*(5517), 686–693. https://doi.org/10.1126/science.1059412

Zaitsev, A. N., McHenry, L., Savchenok, A. I., Strekopytov, S., Spratt, J., Humphreys-Williams, E., Sharygin, V. V., Bogomolov, E. S., Chakhmouradian, A. R., Zaitseva, O. A., Arzamastsev, A. A., Reguir, E. P., Leach, L., Leach, M., & Mwankunda, J. (2019). Stratigraphy, mineralogy and geochemistry of the Upper Laetolil tuffs including a new tuff 7 site with footprints of *Australopithecus afarensis*, Laetoli, Tanzania. *Journal of African Earth Sciences, 158*. https://doi.org/10.1016/j.jafrearsci.2019.103561

Zanolli, C., Davies, T. W., Joannes-Boyau, R., Beaudet, A., Bruxelles, L., de Beer, F., Hoffman, J., Hublin, J. J., Jakata, K., Kgasi, L., Kullmer, O., Macchiarelli, R., Pan, L., Schrenk, F., Santos, F., Stratford, D., Tawane, M., Thackeray, F., Xing, S., Zipfel, B., & Skinner, M. M. (2022). Dental data challenge the ubiquitous presence of Homo in the cradle of humankind. *Proceedings of the National Academy of Sciences of the United States of America, 119*(28), e2111212119. https://doi.org/10.1073/pnas.2111212119

Zeberg, H., Kelso, J., & Paabo, S. (2020). The Neandertal progesterone receptor. *Mol Biol Evol, 37*(9), 2655–2660. https://doi.org/10.1093/molbev/msaa119

Zeberg, H., & Paabo, S. (2020). The major genetic risk factor for severe COVID-19 is inherited from Neanderthals. *Nature, 587*(7835), 610–612. https://doi.org/10.1038/s41586-020-2818-3

Zeininger, A., Schmitt, D., & Wunderlich, R. E. (2020). Mechanics of heel-strike plantigrady in African apes. *Journal of Human Evolution, 145*, 102840. https://doi.org/10.1016/j.jhevol.2020.102840

Zeller, E., Timmermann, A., Yun, K. S., Raia, P., Stein, K., & Ruan, J. (2023). Human adaptation to diverse biomes over the past 3 million years. *Science, 380*, 604–608.

Zhou, S., Butler-Laporte, G., Nakanishi, T., Morrison, D. R., Afilalo, J., Afilalo, M., Laurent, L., Pietzner, M., Kerrison, N., Zhao, K., Brunet-Ratnasingham, E., Henry, D., Kimchi, N., Afrasiabi, Z., Rezk, N., Bouab, M., Petitjean, L., Guzman, C., Xue, X., Tselios, C., Vulesevic, B.,

Adeleye, O., Abdullah, T., Almamlouk, N., Chen, Y., Chasse, M., Durand, M., Paterson, C., Normark, J., Frithiof, R., Lipcsey, M., Hultstrom, M., Greenwood, C. M. T., Zeberg, H., Langenberg, C., Thysell, E., Pollak, M., Mooser, V., Forgetta, V., Kaufmann, D. E., & Richards, J. B. (2021). A Neanderthal OAS1 isoform protects individuals of European ancestry against COVID-19 susceptibility and severity. *Nature Medicine, 27*(4), 659–667. https://doi.org/10.1038/s41591-021-01281-1

Zihlman, A. L., & Bolter, D. R. (2015). Body composition in *Pan Paniscus* compared with *Homo sapiens* has implications for changes during human evolution. *Proceedings of the National Academy of Sciences of the United States of America, 112*(24), 7466–7471. https://doi.org/10.1073/pnas.1505071112

Zimmer, C., & Emlen, D. J. (2013). *Evolution: Making sense of life*. Roberts and Company Publishers, Inc.

Zirkle, D., & Lovejoy, C. O. (2019). The hominid ilium is shaped by a synapomorphic growth mechanism that is unique within primates. *Proceedings of the National Academy of Sciences of the United States of America, 116*(28), 13915–13920. https://doi.org/10.1073/pnas.1905242116

Zohar, I., Alperson-Afil, N., Goren-Inbar, N., Prévost, M., Tütken, T., Sisma-Ventura, G., Hershkovich, I., & Najorka, J. (2022). Evidence for the cooking of fish 780,000 years ago at Gesher Benot Ya'aqov, Israel. *Nature Ecology & Evolution*. https://doi.org/10.1038/s41559-022-01910-z

Zollikofer, C. P., Ponce de Leon, M. S., Lieberman, D. E., Guy, F., Pilbeam, D., Likius, A., Mackaye, H. T., Vignaud, P., & Brunet, M. (2005). Virtual cranial reconstruction of *Sahelanthropus tchadensis* [Historical Article, Research Support, Non-U.S. Gov't]. *Nature, 434*(7034), 755–759. https://doi.org/10.1038/nature03397

Zollikofer, C. P. E., Beyrand, V., Lordkipanidze, D., Tafforeau, P., & Ponce de Leon, M. S. (2024). Dental evidence for extended growth in early *Homo* from Dmanisi. *Nature, 635*(8040), 906–911. https://doi.org/10.1038/s41586-024-08205-2

Zywiczynski, P., Wacewicz, S., & Lister, C. (2021). Pantomimic fossils in modern human communication. *Philosophical Transactions of the Royal Society of London. Series B, Biological Sciences, 376*(1824), 20200204. https://doi.org/10.1098/rstb.2020.0204

# Index

**A**

*ABCC11* gene, 210
Aborigines, 150, 166, 178, 195
*Acanthostega*, 23, 26
Acheulean technology, 166–167, 180, 200–201
Achilles tendon, 56, 83, 111, 143
*Acromion*, 53
Adipose tissue, 137, 165, 192–193
Affenspalte, 125–127
Africa, 15, 19, 35, 37, 40–42, 53, 131, 133, 134–135, 169, 178, 179, 182, 196, 205, 212
African elephant, 66
African erectus, 140, 144, 166, 170
African great apes, 39, 42, 46, 59, 61, 62, 63
African sapiens, 171, 176, 177, 178, 183
*Afropithecus*, 41
Age determination of fossils, 9
Air breathing, 22, 25
Allen's Law, 186, 192
Altai Mountains, 171, 175
Altai Neanderthals, 175
*Alu element*, 55
American Museum of Natural History in New York, 108, 109
Amino acids, 122, 137, 139, 199
Amniota, 26–27
Amniote, sapiens as, 26–27
Amniote eggs, 32
Amniotes, 19, 20, 25
Amphibians, 23, 26, 27
Ancient Greeks, 13
Angular gyrus, 73
Animal, sapiens as, 18–20
Animal characteristics, 17
Ankle bones (tarsalia), 24
Ankle joints, 54, 56, 111, 144
Ankles, 74, 83
Annulus fibrosus, 20, 26
Antelope, 137, 139
Anthropoidea, 37
Anvil (*incus*), 30
Apes, 1, 19, 35–42, 45, 53; *see also* Great apes
    African great apes, 39, 42, 46, 59, 61, 62, 63
    Asian apes, 41

brain, 63
    folds, 71–74
    neurons, 67–70
    olfaction, 70–71
    primate brain, 63–67
European apes, 41–42
as fruit eaters, 42–45
glory days of, 41–42
man and man-like apes, 14
Miocene apes, 58
primitive apes, 41
scientific names, 35–40
from the South, 88
    'the child of Taung', 89–91
    'Lucy in the sky with diamonds', 91–92
Apocrine sweat glands, 62
Aquatic breathing, 22
Arabian Peninsula, 41, 200
Arambourg, Camille, 93
Arboreal, 51, 58
*Archaefructus*, 42
Archaic sapiens, 170, 171
Ardi, 80, 81, 83, 85, 88, 117, 188
*Ardipithecus*, 77, 82, 84, 85, 95
*Ardipithecus kadabba*, 80
*Ardipithecus ramidus*, 79
Argon-argon dating, 9
*ARHGAP11B* genes, 159
Aristotle, 13
Arm, 24, 52, 53, 59, 74, 103–104, 116, 201
Armadillo, 26
Arm length, 24, 48–49
Arm swinging, 55, 104, 148
*Arrector pili* muscles, 31
Artiodactyl mammals, 166
Ascorbic acid, 45
Asia, 15, 19, 37, 53, 113, 133
Asian apes, 41
Atavism, 22
Atlantic Ocean, 37
Attenborough, David, 211
Auditory cortex, 73
Aurignacian technology, 201

Australopithecines, 87, 95, 97–98, 101, 102–103, 104, 106, 110, 111, 112, 113, 118, 122, 131, 135, 140, 141–142, 155, 202
*Australopithecus*, 4, 11, 14, 77, 78, 79, 80, 87, 88, 94, 99, 125, 166
*Australopithecus afarensis*, 88, 91
*Australopithecus africanus*, 88, 89, 90, 91, 92
*Australopithecus anamensis*, 88, 92
*Australopithecus bahrelghazali*, 7, 8, 77, 88, 92
*Australopithecus deyiremeda*, 89, 92
*Australopithecus garhi*, 88
*Australopithecus sediba*, 89, 93, 127
Awash River, 79

**B**

Baboons, 19, 97, 117, 136
Bacteria, 13, 17–18
Balance, 23, 30, 39, 49, 54, 110, 112, 117, 148
Barefoot walking, 109
Barley (*Hordeum*), 181
Basal metabolism, 140, 141, 163–164
Behavior, survival through, 197
    big brains, 197–198
    globular brains, 198–200
    hands and feet, talking with
        from tools to language, 203–205
        tools to talk, 205–207
    healthy brains, 200
    technology
        Acheulean technology, 200–201
        being creative with materials, 202–203
        Handy Harry's, 201–202
Belyaev, Dmitry, 209
Bent-hip-bent-knee, 58, 109, 115, 121
Bergmann's Law, 185, 192
Big brains, 123, 156, 197–198
    Affenspalte, 125–127
    changing in different ways, 159–162
    energy guzzler, 162–166
    intelligence, 156–159
    Taung's child, 124–125
    tools, 166–167
Big brains, walking with, 184
    birth canal, 186–191
    running economy, 186
    walking and running in the cold, 184–186
Bipedal locomotion, 82, 186
Birds, 4, 9, 28, 64, 117, 122, 201
Birth canal, 186–191
Black, Davidson, 8, 133

Black-brown pigments, 63; *see also* Eumelanin
Blood, 24, 31, 163–164, 166, 192
    de-oxygenated, 23
    oxygenated, 23
    oxygen-poor, 23
    warm, 31, 166
Blood hemoglobin, 210
Blue nostrils, 38
Blue scrotum, 38
Blue whale, 68
*Bmp* genes, 152
Bones, 22, 24, 25, 28, 97, 174
    ankle, 24
    heel, 56
    jaw, 22, 29
    lower jawbone, 30
    metacarpal, 57, 202
    middle ear, 25, 29
    midfoot, 24
    palm, 24
    pinky, 171, 173
    skull, 28, 124, 127, 140, 166, 187
    wrist, 24, 57
Bonobos, 39, 40, 41, 42, 43, 56, 57, 72, 193
Bony fish, 20, 22, 28, 29
Brachiation, 101
Brachiators, 39, 40, 48, 52
Brain, 3, 4, 10, 21, 30, 51, 65, 66, 69, 85, 131, 157, 160, 163, 166, 204; *see also* Big brains
    big, 197–198
    Dmanisi, 162
    globular, 198–200
    gyrencephalic, 64
    healthy, 200
    large, 30, 64
    lissencephalic, 64
    mammalian, 21, 64
    primate, 63–67
    reptile, 21
    small, 66, 77, 87, 132, 164, 187
    special brain twister, 71
    triune, 21
Brain cells, 4, 18, 67, 190
Brain-derived neurotrophic factor (BDNF), 200
Brain-gut connection, 18
Brain mosaic look, 66
Brain size, 4, 66, 69, 78, 85, 141, 156–159, 161, 187, 190, 199
Brainstem, 22–23, 64, 67
Brain-to-body ratio, 64

Breathing
    air, 22, 25
    aquatic, 22
    rhythmic, 23
'Broad-nosed', 38
Broca, Paul, 73
Broca's center, 73, 126, 127, 128, 204
Brodmann, Korbinian, 72
Broom, Robert, 92, 93, 99
Brunet, Michel, 11
Buffalo, 137
Bulgarian Bacho Kiro cave, 178
Bulla tympanica, 30
Bush babies, 37

**C**

Calcaneus, 56
Calf muscles, 56, 144, 147
Canine, 28, 29, 39, 45–46, 81
Capitatum, 57
Capuchin monkeys, 2, 38, 49, 97, 136, 202
Car-bodies, 16
Carbohydrates, 43, 136, 139, 164, 165, 182, 212
Carbon-12, 95
Carbon-13, 95
Carbon dating, 9
Carpalia, 24
Carry hypothesis, 102
Cartilages, 21, 155
Cartmill, Matt, 2
Catarrhini, 38
Cellulose, 38, 47, 139
Center of gravity, 52, 54, 58, 111, 113, 114, 142,
    145, 148
Central Africa, 7
Central African chimpanzee, 63
Cerebellum, 64, 66–67, 70
Cerebral rubicon, 132
Cerebrum, 30, 64, 71
Cetaceans, 30
Chad, 77–78
Chatelperronian technology, 201
Cheddar man, 196
Cheek teeth, 28
Chewing, 28, 30, 45–47, 97
'The child of Taung', 89–91
Chimpanzees, 2, 3, 14, 15, 18–20, 39–43, 47, 50,
    53–57, 61, 63, 97, 110, 114, 116, 120,
    136–138, 149, 159–160, 162
China, 41, 133, 135, 181

Chinese erectus, 144, 158
Chinese skulls, 197
*Choanae*, 25
Chorda, 20
Chordates, 20, 21; *see also* Sea squirts (tunicates)
*Chororapithecus*, 42
Clacton-on-Sea, 201
Clambering with hands and feet, 48
    climbing and walking in trees, 58–59
    playing the monkey, 48–51
    swinging, 52–57
Claw-like nails, 38
Claws, 49, 54
Clay, 1
Climbing, 51, 55, 56, 58–59, 103–107
Cold, walking and running in, 184–186
Cold periods, surviving, 192
Collagen, 45
Colors of the rainbow, 193–197
Color vision, 71
Communication, 2, 66, 68, 73, 74, 206
    non-chemical, 72–73
    non-visual, 204
    through sign language, 204
    vocal, 74, 204
Cooked plant foods, 138–140
Cooking food, 139
Cooling, 31, 61–63, 84, 120, 153–156, 191–193
Coppens, Yves, 93
Copper Age, 203
COVID-19 pandemic, 176
C3 plants, 95
C4 plants, 95
Crassulacean acid metabolism (CAM) plants, 95

**D**

*Danuvius*, 41
*Danuvius guggenmosi*, 59
Dark fetal hair, 15
Dark skin, 7, 63, 122, 151, 154, 193–194
Dart, Raymond, 11, 77, 88, 89, 91, 92, 127
Darwin, Charles, 1, 13–15, 21, 42, 58, 102, 122,
    138, 156, 210
Darwin tubercle, 15
*DDB1* genes, 195
Decay rate, 9
De Chardin, Pierre Teilhard, 35
Degenerative spondylolisthesis, 112
Democratic Republic of Congo, 40
Dendrites, 161–162

Denisova Cave, 173, 175
Denisova humans, 10, 171–177, 179, 191, 192
Denisovans, 173, 174
Dentary, 28
Dentin, 22
De-oxygenated blood, 23
Depth perception, 51
Dermatoglyphs, 50, 54
Dermis, 26, 51, 60
*Descent of man* (Darwin), 1, 14
Desiccation, 94
Dexterity, 128, 167, 197, 201, 203
Digestive system, 38, 47–48, 138
Digitigrades, 57, 59
Dinosaurs, 30, 31, 68
Display hypothesis, 102
Diurnal locomotion, 55
*Dkk2* genes, 152
Dmanisi brains, 162
Dmanisi erectus, 158, 161, 164, 166
DNA, 8, 10, 14, 16–19, 32–33, 119, 122, 171–173,
    175, 179
  herpes DNA, 18
  mammalian DNA, 32
  mitochondrial DNA (mtDNA), 17, 170
  Neanderthal DNA, 10, 175–178
  nuclear DNA (nDNA), 16, 17, 175
  viral DNA, 32
DNA methylation, 173
Dobzhansky, Theodosius, 1
Downward-facing nostrils, 19
Downward-nosed monkeys, 20, 38
Dragon skull, 171, 173
*Dryopithecus*, 41, 53
Dubois, Eugene, 8, 133
Düsselbach River, 174

**E**

Ear
  inner ear, 23, 25, 73, 105, 148, 184
  middle ear, 25, 29, 30
Eardrum, 25
Early Neanderthals, 171, 175, 181
Earth, 13
Earthworm, 18, 21
East African mammalian biodiversity, 9
East African savannah, 63
Eastern African rift valley, 9
Eccrine sweat glands, 62, 154
Echidna, 32

Ectoderm, 27
*EDAR* gene, 210
Edinger, Ludwig, 21
Egg-laying mammals, 32
Eggs, 27
Eggshell, 33
*EGLN1* genes, 212
Elbow, 24
Elbow joint, 25
Electromagnetic radiation, 120
Embryonic development, 21
Embryos, 33
Enamel, 22, 46, 78, 80, 88, 95, 97, 99
Encephalization quotient, 156, 158
Endocasts, 125
Endoderm, 27
Endogenous retroviruses, 32
Endogenous viruses, 33
Endurance running, 145, 184, 185
Energy, 43, 44, 47, 56, 96–98, 114–117
  consumption, 98, 101, 113, 114, 116, 148,
    149–150, 186, 193
  kinetic energy, 114–115, 145, 146
  management, 121
  potential energy, 114–115, 145–146
  in an unstable environment, 140–141
Energy-consuming organs, 140, 141
Energy guzzler, 162–166
Engis, Belgium, 14
Envelope glycoprotein-encoding genes (*env*)
    genes, 32
*EPAS1* genes, 212
Epidermis, 26, 31, 60, 193
Equatorial forest, 35, 100
  ape brain, 63
    folds, 71–74
    neurons, 67–70
    olfaction, 70–71
    primate brain, 63–67
  apes, 35–42
    glory days of, 41–42
    scientific names, 35–40
  fruit, 42
    apes as fruit eaters, 42–45
    jaws and other adaptations, 45–48
  hands and feet, clambering with, 48
    climbing and walking in trees, 58–59
    playing the monkey, 48–51
    swinging around, 52–57
  tropical paradise, 60
    fur, cooling down with, 61–63

largest body part, 60–61
pale chimpanzee with hair, 63
Erectus skeleton, 143
Esophagus, 21, 22, 47
Ethiopia, 11, 79, 92, 93, 167
Eukaryotic cells, 16, 18, 19
Eumelanin, 193
Eurasia, 35, 41, 169–170, 171, 184, 191
Europe, 15, 41–42, 170, 175, 178, 180, 200
European apes, 41–42
Eustachian tube, 21
Eve hypothesis, 170
Evolutionary adaptations, 13
Evolutionary differences between species, 3
Evolutionary tree, 19, 75, 76, 78
External nasal openings (the *nares*), 25
Eye, 28, 30, 37, 51

**F**

Fallback food, 94
food, changes in, 95–97
lions that feed on grass, 98–100
plants, adaptation of, 94–95
relying on, 97–98
Farmers, hunter-gatherers becoming, 179–181
Fat, 165, 192–193
Fat metabolism, 193
Fatty tissue, 31, 61
Feeling, 51
Feet, 3, 23, 53, 74; *see also* Hands and feet
Feldhof fossils, 174
Femur, 24, 78, 82, 112, 113, 144, 185
Fiber-poor food, 47
Fibula, 24
Filamentous fungi, 44
Filter feeding, 20, 21, 22
Fin, 3, 24
Fine motor skills, 167
Fingers, 24, 49–50, 54, 57, 74
Fin rays, 23, 24
Fins, 23, 24
Fire use, 138
Firm grip, 50–51
First family, 8
First hominins, 74, 75, 77, 80, 81, 82, 85, 86, 100, 110, 112
First man, rise of, 131–133
Fish, 20, 22, 24
bony fish, 20, 22, 28, 29
lungfish, 24, 25

sapiens as, 20–23
Fish ancestors, 23, 44
Fleshy nose, 155
Flower-forming plants, 42
Folate, 122, 123
Folds, 64, 65, 71–74
Food; *see also* Fallback food
changes in, 95–97
cooked plant foods, 139
cooking food, 139
fiber-poor food, 47
softer food, 139, 140, 207
Foot, 24, 49, 59, 109, 110, 111, 144–145
Foot movement, 56
Foot skeleton, 110, 111, 131
Foraging hypothesis, 102
Foramen magnum, 82
Forbes mine in Gibraltar, 14
Forearms, 15, 55
Forests, 87, 100–101
Fossilization, 7, 105
Fossilized skull, 89, 92
Fossils, 2, 10, 11, 16, 23, 35, 42, 59, 134, 177, 178
age determination of, 9
as direct evidence, 7–9
Feldhof, 174
iconic, 174
mammalian, 9
Piltdown, 15
subfossil, 7
'true' fossil, 7, 8
Fossil skulls, 11, 77, 88, 89, 90, 132
Four-footed ancestors, 25
Four-footed monkeys, 49
*FOXP-2* gene, 206
Fragments of existence of humans, 7
fossils as direct evidence, 7–9
indirect traces, 9–10
opinions, battle of, 10–12
Frog tadpoles, 23
Frugivores, 42
Fruit, 42
apes as fruit eaters, 42–45
jaws and other adaptations, 45–48
no longer central in diet, 80–81
Fruit-eating primates, 136
Functional jaw joint, 29
Fur, 15, 30–31
cooling down with, 61–63

## G

Gait pattern, 48, 146, 148–149
Galactose, 183
Galagos, 36, 37
Gelada baboons, 99
Genetic traits, 123, 173
Georgia, 7
Germany, 41
Gibbons, 39, 41, 45, 52–53, 56, 57
*Gigantopithecus*, 41
Gills, 21, 22, 23
Girdles, 26
Glacial periods, 191
Gladiators, 52
*Glandulae mammariae*, 32
Glial cells, 67, 69, 70, 164, 190
Globular brains, 198–200
Gloger, Constantin, 122
Gloger's law, 122
Glottis, 23
Glucose, 183
Gluteus maximus muscles, 144
Gluteus muscle, 147
Gokham, David, 173
Gondwana primordial continent, 35
Gorham Cave, 175
Gorillas, 14, 18, 20, 39, 40, 42, 47, 50, 53, 54, 56, 57, 72, 119, 158, 164
Gottschall, J., 2
Goyet Cave, 175, 180
*Graecopithecus*, 41
Grasping feet, 48, 49, 59, 85, 108
Grasping hand, 43, 48, 49, 54
Grasping system with arms, 53, 54
Grasses, 94, 95
Grassland/savannah, 80, 84, 87, 100–101, 136–137
  big brains, 123
    Affenspalte, 125–127
    Taung's child, 124–125
  climbing and walking, 103–107
  energy, 114–117
  fallback food, 94
    food, changes in, 95–97
    lions that feed on grass, 98–100
    plants, adaptation of, 94–95
    relying on, 97–98
  pioneers of becoming human, 87
    ape from the South, 88–92
    lions among hominins, 92–94

strolling in Laetoli, 107–109
sun, 117
  less heat being produced, 121
  less heat coming in, 117–119
  more heat going out, 120–121
  UV protection, 121–123
tree, 100–117
two points of support, 110–114
Great apes, 14, 15, 18–20, 35, 39–50, 52, 55, 59, 72, 74, 75, 77–78, 81–83, 90, 97, 102, 105, 112, 127, 155, 187, 201, 202
Greece, 41
Greenland, 194
Green vervet monkeys, 138, 205
Grip, 48, 50–51
Grooves, 64, 66
Gyrencephalic brains, 64
Gyrus, 64, 73

## H

*HACNS1* gene, 202
Haeckel, Ernst, 15, 21
Hair, 15, 17, 30–31, 61, 63, 119, 152
Hair cells, 23
Hair color, 119, 193, 197
Hair follicles, 152
Hair louse, 119
Hallux, 49
Hammer (*malleus*), 30
Hand movements, 167, 204
Hands, 23, 24, 53
Hands and feet
  clambering with, 48
    climbing and walking in trees, 58–59
    playing the monkey, 48–51
    swinging around, 52–57
  talking with
    from tools to language, 203–205
    tools to talk, 205–207
Haplorrhini, 37
*HCA3* gene, 44
Head hair, 119
Head louse, 119
Healthy brains, 200
Hearing, 51
Heart, 23
Heat, 31, 61, 62, 117–121
Heel bone (*calcaneus*), 56
Heidelbergensis, 171
Hemoglobin, 210

Herbivores, 43, 96, 99, 136, 137, 179
Herculano-Houzel, Suzana, 21, 69
Hernia, 21
Herpes DNA, 18
*Herpes simplex* virus, 18
Herpes virus, 18
Heterotrophic animal, 18
Hiccups, 22, 23
*Hispanopithecus*, 41
Hobbit, 7, 12
Hominidae, 39
Hominins, 9, 14, 16, 19, 40, 75, 77, 78, 82, 86,
        138, 188, 202
    evolutionary tree, 76
    evolution of, 2
    lions among, 92–94
Hominoidea, 39
*Homo*, 132, 158, 159, 164, 166, 186, 187
*Homo altaiensis*, 173
*Homo antecessor*, 170
*Homo denisensis*, 173
*Homo denisovensis*, 173
*Homo destructus*, 214
*Homo erectus*, 8, 10, 93, 98, 112, 115, 133
*Homo ergaster*, 133
Homo fictus, 2
*Homo floresiensis*, 7, 11, 113, 135
*Homo georgicus*, 133
*Homo habilis*, 98, 128, 132, 157
*Homo heidelbergensis*, 4, 171
*Homo helmei*, 178
*Homo juluensis*, 11
*Homo longi*, 11, 171
*Homo luzonensis*, 12, 135
*Homo naledi*, 8
*Homo neanderthalensis*, 174
*Homo rhodesiensis*, 135, 171
*Homo sapiens*, 2, 4, 7, 10, 11, 13, 15–19, 24, 39,
        82, 87, 170
Howler monkeys, 49
*Hox* genes, 106
Human cradle, 75
    *Ardipithecus*, 79–80
    brain, body to carry around, 85
    Chad, 77–78
    first hominins, 75
    fruit no longer central in diet, 80–81
    legs gaining importance, 81–83
    *Orrorin*, 78–79
    skin, feeling good in, 84
    split, 76–77

Hungary, 41
Hunter-gatherers, 96, 97, 141, 177, 179–181,
        184, 193
Hunting, 118, 140, 150, 162, 179, 181
Huxley, Thomas, 1, 14
Hypodermis, 26, 61

I

Iberian Peninsula, 184
*Ichthyostega*, 23
Iconic fossils, 174
IGF-1, *see* Insulin-like Growth Factor 1
Iliopsoas muscle, 144
India, 41
Indirect traces, 9–10
Inner ear, 23, 25, 73, 105, 148, 184
Insect, 18
Insulin-like Growth Factor 1 (IGF-1), 200
Intelligence, 90, 156–159
Interglacials, 177, 184, 191
Intermembral index, 49
Internal nasal openings (the *choanae*), 25
Intervertebral discs, 20
Inuit, 7, 166, 194
Inverted pendulum, 145, 146,
        147, 153
Island of Flores, 7, 12, 135
Isotopes, 9, 95, 99, 180
Italy, 41

J

Java man, 8, 9, 131
Jaw bones, 22, 29
Jawed animals, 20, 22
Jawless animals, 22
Jaw muscles, 28, 45, 46, 47, 81
Jaws, 21, 22, 28, 45–48
    functional jaw joint, 29
    lower jaw, 8, 28–30, 46, 78, 81, 90, 135,
        207
    upper jaw, 22, 28, 29, 45, 81, 89, 93
Jerrison, Harry, 156
Johanson, Donald, 8, 11, 87, 91
Joints
    ankle, 54, 56, 111, 144
    elbow, 25
    functional jaw joint, 29
Jumping galagos, 49

## K

Keith, Arthur, 81, 90
Kenya, 41
Kenyan National Museum, 79
*Kenyanthropus*, 87
*Kenyanthropus platyops*, 92
*Kenyapithecus*, 41
Keratin, 26–27, 30, 50, 193–194
Kidneys, 180
Kilocalories, 137
Kinetic energy, 114–115, 145, 146
King, William, 174
*KITLG* gene, 193, 196
Knee, 24, 58, 88, 113, 143
Knuckle walking, 40, 42, 53, 57, 58, 81, 83, 85, 100

## L

Lactase, 183
Lactase persistent, 183
Lactose, 182–183
Lactose tolerant, 183
Laetoli
    impressions, 108–109
    strolling in, 107–109
Lancelet, 20
Landau, Misia, 2
Language, 30, 72
Lanugo hair, 15, 61
Large brain, 30, 64; *see also* Cerebrum
Larynx, 21, 32, 206
Lassa virus, 213
Last Glacial Maximum, 178, 192
Lateralization, 204
*LCT* gene, 183
Leaf eaters, 38, 136
Leaf-eating primates, 38, 137
Leaf monkeys, 38, 43
Leakey, Louis, 42, 79, 90, 99, 131
Leakey, Mary, 107, 131
Legs, 4, 23, 24, 26, 48–49, 53, 81–83
Lemurs, 36
Levallois technique, 167, 180, 200, 203
Levant, 8, 191
Limbs, 23, 24, 59, 105, 185
Linnaeus, Carolus, 13, 131
Lions
    among hominins, 92–94
    that feed on grass, 98–100
Lissencephalic brain, 64

Lomekwi technology, 128
Longitudinal arch, 110
Lordosis, 112
Lorises, 36, 37
Lower back, 55, 85, 112, 184
Lower jaw, 8, 28–30, 46, 78, 81, 90, 135, 207
Lower jawbone, 30
Lower temporal window, 28
Lucy, 4, 8, 11, 79, 80, 85, 87, 98, 103–107, 110, 113–117, 121, 123, 142, 164, 188
'Lucy in the sky with diamonds', 91–92
Lumbar lordosis, 112
Lunate sulcus, 124, 125, 126, 127
Lungfish, 24, 25
Lungs, 22, 23, 25

## M

MacLean, Paul, 21
Madagascar, 37
Magnetic resonance imaging (MRI), 72
Maillard reaction, 139
Mammalian brain, 21, 64
Mammalian DNA, 32
Mammalian fossils, 9
Mammalian neocortex, 71
Mammals, 19, 21, 35, 62, 66, 117
    artiodactyl, 166
    egg-laying, 32
    sapiens as, 27–33
Mammary glands, 32
Man and man-like apes, 14
Mandrin Cave, 175
Manipulations, 202
Marine Isotope Stage (MIS), 191
Masseter muscle, 46
Maxilla, 22
Mayr, Ernst, 133
McGrew, William, 102
*MCM6* gene, 183
*MC1R* gene, *see Melanocortin-1-Receptor* gene
Meat eating, 43, 98, 136–137, 139, 140, 141, 168, 181
    steak and fries, 135
        energy in an unstable environment, 140–141
        grasslands full of steaks, 136–137
        steak well done with cooked potatoes, 138–140
Meissner's corpuscles, 60
Melanesia, 172

Melanin, 63, 122, 123, 152, 193
*Melanocortin-1-Receptor (MC1R)* gene, 123, 195–197
Melanocytes, 63, 193, 194, 195
Melanophores, 122
Mental traits, 14
Mesoderm, 27
Metabolism, 10, 121, 140, 141, 155
Metacarpal bones, 57, 202
Metacarpalia, 24
Metatarsalia, 24
Microorganisms, 17
Middle ear, 25
Middle ear bones, 25, 29, 30
Middle East, 8, 135, 151, 178, 180, 184, 191, 195, 202
Middle Miocene, 94
Midfoot bones (metatarsalia), 24
Migration, 135, 175, 178, 180
Milk, 3, 182–183
Milk glands, 32
Miocene, 41
Miocene apes, 58
Mirror neurons, 204
MIS, *see* Marine Isotope Stage
Misliya, Israel, 178
Mismatch diseases, 213
Mitochondria, 10, 17
Mitochondrial DNA (mtDNA), 17, 170
M/L-opsin, 44
Modern man, 15
Mojokerto child, 161
Monkeys, 1–2, 11, 19, 37–38, 48–51, 53
    capuchin, 2, 38, 49, 97, 136, 202
    downward-nosed, 20, 38
    four-footed, 49
    green vervet, 138, 205
    howler, 49
    leaf, 38, 43
    New World monkeys, 37, 49, 66
    Old World monkeys, 37, 38, 39, 47, 50, 51, 72
    prosimian, 64
    spider, 38, 49
    woolly, 49
Morocco, 177, 201
Mother's milk, 32
Motor skills, 167, 201
Mousterian technology, 201
MRI, *see* Magnetic resonance imaging
*MSFD12* genes, 195
mtDNA, *see* Mitochondrial DNA

Mucous glands, 26
Multiregional model, 169
Muscle attachment sites, 206
Muscle spasms, 23
Mutations, 139, 183, 195, 196, 202, 210–211, 212–213
*Myosin Heavy Chain 16* gene, 139

**N**

Nails, 54
*Nares*, 25
Nariakotome River, 134
Natural fat, 137
Natural selection, 3, 10, 13, 16, 84, 122, 147, 193, 200, 207, 211, 212, 213
*Nature*, 11
nDNA, *see* Nuclear DNA
Neander, Joachim, 174
Neanderthal DNA, 10, 175–178
Neanderthal evolutionary line, 175
Neanderthal fossils, 174
Neanderthals, 10, 14, 131, 167, 170–175, 179, 180, 184–185, 191, 192, 197–198
Neck, 25–26, 82, 111, 144, 148
Neocortex, 30, 64, 65, 66, 71–73, 124–126, 157–159, 161
Neocortical neurons, 64
Neolithic, 182, 183, 211
Nervous system, 18, 122, 123, 200
Neurons, 4, 64, 67–70, 156, 161, 162, 164, 204
    mirror, 204
    neocortical, 64
    Von Economo neurons, 68, 156
Neurotransmitters, 67–68, 122, 164
New Stone Age, 182
New World monkeys, 37, 49, 66
Nipples, 32
Nocturnal, 31, 37, 38, 70
Non-chemical communication, 72–73
Non-visual communication, 204
Nose, 19, 37, 38, 66, 155
*NOTCH2NL* genes, 159
Nuclear DNA (nDNA), 16, 17, 175
Nucleus, 19
Nucleus pulposus, 20, 26

**O**

Oblique iliac wings, 112
Obstetrics dilemma, 186

OCA2 gene, 196
Occipital lobe, 66, 71, 72, 205
Oldowan technology, 128, 133, 134, 135, 166
Old World monkeys, 37, 38, 39, 47, 50, 51, 72
Olfaction, 70–71
Olfactory lobe, 71
Olfactory memory, 70
Omnivorous, 81, 99, 179
'On the origin of species' (1859), 14
Opposable thumb, 50, 53, 202
Orangutans, 20, 39, 50, 53, 55–57, 72
Orbitofrontal cortex, 71
Oreopithecus, 41
Orrorin, 11, 77, 78–80, 82, 83, 85
Orrorin tugenensis, 11, 79
Orthograde, 58
Osborne, Charles, 23
Ouranopithecus, 41, 42
'Out of Africa' model, 134–135
Owen, Richard, 15
Oxygenated blood, 23
Oxygen-poor blood, 23

P

Pääbo, Svante, 10, 171, 173
Paleoanthropologists, 8, 10–12
Paleoanthropology, 2, 12
Paleocene-Eocene Thermal Maximum, 36
Paleolithic, 9
Paleontologists, 7
Palm, 51, 57, 152
Palmaris longus muscle, 15
Palm bones (metacarpalia), 24
Pan, 42
Panting, 120
Paranthropus, 87, 93, 94, 98–99, 124, 125, 129, 131, 132, 166
Paranthropus aethiopicus, 93
Paranthropus boisei, 93
Paranthropus robustus, 93, 123
Paraustralopithecus aethiopicus, 93
Peas (Pisum), 181
Pediculus humanus capitis, 119
Pediculus humanus humanus, 119
Peking man, 8, 90, 134
Pelvic girdle, 26, 107, 112, 164
Pelvic stabilization, 82, 112
Pelvis, 55, 58, 82, 86, 174
Phalanges, 24, 85
Pharynx, 21, 22, 23, 206

Pheomelanin, 193, 195
Phrenic nerve, 22
Pickford, Martin, 78–79
Pierolapithecus, 41
Pigmented hairs, 31
Pilot whales, 158
Piltdown forgery, 133
Piltdown fossil, 2, 11, 15, 90
Pinky bone, 171, 173
Pinky finger, 57
Pithecanthropus erectus, 133
Placenta, 33
Plant-eating australopithecines, 135
Plantigrades, 59
Plants, adaptation of, 94–95
Platypus, 32
Platyrrhini, 38
Pleistocene, 131
Plesiadapiformes, 35
Plesianthropus transvaalensis, 93
Pliocene, 75, 87
PLPP1 gene, 193
Polar bears, 31
Polar foxes, 31
Pollex, 50
Porridge, 181–182
Potential energy, 114–115, 145–146
Predators, 51, 100, 103, 136, 138, 205
Prefrontal cortex, 68, 72, 125, 126, 127, 161, 198, 199, 200
Prehensile tail, 38, 49, 50, 51
Premaxilla, 22
Previtamin $D_3$, 194
Primate brain, 63–67
Primates, 14, 19, 20, 26, 30, 35, 36, 42, 44, 48–50, 55, 58, 69, 136, 158, 164, 166
    fruit-eating, 136
    leaf-eating, 38, 137
    true primates, 35
Primitive apes, 41
Pritchard, Jonathan, 211
Proconsul, 41, 53, 54
Pronograde, 58
Prosimians, 51, 64, 70, 201
Proto-languages, 207
Pthirus gorilla, 119
Pthirus pubis, 119
Pubic hair, 119
Pubic louse, 119
Purkinje cells, 68
Pyrenees, 41, 171

**Q**

Quadrosphenic molars, 29
Quadrupeds, 19, 26, 105, 118, 154

**R**

Radioactive decay, 9
Radius, 24
Red noses, 38
Red penis, 38
Reptile brain, 21
Reptiles, 25, 26, 28
Reptilian ancestors, 32
Respiration, 107
Retina, 37
Rhodesiensis, 171
Rhythmic breathing, 23
Rib cage, 23, 134, 164, 206
Rickettsia, 174
Ring-tailed lemur, 37
Ripe (wild) fruits, 43
Rising Star cave, 8
RNA, 18, 32
Rodents, 30
Rotting fruit, 43
*Rudapithecus*, 41
*Rudapithecus hungaricus*, 59
Ruffini corpuscles, 61
Running, 146–151, 184–186
Running economy, 186
Running paradox, 150

**S**

Sacral vertebrae, 26
*Sahelanthropus*, 9, 11, 75, 77, 78, 80, 82, 83, 85,
    95, 157
*Sahelanthropus tchadensis*, 7, 11, 77
*Sapiens; see also Homo sapiens*
    as an amniote, 26–27
    as an animal, 18–20
    as a fish, 20–23
    as a mammal, 27–33
    as a tetrapod, 23–26
*Sapiens* track, 169
    closest friend, 174–177
    track splits, 169–171
    wise one, 177–179
Savannah theory, 101
*Scala naturae*, 13

Scales, 22
Scaphoid, 57
Scapula, 53
Schaafhausen, Hermann, 174
Sedges, 95
Seeing, 51
Sensory function, 30
Sensory organs, 46, 51
Senut, Brigitte, 11, 79
Sexual reproduction, 18
Sexual selection, 15, 193
Shanidar Neanderthals, 181, 185
Shoulder, 53, 55, 187, 188
Shoulder girdle, 26, 103–104, 148
Shubin, Neil, 20
Signals, 68
Sign language, 10, 74
    communication through, 204
Silberberg Grotto, 88
'Simple-nosed', 37
*Sinanthropus pekinensis*, 133
*Sivapithecus*, 41, 53
Skeletal-muscle configuration, 144
Skeleton, 7, 8, 21, 60, 80, 84
    erectus, 143
    foot, 110, 111, 131
Skin, 3, 4, 26, 31, 51, 60, 63
    color, 122, 123, 193, 196
    dark skin, 7, 63, 122, 151, 154,
        193–194
    feeling good in, 84
    fossilized, 89, 92
Skinny dipping, 151–153
Skull, 11, 14, 22, 25–26, 28, 82
Skull bones, 28, 124, 127, 140, 166, 187
Skull windows, 28
*SLC24A5* gene, 196, 210
Small brain, 66, 77, 87, 132, 164, 187; *see also*
    Cerebellum
Smell, sense of, 70–71
Smith, Grafton Elliot, 126, 127
Smith, Thierry, 35
Snout, 70
Social communication, 99
Softer food, 139, 140, 207
S-opsin, 44
Sound vibrations, 25, 206
Sound waves, 25
South, ape from, 88
    child of Taung, 89–91
    Lucy in the sky with diamonds, 91–92

South Africa, 8, 11, 77, 89, 91, 92, 93, 127, 132, 135, 138, 177
South America, 37–38
Southeast Asia, 11, 41, 170, 172, 213
Spain, 41, 170, 181, 197
Spatulate incisors, 46
Speech center, 73
Speech movements, 204
Spider monkeys, 38, 49
Spinal cord, 21, 67, 68, 104
Spine, 3, 26, 74, 85, 105–106
Split, human ancestor from ape ancestor, 76–77
Squirrels, 51
*SRGAP2* gene, 10
*SRY* gene, 214
Stabilization of pelvis, 82, 112
Starch, 97, 139, 210
Stargazers, 188
Stepping movements, 24
Stereoscopic vision, 51
Stomach, 22, 38, 43, 47, 60, 68, 137
Strepsirrhini, 37
Stretched legs, 108, 109, 114, 115, 145
Strolling in Laetoli, 107–109
Subcutaneous hypodermis, 31, 60
Subfossil, 7
Subterranean roots, 95
Sucking, 22, 23
Sumatra, 39
Sun, 117
    less heat being produced, 121
    less heat coming in, 117–119
    more heat going out, 120–121
    UV protection, 121–123
Super-archaic humans, 173
Support, two points of, 110–114
Support phase, 145
Sweat, 31–32, 62
Sweat glands, 31–32, 61, 152
    apocrine, 62
    eccrine, 62, 154
Sweating, 120–121, 153–156
Swimming, 20
Swinging, 52–57, 148
Swing phase, 145
Synapse, 68, 157, 161–162
Synapsida, 27–29
Synapsid skull, 28
Synaptic vesicle, 67
Syncytiotrophoblast, 33
Syntax, 205

**T**

Tactile organs, 51, 60, 61, 68
Tail, 18, 49, 54
    long-plumed, 38
    prehensile, 38, 49, 50, 51
Tail fin of shark and whale, 3
Tarsalia, 24
Taste, 30, 43, 44–45
Taung region, 11
Taung's child, 90, 124–125, 127
Teardrop-shaped hand axes, 200
Teeth, 8, 22, 28, 90, 99, 207
*Teilhardina*, 35, 36, 37
Temporalis muscle, 28, 46
Temporal lobe, 66
Terminal hair, 152
Testosterone levels, 2
Tethys Sea, 41
Tetrapods, 22, 30
    sapiens as, 23–26
Thermoregulation, 61
Tibet, 194
Tibia, 24, 88
*Tiktaalik*, 24, 26
Time window, 8–9, 78, 92
*TKLT1* gene, 199
Tobias, Phillip, 131
Toe, 3, 24, 49, 54, 59
Tongue, 62
Tool making, 128, 202
Tool use, 103, 205
Toumai, discovery of, 77–78
Traces, indirect, 9–10
Trachea, 23
Transitional vertebra, 106
Transverse arch, 110
Trees, 100
    climbing and walking in, 58–59
    living in, 48
Tribosphenic molars, 29
Triconodonta, 28
Triune brain, 21
Tropical paradise, 60
    fur, cooling down with, 61–63
    largest body part, 60–61
    pale chimpanzee with hair, 63
'True' fossil, 7, 8
True primates, 35
Tugen Hills, 79
Tunicates, 20

Turkana boy, 133–134
Turkey, 41

**U**

Uganda, 41
Ukraine, 171
Ulna, 24
Ungulates, 30
Uniqueness of humans, 4, 13, 209–211
    amniote, sapiens as, 26–27
    animal, sapiens as, 18–20
    fish, sapiens as, 20–23
    mammal, sapiens as, 27–33
    sapiens body, 15–18
    tetrapod, sapiens as, 23–26
Upper jaw, 22, 28, 29, 45, 81, 89, 93
Upright locomotion, 59
Uranium-lead dating, 9
Uranium-thorium dating, 9
Urethra, 17
UV protection, 121–123
UV radiation, 7, 63, 191, 194
    UV-A rays, 121
    UV-B rays, 121–122, 193
    UV-C rays, 122

**V**

Vagina, 17
Vagus nerve, 22
Valgus angle, 113
Van Straalen, Nico, 211
Varus angle, 113
Vater-Pacini corpuscles, 61
Verraes, Walter, 3
Vertebrae, 20–21, 85, 106–107, 112
Vertebrates, 3, 20, 22, 28
Vertical climbing, 55, 56, 57, 74, 81
Vestigial organs, 15
Vibrations, 25, 60, 61
Videan, Elaine, 102
Vigilance hypothesis, 102
Viral DNA, 32
Virunga Mountains, 40
Viruses, 13, 18

Vision, 30, 37, 43, 51, 70, 102, 148
Vitamin C, 45
Vitamin D$_3$, 193, 194, 196
Vocal communication, 74, 204
Von Economo neurons, 68, 156
Von Haller, Albert, 156

**W**

Walking, 4, 21, 48, 56, 58–59, 103–107,
        141–146
    barefoot walking, 109
    with big brains, 184
        birth canal, 186–191
        running economy, 186
        walking and running in the cold, 184–186
    in the cold, 184–186
    knuckle walking, 40, 42, 53, 57, 58, 81, 83, 85,
        100
    in trees, 58–59
Wallace, Alfred Russel, 13
Warm blood, 31, 166
Wernicke, Carl, 73
Wernicke's center, 73, 204
West African chimpanzee, 63
Wheat (*Triticum*), 181
White Sands National Park, 178
Wilberforce, Bishop Samuel, 1
Wildebeest, 137
Wisdom teeth, 3, 207
Woolly monkeys, 49
Wrist, 24, 54
Wrist bones (carpalia), 24, 57

**Y**

Y chromosome, 10, 170, 176, 214
Yoruba population in Nigeria, 213
Young Dryas, 181

**Z**

Zagros mountain region, 176
Zaire River, 40
Zebra, 137
*Zinjanthropus boisei*, 93